Green Energy and Technology

Climate change, environmental impact and the limited natural resources urge scientific research and novel technical solutions. The monograph series Green Energy and Technology serves as a publishing platform for scientific and technological approaches to "green"—i.e. environmentally friendly and sustainable—technologies. While a focus lies on energy and power supply, it also covers "green" solutions in industrial engineering and engineering design. Green Energy and Technology addresses researchers, advanced students, technical consultants as well as decision makers in industries and politics. Hence, the level of presentation spans from instructional to highly technical.

Indexed in Scopus.

Indexed in Ei Compendex.

Amos Madhlopa · Richard Nkhoma

Principles of Solar Gas Turbines for Generating Electricity

Second Edition

Springer

Amos Madhlopa
Department of Engineering
Malawi University of Science
and Technology
Thyolo, Malawi

Richard Nkhoma
Department of Engineering
Malawi University of Science
and Technology
Thyolo, Malawi

ISSN 1865-3529 ISSN 1865-3537 (electronic)
Green Energy and Technology
ISBN 978-3-031-84095-1 ISBN 978-3-031-84096-8 (eBook)
https://doi.org/10.1007/978-3-031-84096-8

This Springer imprint is published by the registered company Springer Nature Switzerland AG
The registered company address is: Gewerbestrasse 11, 6330 Cham, Switzerland

Preface

Since the industrial revolution in the eighteenth century, fossil fuels (natural gas, oil, and coal) have played a critical role in socio-economic growth. They fuel many technologies, ranging from motor vehicles to power plants. Nevertheless, environmental degradation is one main concern about their exploitation. It is perceived that greenhouse gas (GHG) emissions from the consumption of fossil fuels are contributing to climate change and other environmental problems. Therefore, international efforts are aiming at protecting the environment while developing economies. Another concern is that fossil fuels occur in finite quantities, implying that they can be depleted, thereby posing a risk to energy security. On the other hand, renewable energy resources replenish themselves through natural mechanisms and are generally cleaner. In this vein, one of the most important renewable energy resources is solar radiation that can be converted to electricity by using a suitable technology.

Conversion of primary energy to electricity requires technologies that usually come in the form of engines. Heat engines (including gas turbines), predominantly fuelled by fossil resources at present, have found wide application in the energy sector. In this context, a gas turbine is one of the commonest heat engines for generating electricity. It operates based on the Brayton cycle and fundamentally comprises a compressor, a heater, and a turbine. The Brayton cycle is the ideal cycle for gas turbine engines, and so it is also known as the gas turbine cycle. A standard gas turbine engine utilizes air as the working fluid. The working fluid is drawn into the compressor, where both its temperature and pressure are elevated. The pressurized fluid flows into the heater for direct or indirect heat addition at constant pressure, yielding a high-temperature fluid. The hot fluid enters the turbine section where it expands and produces mechanical power. After exiting the turbine, the fluid is cooled before it re-enters the compressor to start a new cycle (in a closed cycle gas turbine), or it is released into the ambient environment (in an open cycle gas turbine). An electric generator is used to convert mechanical power to electricity.

Natural gas is often used as a fuel for the conventional gas turbine engine. Nevertheless, concerns about sustainable development are driving changes in national and international laws and policies that support transition to renewable energy. These developments are affecting the direction of research on gas turbines and other technologies. In this connection, solarization of the gas turbine engine is gathering momentum as an important option for achieving sustainable energy systems. A solar gas turbine (SGT) is a device in which concentrated solar radiation is used to heat up a working fluid before it expands through the turbine section.

Generally, a gas turbine operates at high turbine inlet temperature (>673 K), and therefore, concentrated solar radiation is a suitable heat source for driving this engine. The intensity of direct normal irradiation is a good indicator of the suitability of a site for exploitation of concentrating solar power (CSP) technologies. Basically, a CSP power plant comprises a solar field (concentrator and receiver) and a power block. The concentrator focuses direct (beam) radiation onto the receiver that heats up a heat transfer fluid circulated through it. Then, the hot fluid is directly or indirectly used to drive an engine cycle in the power block. The recommended minimum annual sum of direct normal irradiation for CSP technology to be economically viable is 2000 kWh/m^2, and many locations within the sunbelt in the world meet this requirement.

The commonest CSP technologies exploited globally are the parabolic trough concentrator (PTC), linear Fresnel reflector (LFR), solar tower (ST), and parabolic dish concentrator (PDC). Studies have shown that the PTC and LFR technologies achieve lower temperatures than the ST and PDC technologies. Nevertheless, at the time of writing this book, most of the operating CSP power plants in the world were based on the steam Rankine cycle driven by the PTC technology, followed by the ST. However, this cycle exhibits lower thermodynamic efficiency than the Brayton cycle. Therefore, one of the strategies for improving the efficiency of CSP technology is the exploitation of the Brayton cycle in the power block. Recent studies show that the third generation of the CSP technology is expected to be based on the Brayton cycle. Moreover, an increase in the system efficiency results in the reduction of the generation cost of electricity and the emission of GHGs. Consequently, solar gas turbines are expected to play a critical role in the sustainable production of electricity.

Principles of Solar Gas Turbines for Generating Electricity brings together pieces of new knowledge on solar gas turbines with adequate illustrations and coherent treatment. At present, the SGT technology is under active research and development, which has necessitated writing a second edition of this book. Simulation is an important tool for designing solar gas turbines. In addition, the environmental performance of a given technology is a critical aspect of sustainable development. Consequently, this new edition includes two separate chapters on simulation and environmental sustainability of solar gas turbines.

This book aims at providing the reader with principles of solar gas turbines and a state of the art. The context of solarising gas turbines is presented in Chap. 1. This chapter also introduces relevant fundamentals of conventional and solar gas turbines. Chapters 2 and 3 focus on fuels and solar radiation, respectively, as sources of heat for the SGT engine. The discussion on fuels includes aspects of emissions, which is important for environmental protection. Various components of an SGT are covered in Chap. 4. It is observed that the solar receiver and the combustor are critical components in the solarization process. Based on the heat transfer medium, solar receivers can be classified into gas, liquid, and solid particle categories. Gas receivers are capable of attaining high temperatures (up to 1773 K) and therefore more suitable for exploitation in solar gas turbines. Liquid and particle receivers can heat up the working fluid in an SGT only via heat exchangers, which would tend to reduce the efficiency of converting solar energy to useful heat. Solar receivers can also be classified based on their geometry: tubular, volumetric, and microchannel receivers. Although each category has advantages and some disadvantages, tubular and volumetric receivers are emerging to be in the most advanced stage of development. These receivers can be linked to combustors in hybrid solar gas turbines. At present, combustors in use are designed for conventional gas turbines. Thus, they cannot just be integrated with solar receivers without modification. For example, the Mercury 50 gas turbine allows a combustor inlet air temperature of 923 K. However, the receiver outlet air temperature may theoretically reach temperatures of 1773 K. This challenge is aggravated by the lack of gas turbines tailor-designed to operate on solar energy. Selected relevant engine cycles have been presented in Chap. 5. Based on the flow path of the working fluid, gas turbines can be classified into closed, open, and semi-closed cycle systems. In the closed cycle, the working fluid is often heated via a heat exchanger, which reduces the thermal efficiency of the system. Combustion gases constitute part of the working fluid at the turbine inlet in an open cycle gas turbine, thereby elevating the thermodynamic efficiency of this cycle. In view of this attractive thermal performance, open cycle gas turbines have found widest application in the electricity industry. However, the use of carbon dioxide (CO_2) as a working fluid in the supercritical phase is improving the thermodynamic performance of the closed cycle. It is also possible to combine the Brayton and steam cycles to obtain a combined cycle with a high thermodynamic efficiency (>50%). Exergy analysis reveals that most of the exergy destruction occurs in the solar receiver due to the low rate of conversion of solar radiation to useful heat. Components of an SGT can be arranged in many different configurations, some of which are presented in Chap. 6. This flexibility in engine layout is an advantage in the development of the SGT engine. System design is one important aspect of the development of the SGT technology. So, Chap. 7 covers principles of gas turbine design. The SGT system involves complex multivariate interactions that require suitable design tools. Simulation enables engineers and scientists to design complex systems. Thus, Chap. 8 focuses on the principles of simulation and their application in the development of solar gas turbines. After designing a system, it is necessary to prototype and test it (see Chap. 9). Even if a solar gas turbine thermodynamically performs well, its rate of diffusion on the market is influenced by the cost of electricity production. Therefore, Chap. 10 examines the economic performance of solar gas turbines. Findings show that the levelized cost of CSP-based electricity has significantly declined over the past decade. Environmental protection is a key consideration in the deployment of energy technologies. Various aspects of the environmental performance of the

SGT technology are presented in the last chapter (Chap. 11). It is evident that solar gas turbines are environmentally more sustainable than fossil-fuelled systems.

Thyolo, Malawi

Amos Madhlopa
Richard Nkhoma

Competing Interests The authors have no competing interests to declare that are relevant to the content of this manuscript.

Contents

Abbreviations

AC	Alternating current
AD	Anaerobic digestion
AI	Artificial intelligence
AMD	Acid mine drainage
ANN	Artificial neural network
AQG	Air quality guidelines
ASHRAE	American society of heating, refrigerating, and air-conditioning engineers
ASME	American society of mechanical engineers
BRT	Backward ray tracing
CCC	Cash conversion cycle
CCGT	Combined cycle gas turbine
CCS	Carbon capture and storage
CCSGT	Combined cycle solar gas turbine
CEA	Cumulative exergy analysis
CFCs	Chlorofluorocarbons
CFD	Computational fluid dynamics
CHP	Combined heat and power
CLCGT	Closed cycle gas turbine
CMC	Ceramic matrix composite
CPC	Compound parabolic concentrator
CR	Concentration ratio
CSP	Concentrating solar power
DC	Direct current
DEA	Diethanolamine
DLN	Dry low oxide of nitrogen (NO_x)
DNI	Direct normal irradiance
DNS	Direct numerical simulation
DPB	Discounted pay-back
EC	External combustion
ED	Electrodialysis

EIA	Environmental impact assessment
ERA	Environmental risk assessment
FAE	Fatty acid ester
FEA	Finite element analysis
FEM	Finite element method
FRT	Forward ray tracing
FVM	Finite volume method
GEF	Global environment facility
GHG	Greenhouse gas
GT	Gas turbine
GWP	Global warming potential
HEX	Heat exchanger
HHV	Higher heating value
HRSG	Heat recovery steam generator
HV	Heating value
HVAC	Heating, ventilation, and air-conditioning
IC	Internal combustion
IEC	International Electrotechnical Commission
IGCC	Integrated gasification combined cycle
IRR	Internal rate of return
ISCC	Integrated solar combined cycle
ISGTCPP	Integrated solar gas turbine cogeneration power plant
ISO	International organization of standardization
LCA	Lifecycle assessment
LCOE	Levelized cost of electricity/energy
LES	Large eddy simulation
LFR	Linear Fresnel reflector
LHS	Latent heat storage
LHV	Lower heating value
LR	Linear regression
MDEA	Methyl diethanolamine
MED	Multi-effect distillation
MGT	Micro gas turbine
MLR	Multiple linear regression
MON	Motor octane number
MSF	Multi-stage flash
MTBE	Methyl-tertio-butyl-ether
NF	Nanofiltration
NG	Natural gas
NPV	Net present value
NREL	National renewable energy laboratory
OCGT	Open cycle gas turbine
ORC	Organic Rankine cycle
PCM	Phase change material
PDC	Parabolic dish concentrator

PID	Proportional-integral-derivative
PM	Particulate matter
PME	Palm methyl ester
PMG	Permanent magnet generator
PTC	Parabolic trough concentrator
PV	Photovoltaic
R&D	Research and development
RANS	Reynolds-averaged Navier-Stokes
RE	Renewable energy
RHE	Recuperative heat exchanger
RO	Reverse osmosis
RON	Research octane number
RSM	Response surface methodology
RTD	Resistance temperature detector
SCLCGT	Semi-closed cycle gas turbine
SCR	Selective catalytic reduction
SGS	Subgrid-scale
SGT	Solar gas turbine
SHS	Sensible heat storage
ST	Solar tower
TC	Technical committee
TES	Thermal energy storage
TIT	Turbine inlet temperature
TRL	Technology readiness level
TRNSYS	Transient system simulation
UFM	Ultrasonic flow metre
UHCs	Unburned hydrocarbons
UHT	Ultra-high temperature
UV	Ultraviolet
VCD	Vapour compression distillation
WRC	World Radiation Centre

List of Figures

List of Tables

Chapter 1
Introduction to Solar Gas Turbines

1.1 Introduction

Energy is a key driver of socio-economic development of nations. Its significance is reflected in the myriad ways it powers industries, sustains households, and fuels technological advances. In this vein, electricity stands out as the major form of modern energy (Opiyo 2016). A gas turbine is one of the major technologies that offer a dynamic and efficient means of converting energy into electricity. It basically comprises connected components: a compressor, a combustion chamber (combustor), and a turbine. In this symbiotic arrangement, a working fluid, often air, is the medium through which energy is harnessed and transformed. This fluid undergoes a sequence of pressurization and heating before expanding through the turbine section, where it produces mechanical power that is ultimately converted to the required electrical output by using an electric generator. However, the conventional landscape of gas turbine power plants has been dominated by the combustion of fossil fuels that do not only contribute to environmental degradation but also raise red flags pertaining to the finite nature of these conventional energy resources (see Chap. 11). The repercussions of this heavy reliance on fossil fuels have become all so apparent in the form of climate change and other environmental problems, and in posing a substantial risk to the global energy security. Consequently, there is an imperative to chart a course towards energy practices that align with the principles of sustainable development.

The call for a paradigm shift in the energy landscape is met with the rising significance of renewable energy sources that offer a more harmonious balance between energy production and environmental preservation. The inexhaustible and environmentally benign nature of renewable resources positions them as beacons of hope in the quest for a cleaner and more sustainable energy future. Solar energy is one of the most promising sources of renewable electricity, not only for its abundance but also for its minimal environmental impact and capacity for self-replenishment through natural processes. In this context, a solar gas turbine (SGT) is one of the key innovations in the electricity generation sector. Solar gas turbines have relatively

A. Madhlopa and R. Nkhoma, *Principles of Solar Gas Turbines for Generating Electricity*, Green Energy and Technology,
https://doi.org/10.1007/978-3-031-84096-8_1

high thermodynamic efficiency, and so, they offer a unique synergy of efficiency and sustainability.

1.1.1 Energy Resources

Energy resources are all forms of fuels utilized for heating, generation of electricity, or for other processes of energy conversion (Rashid 2016). These resources can be classified into two main categories: renewable and non-renewable. Renewable energy (RE) resources possess the remarkable ability to replenish themselves through natural processes. The sustainability of these resources hinges on the delicate balance between their rate of replenishment and consumption. Within the realm of renewable energy, a fundamental distinction exists between solar and non-solar resources. Direct harnessing of solar energy is achievable through cutting-edge technologies such as solar photovoltaic (PV) or thermal systems. Alternatively, indirect forms of solar-based resources, such as biomass, hydro, and wind energy, contribute significantly to the landscape of renewable energy. Biomass is a product of photosynthesis facilitated by solar radiation in plants, and it reveals the intricate relationship between solar energy and organic matter. The hydrological cycle is driven by the sun's heat that evaporates water from the earth's surface. This water vapour then condenses in the atmosphere, ultimately returning to the earth as precipitation. The ensuing flow of water in rivers becomes a valuable resource for generating hydroelectric power. Moreover, the manipulation of water resources extends beyond immediate energy generation needs. Water can be strategically pumped and stored in elevated reservoirs during periods of low electricity demand, creating a reservoir of potential energy (Vahidinasab and Habibi 2021). Subsequently, when the demand for electricity peaks, this stored water is released to drive hydroturbines, thereby meeting the increased demand for electricity. A fraction of the solar radiation incident on the earth's surface is transformed into heat, initiating a cascade of atmospheric phenomena. As the earth's surface warms, the surrounding air rises through natural convection, thereby generating wind. The complex interaction among solar radiation, heat conversion, and atmospheric dynamics highlights the diverse avenues through which solar energy manifests itself, contributing to the renewable energy portfolio. In essence, understanding the multifaceted nature of energy resources, particularly those intertwined with solar influence, becomes necessary during the transition to sustainable energy production and consumption.

In contrast to renewable resources, non-renewable resources are depleted at a rate far exceeding the pace at which they are naturally formed. These resources can be broadly categorized into fossil and non-fossil types, with fossil fuels (including coal, oil, and natural gas) dominating the non-renewable energy landscape. In addition to fossil fuels, non-fossil resources such as nuclear energy contribute to the non-renewable energy mix, albeit with unique considerations. Figure 1.1 provides a comprehensive overview of the classification of energy resources, distinguishing between the renewable and non-renewable categories. While renewable resources are replenished through natural mechanisms, non-renewable resources stand on the precarious edge of finite availability. On a global scale, fossil fuels have long held the position of primary contributors to the global energy mix. However, this dominance comes at a considerable cost, with growing apprehensions about the social, economic, and environmental ramifications associated with the extraction and utilization of these fuels. In the light of these challenges, research and development efforts are increasingly focusing on cultivating sustainable strategies for both energy supply and consumption. In this context, renewable energy is globally perceived as one of the pillars of sustainable energy systems (see Chap. 11).

A key advantage of RE resources is their ability to replenish themselves through natural mechanisms. It is possible to exploit these resources unsustainably (this occurs when the rate of resource utilization exceeds that of replenishment as may be the case with biomass production and consumption). Nevertheless, renewable energy

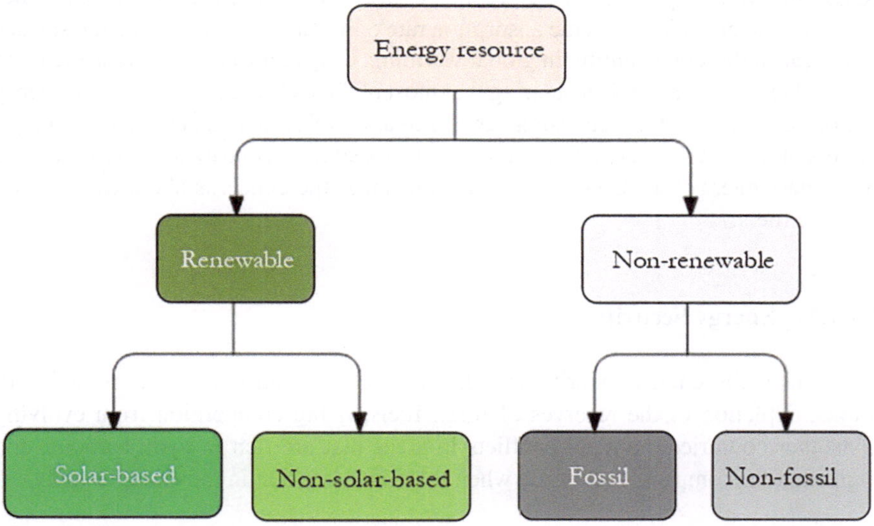

Fig. 1.1 Classification of energy resources

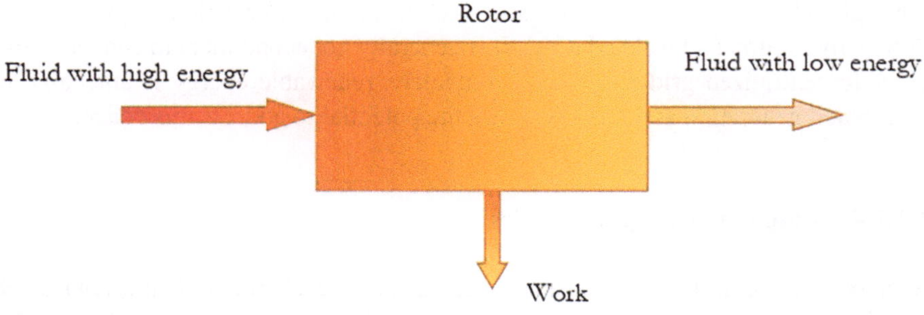

Fig. 1.2 Schematic diagram of a power producing turbomachine

is regarded as beneficial to society, economy and environment as reflected in the following drivers of RE resource deployment (Bhattacharyya 2011): (a) Reduction of carbon dioxide emissions and mitigation of climate change, (b) energy security, (c) increasing energy access rate, (d) employment opportunities, and e) other spill-over benefits.

1.1.1.1 Reduction of Carbon Dioxide Emissions and Mitigation of Climate Change

Anthropogenic activities are generating greenhouse gas (GHG) emissions, and energy consumption is the main contributing factor to this environmental problem. Greenhouse gases absorb and emit radiation in the infrared band, and they are responsible for causing climate change. These gases include carbon dioxide (CO_2), methane (CH_4) and water vapour (H_2O). Thus, an increase in the concentration of these gases in the atmosphere augments the absorption rate of infrared radiation from the surface of the earth, thereby resulting in global warming. CO_2 in the atmosphere is the main cause of climate change. Considering that most of the CO_2 is generated by the energy sector, reduction of CO_2 emissions and mitigation of climate change are the major drivers of renewable energy. On a global scale, fossil fuels contribute the largest share of primary energy (see Chap. 11), which enhances the concerns about exploitation of these fuels.

1.1.1.2 Energy Security

Concerns about energy security arise from the consideration of fossil price fluctuations, depletion of the reserves of fossil fuels, rising competition from evolving consumer countries, political conflicts in areas that are rich in hydrocarbons, and high economic impacts that ensue when there is disruption in the energy supply.

1.1.1.3 Increasing Energy Access Rate

On a global scale, there is limited access to clean and affordable energy supplies, especially in rural and isolated areas. It is generally not economical to connect these areas to centralized grid networks. Fortunately, renewable energy is amenable to distributed generation and, therefore, has the potential to address this concern.

1.1.1.4 Employment Opportunities

Society is one of the three dimensions of sustainable development. In this regard, job creation is one of the common social indicators of energy sustainability. Renewable energy can directly create jobs due to its decentralized and modular structure of the technologies and implementation of projects at local level. For instance, it has been found that concentrating solar power has a high potential for job creation through manufacturing of components in developing countries (Pierce et al. 2013).

1.1.1.5 Other Spill-Over Benefits

Dependence on renewable energy can assist in bettering macro-economic stability. This is due to: (a) promotion of renewable energy can reduce reliance on imports such as petroleum products, thereby preserving the foreign reserves of a country, (b) fossil fuels constitute an important component of international trade, and so replacing these fuels with renewable energy would result in a diminished international trade balance, and (c) decrease in international trade helps to reduce the likelihood of economic shocks that are created by external factors.

1.1.2 Energy Conversion

The conversion of primary energy to electricity requires technologies that usually come in the form of engines. An engine is a mechanical device designed to transform some form of energy into mechanical energy. The input energy to the engine is often in a chemical form (fuels), solar, and wind. Engines that convert fuel to mechanical energy, via thermal energy, are referred to as heat engines. In these engines, the engine-fuel interaction passes through various stages from one thermodynamic state of the engine to another and back to its initial state, forming an engine cycle.

A heat engine is a closed cycle device that obtains heat from a hot reservoir, does useful work, and rejects heat to a cold reservoir. Energy transformation in a heat engine is exothermic, where heat is given out and utilized to produce mechanical work (shaft work). For combustion engines, the burning of fuel yields the required thermal energy to drive the engine, and these engines are of two types: internal combustion (IC) and external combustion (EC). An IC engine is a heat engine in which the combustion of fuel takes place with an oxidizer (usually air) in a combustor that forms part of the flow circuit of the working fluid. Examples of IC engine cycles include the Brayton, Otto and diesel cycles. In an EC engine, the working fluid is heated by fuel combustion in an external combustor, and heat is transferred to the working fluid through a heat exchanger. Some of the EC cycles are the Rankine (steam) and Stirling cycles. Detailed information on the various engine cycles is reported in many books on classical thermodynamics such as Çengel and Boles (2006) and Eastop and McConkey (2009). For electricity generation, the Brayton cycle (the ideal cycle for gas turbine engines) is gaining popularity. A gas turbine belongs to a group of machines known as turbomachines.

A turbomachine is a fundamental device for the transfer of mechanical energy between a continuously flowing fluid and a rotor system, achieved through the dynamic interaction of rotating blades arranged in rows. This dynamic action leads to the transfer of energy, resulting in an enthalpy change in the fluid as it passes through the turbomachine. Based on the direction of energy transfer (whether it is from the fluid to the rotor system or vice versa), turbomachines are broadly classified into two categories: power producing and power absorbing.

1.1.2.1 Power Producing Turbomachines

In a power producing turbomachine, energy is transferred from a flowing fluid with high energy to the rotor system, ultimately producing shaft work (Fig. 1.2). Power producing turbomachines play a crucial role in various industries and applications, contributing significantly to power generation and conversion. Several types of power producing turbomachines include: gas, hydraulic, steam, and wind turbines. Gas turbines are widely used in aviation, power plants, and various industrial applications. They operate on the basis of the Brayton cycle: compressing air, combusting it with fuel, and then expanding the resulting high-temperature, high pressure gases to produce mechanical work. Hydraulic turbines harness the kinetic and potential energy of flowing water to generate mechanical work, which is then converted into electrical energy by connected generators in hydropower plants. Steam turbines are integral components in steam power plants, where they convert the thermal energy of steam into mechanical work by expanding the steam through a series of stages. Wind turbines convert the kinetic energy of the wind into mechanical energy, which is then used to generate electricity through a generator. They are pivotal in harnessing wind power for renewable energy. These power producing turbomachines are critical elements in diverse energy conversion processes, contributing to the generation of electricity and the propulsion of various vehicles and machinery.

1.1.2.2 Power Absorbing Turbomachines

The primary function of a power absorbing turbomachine is to transfer energy from shaft work to a fluid with lower energy, typically accomplished through the action of an impeller. This process results in the production of a fluid with higher energy. Figure 1.3 illustrates the general configuration of this type of turbomachine. Power absorbing turbomachines play a crucial role in various applications where mechanical energy needs to be imparted to a fluid for specific purposes. Examples of power absorbing turbomachines include: blowers, compressors, fans, propellers, and pumps. Blowers are devices that absorb mechanical energy and transfer it to a fluid, typically air or gas, increasing its pressure. They find applications in ventilation systems, air circulation, and industrial processes where controlled air movement is essential. Compressors are utilized to increase the pressure of a gas, resulting in a higher-energy fluid. They are essential components in various industries, such as refrigeration, air conditioning, and gas compression for industrial processes. Fans are similar to blowers but are designed to move larger volumes of air at lower pressures. They are commonly used in heating, ventilation, and air-conditioning (HVAC)

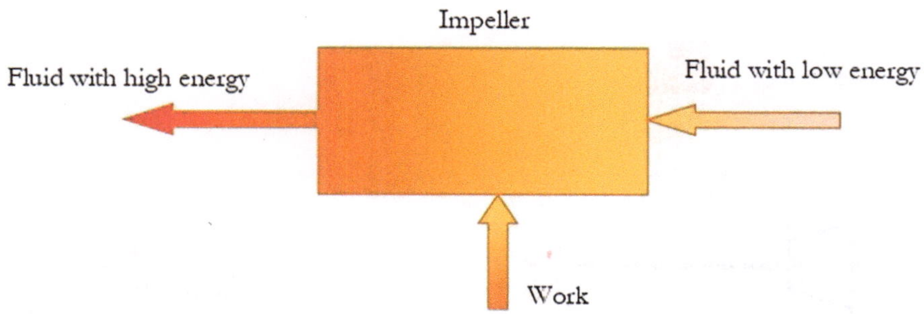

Fig. 1.3 Schematic diagram of a power absorbing turbomachine

systems, cooling applications, and electronics to dissipate heat. Propellers are power absorbing turbomachines used in marine and aerospace applications. They absorb energy from the engine shaft and transfer it to a fluid, usually air or water, to produce thrust or propulsion. Pumps are devices that absorb mechanical energy and transfer it to a liquid, increasing its pressure and facilitating its movement. Pumps are crucial in various industries, including water supply, wastewater treatment, and chemical processing.

These power absorbing turbomachines are integral components in systems where the transfer of mechanical energy to a fluid is required for specific functions. The design and operation of each type of a power absorbing turbomachine are tailored to the characteristics and requirements of the fluid involved, emphasizing efficiency, reliability, and performance in diverse applications.

1.1.3 Basic Gas Turbine for Electricity Generation

A gas turbine (GT) is an internal combustion engine (IC) that can be used to efficiently generate electricity, relying on air as the primary working fluid. The fundamental operation of a GT involves a sequence of processes that facilitate the conversion of fuel into mechanical power, ultimately producing electrical energy. Major components of the GT system are compressor, combustion chamber (combustor), turbine, and exhaust (Fig. 1.4). Based on flow path, gas turbines are classified into three basic types: (a) closed cycle gas turbine, (b) open cycle gas turbine and (c) semi-closed gas turbine. In the closed cycle GT, the working fluid can be a gas such as air, supercritical carbon dioxide (s-CO_2), helium or nitrogen (Olumayegun et al. 2016). This fluid is heated via a heat exchanger, expanded through a turbine, and then cooled before being recompressed and re-circulated. In the basic open cycle GT systems, ambient air enters the compressor where it is compressed. Now at an elevated pressure, the compressed air enters the combustion chamber where it combines with fuel. The introduction of fuel triggers the combustion process, generating gases at high temperature and pressure. This phase is critical as it marks the conversion of

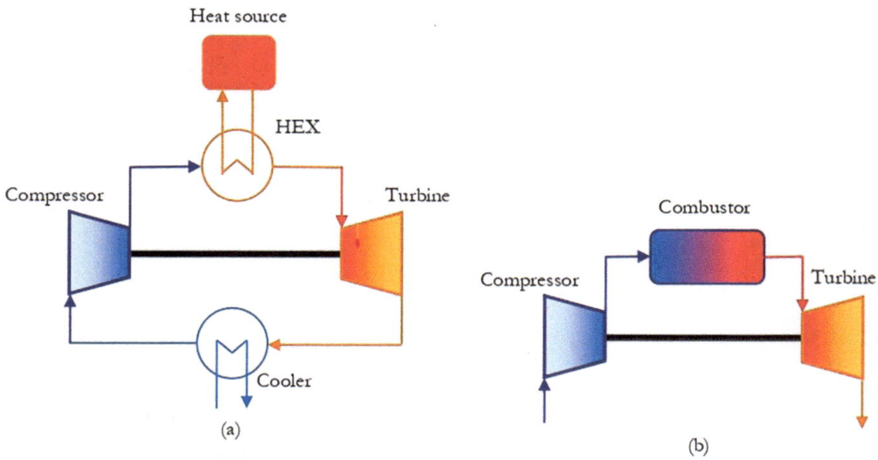

Fig. 1.4 Schematic representation of a simple gas turbine: **a** closed cycle and **b** open cycle. HEX = heat exchanger

chemical energy from the fuel into thermal energy. These gases enter the turbine section where they expand rapidly, undergoing a controlled release of energy. This expansion propels the turbine blades, converting the thermal energy into mechanical energy. It is in this phase that the turbine generates the power needed to drive the compressor and other auxiliary systems. The final stage involves the release of exhaust gases from the turbine, completing one cycle of the gas turbine operation. The expelled gases may be released into the atmosphere, depending on the overall system design. Semi-closed cycle gas turbine systems combine features of both the closed and open cycle gas turbines. At present, there is no power plant based on this cycle. Thus, there are two types of practical gas turbines: closed and open cycle gas turbines.

A gas turbine operates with the working fluid consistently in a gaseous phase throughout the entire cycle, unlike some other power cycles such as the Rankine cycle used in steam turbines. This continuous gaseous phase is a distinguishing feature of gas turbines, enabling efficient and rapid energy conversion. Moreover, a single gas cycle can be combined with other cycles to obtain combined cycles with higher thermodynamic efficiency (see Chap. 5). Gas turbines are known for their high power-to-weight ratios and their ability to provide a quick response to changes in the electricity demand. These characteristics make them suitable for aviation, power plant, industrial and other applications. In the electricity sector, gas turbines are often employed in standalone power plants or as part of combined cycle systems where they commonly work in conjunction with steam turbines.

1.1.4 *Classification of Gas Turbines by Application*

Gas turbines, versatile in their applications, can undergo classification based on their intended use. This categorization primarily distinguishes between aero-engines, designed for aviation purposes, and land-based (industrial) gas turbines, tailored for various terrestrial applications. The complexity of this classification is further enriched by the emergence of aero-derivative engines, a subset of aero-engines adapted for land-based applications. The classification of gas turbines based on application highlights the diversity and adaptability of this technology. From aviation to power generation, gas turbines, in their various forms, play crucial roles in meeting the dynamic energy needs of modern society. The continuous evolution of gas turbine design and technology ensures their relevance across a spectrum of applications, contributing to the efficiency and sustainability of energy systems worldwide.

1.1.4.1 Aero-Engines

Some aero-engines have successfully transitioned from their original aviation context to find utility in land-based applications. Notable examples include Pratt & Whitney Aircraft J75/FT4, General Electric J75/LM1500, and Rolls Royce RB211 (Giampaolo 2003). This adaptability is driven by their advantageous characteristics.

1.1.4.2 Aero-derivative Engines

Aero-derivative engines, born from aero-engines, retain their lightweight design and are well-suited for various land-based applications. Their ability to start quickly, efficiently handle load variations, and their lightweight nature make them particularly attractive for power generation and marine industry use where weight consideration is crucial.

1.1.4.3 Industrial Gas Turbines

Many industrial gas turbines, exemplified by models like the General Electric Frame 7001 and the Solar MARS, have historical ties to steam engines. Derived from the steam turbine technology, these industrial gas turbines share similarities with their steam-driven counterparts, often featuring larger and heavier designs.

Heavy industrial gas turbines, owing to their robust design and power capabilities, find applications in various sectors, most notably in power generation. Their capacity to generate substantial power makes them indispensable in meeting the electricity demands of large-scale industrial processes and communities.

1.1.5 Selection and Development of Gas Turbines

The weight of a gas turbine often plays a pivotal role in its suitability for specific applications. Aero-derivative engines, being lightweight, are favoured in scenarios where weight is a critical factor. Conversely, heavy industrial gas turbines, despite their weight, excel in scenarios where high-power output is paramount.

Ongoing research and development initiatives continue to refine the gas turbine technology. Innovations in materials, design, and control systems aim to enhance efficiency, reduce environmental impact, and expand the range of fuels that can be utilized. These efforts contribute to the evolution of gas turbines across all categories. Due to environmental concerns, research attention has also been given to solarization of gas turbine systems.

1.2 Solar Gas Turbines

A solar gas turbine (SGT) is a device that uses concentrated solar radiation to heat up the working fluid before it expands through the turbine section. To augment the solar power, the principles of reflection or refraction are used to concentrate solar radiation on a small area (line or point). Smooth surfaces are used to reflect solar radiation onto a target small area (receiver). Convex lenses utilize the concept of refraction to focus solar radiation. In this vein, a basic SGT for generating electricity comprises a solar field and a gas turbine engine (Fig. 1.5). The solar field is made up of a solar concentrator and a solar receiver. A solar concentrator is a device that focuses incoming solar radiation onto a smaller surface (solar receiver). The receiver absorbs and converts part of the concentrated solar radiation to heat. A working fluid is circulated through the receiver to evacuate the heat and distribute it to other components of the SGT system. For direct heat addition, pressurized solar receivers can be used because they operate at high pressure and temperature (Montes et al. 2023). Non-pressurized solar receiver can also be used to indirectly heat the working fluid. The use of s-CO_2 is promising to significantly improve the thermal efficiency of the closed cycle SGT systems. More details about basic and advanced SGT cycles are reported in Chap. 5.

Conventional gas turbines are designed to be driven by conventional fuels, and, therefore, they need modification before solar energy can be integrated on the inlet side of the turbine. Some aspects of the conventional gas turbine that need consideration are (Stein and Buck 2017): (a) flow path of the working fluid: extraction of compressed working fluid, and reintroduction of the preheated working fluid, (b) connection configuration of combustor and solar heater: whether the combustor and solar heater should be connected in parallel or serial configuration, (c) optimization for solar operation: results in reduction of turbine inlet temperature, (d) adapted safety and emergency measures, and (e) adapted control system.

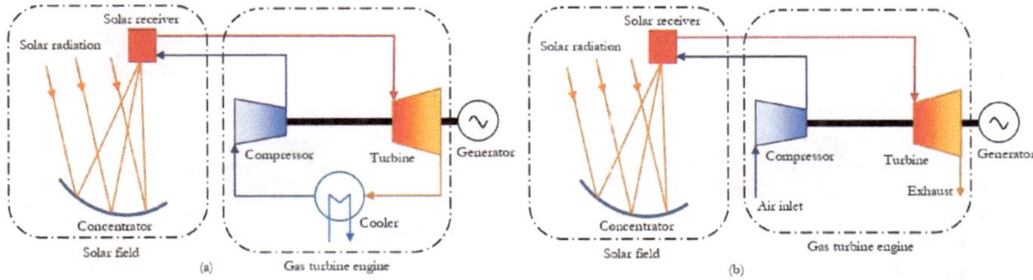

Fig. 1.5 Basic solar gas turbine: **a** closed cycle, **b** open cycle

Solar radiation can be the only source of thermal energy, or it can be hybridized with a fuel (Fig. 1.6). In a solar-only gas turbine system, hot air from the solar receiver is allowed to expand through the turbine, thereby driving the gas turbine engine. The hot fluid from the solar receiver can further be heated by an external heater or a combustor in the hybrid system. Another important strategy for enhancing the capacity factor of SGT systems is thermal energy storage. Both hybridization and thermal energy storage are discussed in more detail in Chap. 6. Solar energy can also be added to the working fluid via steam injection into the combustion chamber (Fig. 1.7). In this configuration, solar radiation generates steam which is then injected into the combustion chamber for the augmentation of output power and for the abatement of NO_x (Poullikkas 2005). Nonetheless, gas turbines are designed to take steam up to a certain limit (5–10% of compressor flow rate) (Selwynraj et al. 2015), which shows that steam injection can only contribute a small fraction of the solar thermal energy to the working fluid. It should be noted that there are many different configurations of SGTs as shown in Chap. 6.

1.3 Concentrating Solar Power

Developments in solar concentrating components have led to the evolution of concentrating solar power (CSP) technology that generates electricity by using thermal energy from solar radiation. In this technology, solar radiation coming from the sun is reflected by a large area (A_1) of mirrors onto the small area (receiver with area A_2) where it is converted to heat, which is then collected by a working fluid and often employed to drive a gas or steam engine cycle for electricity generation.

The amount of solar radiation intercepted by the receiver increases with the concentration ratio (A_1/A_2). The concentration of solar radiation on a small area enables the achievement of high temperatures of the working fluid, thereby rendering the CSP technology to be thermodynamically comparable with conventional power plants. The highest possible values of the concentration ratio are 45,000 and 212 for point (circular) and linear concentrator, respectively (Duffie and Beckman 2013). Clearly, point concentration has greater capability of achieving high thermodynamic performance. According to a ranking of different renewable energy technologies

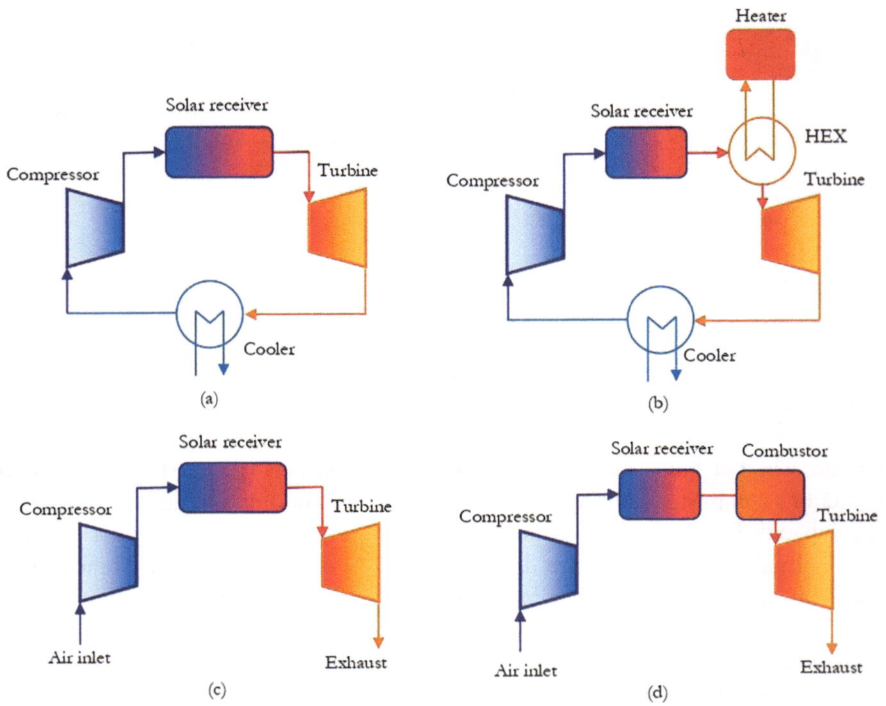

Fig. 1.6 Schematic representation of a solar gas turbine: **a** solar-only closed cycle, **b** hybrid closed cycle, **c** solar-only open cycle, and **d** hybrid open cycle

Fig. 1.7 Addition of solar thermal energy via steam injection

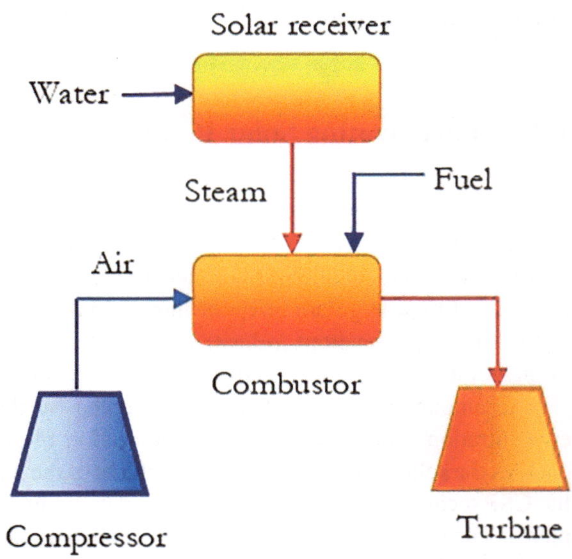

reported by Jacobson and Delucchi (2011), CSP technology exhibits the second highest potential (after wind technology) for power generation worldwide. Based on the type of solar concentrator, CSP technologies are classified into four main types: parabolic trough concentrator (PTC), linear Fresnel reflector (LFR), solar tower (ST), and parabolic dish concentrator (PDC) (Alami et al. 2023). All these technologies are imaging, but the LFR and PTC are linear concentrators while the PDC and ST concentrate solar radiation at a point.

1.3.1 Parabolic Trough Concentrator

The parabolic trough concentrator is one of the most widely used CSP technologies. It employs long, curved mirrors to focus sunlight onto a tubular solar receiver positioned along the focal line of the trough (Fig. 1.8). By mid-2015, there were 63 operational solar thermal electric plants utilizing parabolic trough technology globally, with 61 of them using oil as the heat transfer fluid (Blanco and Santigos 2017). This widespread usage of oil indicates the high maturity of oil-based PTC technology. Thermal oils offer several advantages: they are cost-effective, have low vapour pressure, exhibit good thermal stability, and provide long lifespans when used within specified limits. For instance, if a manufacturer recommends a maximum operating temperature of 573 K, the oil should be used below this threshold to maximize its longevity. However, due to the inherent temperature limitations of thermal oils, researchers are exploring alternative heat transfer fluids. Three notable alternatives gaining attention are water/steam for direct steam generation, pressurized gases such as CO_2, N_2, or air, and molten salts. While PTCs are relatively simple and cost-effective to produce, they have a limited concentration ratio ranging from 15 to 45. As a result, they typically generate low to medium temperatures of 293–673 K (Zhang et al. 2013), which is sufficient for driving steam turbines in most CSP plants. Nevertheless, the most widely deployed technology is currently the PTC (with about 72% of the total global installed capacity of 7889 MW), followed by the ST (23%) and LFR (5%) (Alami et al., 2023). This trend may change in future due to the ongoing research activities.

1.3.2 Linear Fresnel Reflector

A linear Fresnel reflector (LFR) system consists of numerous slightly curved mirror strips arranged to focus solar radiation onto a stationary tubular solar receiver positioned above the mirrors (Fig. 1.9). These mirrors are typically aligned in a trough-like configuration and can be mounted on either single-axis or dual-axis tracking systems to improve solar collection efficiency. The mirrors reflect solar radiation onto the solar receiver, which absorbs and converts the radiation to heat. A heat transfer fluid evacuates the heat from the solar receiver to other components of the CSP system.

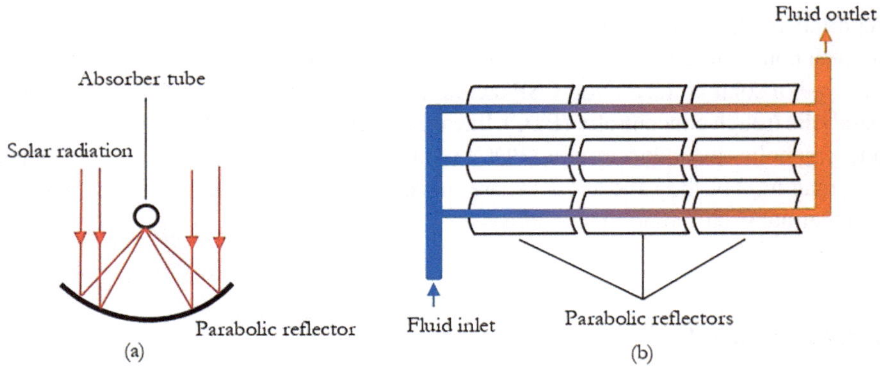

Fig. 1.8 Schematic representation of a parabolic trough concentrator **a** cross section of a single collector and **b** array of collectors

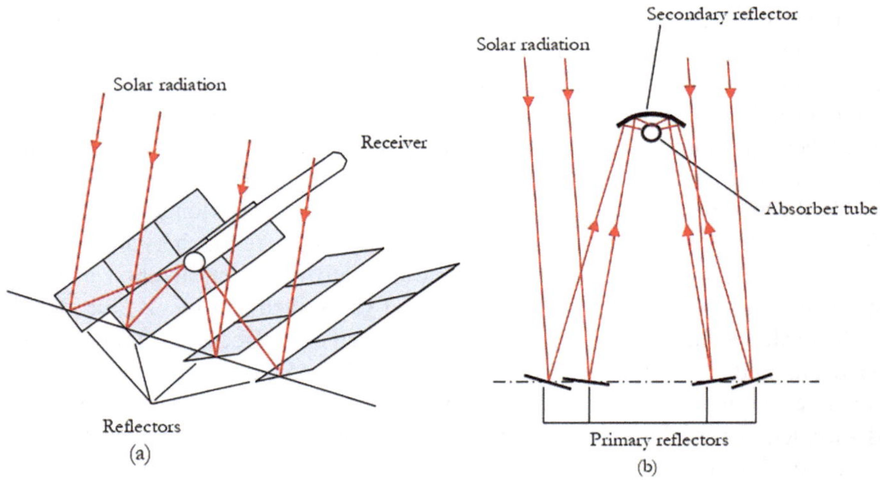

Fig. 1.9 Schematic representation of a linear Fresnel concentrator: **a** perspective view and **b** cross section of receiver cavity around the absorber tube

LFR systems require a large space of land to deploy the mirror rows necessary for capturing sufficient solar radiation. However, the use of two parallel receivers for each row of mirrors generally requires less land than the parabolic trough concentrator for the same output power (Zhang et al. 2013; Islam et al. 2018). The use of secondary mirrors can enhance performance by creating a cavity around the absorber tube (Fig. 1.9b). This configuration, often referred to as an inverted air receiver, helps suppress convective heat loss during operation, thereby improving the system's thermal efficiency. The simplicity and reduced land requirement of LFRs, combined with their ability to minimize heat losses through innovative designs like secondary mirrors, make them a viable alternative for CSP applications, particularly in locations with space constraints.

LFR technologies have concentration ratios ranging from 10 to 40, and they typically operate in a temperature range of 323–573 K (Zhang et al. 2013). These temperature levels are sufficient for driving vapour cycles but they are not high enough to support gas turbine operations that require temperatures exceeding 673 K. To integrate LFR-CSP technology with gas turbine systems, a backup heater, such as a combustor, would be necessary. In this case, the LFR system could serve as a preheater for the working fluid before it mixes with fuel in the combustor, thereby enabling higher-temperature operations and consequently improving efficiency.

Several CSP power plants utilizing LFR technology have been constructed globally. The first operational LFR plant, Puerto Errado 1 (PE 1), was commissioned in March 2009 in Germany, with a capacity of 1.4 MW. Building on the success of this initial project, Puerto Errado 2 (PE 2) was established in Spain, operating at a temperature of 543 K and generating 30 MW of power. When it began operations in August 2012, it was the largest LFR-based CSP plant in the world. Since then, additional LFR-CSP plants have been constructed, further demonstrating the viability of this technology for various applications. The LFR technology contributed about 5% of the total installed CSP capacity in 2023.

1.3.3 Solar Tower

A solar tower (ST), also known as a central receiver system, focuses solar radiation onto a central point (receiver) located at the top of a tower surrounded by a field of heliostats (Fig. 1.10). These heliostats, which can be flat or slightly curved, track the sun's position along two axes to reflect sunlight onto the stationary receiver. A heat transfer fluid, such as molten salt, circulates through the solar receiver to transfer the concentrated solar energy.

With concentration ratios ranging from 150 to 1500, the ST technology operates at temperatures between 573 and 1273 K (Zhang et al. 2013; Stein and Buck 2017). These high temperatures are ideal for driving gas turbine engines, similar to parabolic dish concentrators. One of the key advantages of the solar tower system is its scalability and ability to efficiently store thermal energy, particularly when molten salt is used as the heat transfer medium. This stored heat can be used to generate electricity even when the sun is not shining, offering a solution to the intermittency problem commonly associated with solar energy. The ST technology has seen more commercial success compared to PDCs, with numerous plants operating globally, particularly in regions like the USA, Spain, and South Africa where large-scale CSP facilities are commercially operating at present.

Commercial ST power plants have been developed worldwide, primarily utilizing the Rankine cycle for electricity generation. Although the ST technology currently represents a smaller fraction of CSP capacity compared with the PTC technology, its potential for integration with advanced power cycles, thermal energy storage,

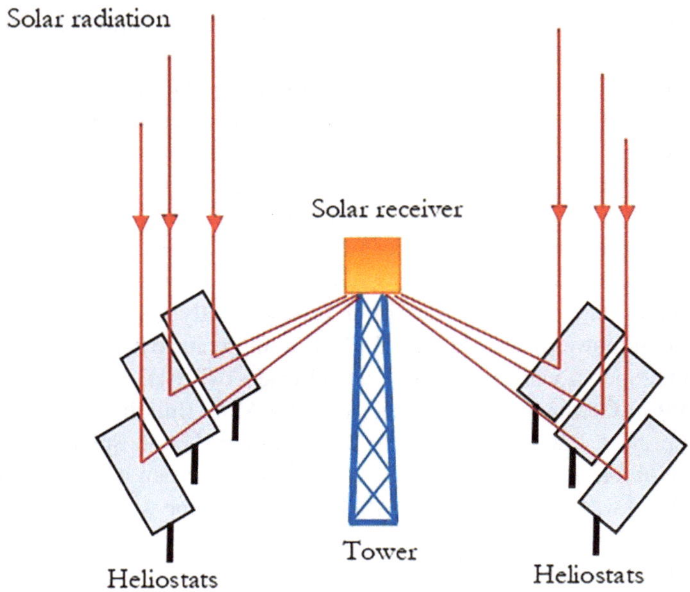

Fig. 1.10 Schematic representation of a solar tower

and hybrid systems has positioned it as a key player in the next generation (third generation) of CSP technology.

1.3.4 Parabolic Dish Concentrator

The parabolic dish concentrator (PDC) focuses solar radiation onto a single point, or receiver, located above the centre of the dish (Fig. 1.11). This technology consists of four primary components: a base support, a concave reflecting dish, a receiver unit, and a dual-axis sun-tracking system. Both the dish and receiver are mounted on the sun-tracking system, allowing them to move together and maintain optimal orientation relative to the sun.

PDC systems boast significantly higher concentration ratios, ranging from 100 to 1000, allowing them to reach operational temperatures between 393 and 1773 K (Zhang et al. 2013). These high temperatures make PDCs suitable for integration with gas turbine engines, which require elevated temperatures for efficient operation. The versatility and high-temperature capability of PDC technology make it a promising solution for generating power in high-performance applications such as solar-driven gas turbines. Furthermore, the PDC's high concentration ratio enhances efficiency, making it a valuable option for CSP plants looking to maximize power output while minimizing land use.

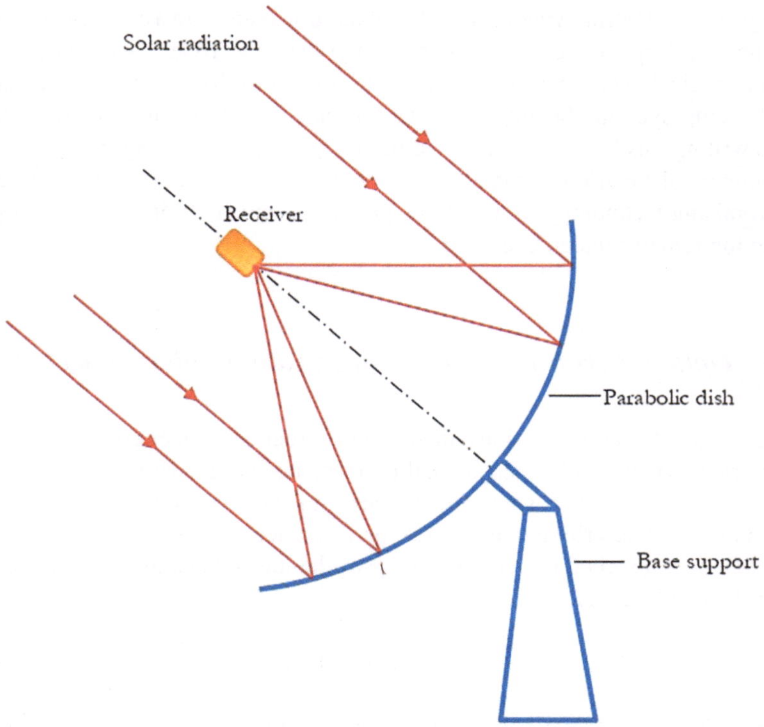

Fig. 1.11 Schematic representation of a parabolic dish concentrator

Despite its high thermodynamic efficiency, the parabolic dish concentrator remains costly and presents significant challenges in terms of hybridization and integration with thermal storage systems. One of the major drawbacks of the PDC technology is its limited power capacity; each dish operates independently with a relatively low output, necessitating the deployment of a large number of dishes to achieve utility-scale power generation. This scale-up introduces additional complications, including the issue of shading between adjacent dishes in large solar fields, which can reduce the overall efficiency.

To mitigate shading problems, a method for optimizing the layout of parabolic dishes in expansive solar fields has been proposed by Dähler et al. (2017). This technique applies principles from the two-dimensional Bravais Lattice, commonly used to study atomic arrangements in crystalline solids, to the spatial arrangement of dishes. While this approach has shown promise in reducing shading effects, the commercialization of PDC technology remains limited.

According to data from the National Renewable Energy Laboratory (2024), only two PDC projects have been initiated globally, underscoring the slow adoption of this technology. A review by Okoroigwe and Madhlopa (2016) also found that, despite its thermodynamic advantages, no PDC-driven power plants were operational at the time of their study. The first PDC project, the Maricopa Solar Project, was launched in

January 2010 in Peoria, Arizona, as a demonstration plant with a capacity of 1.5 MW, utilizing a Stirling engine cycle. The second project, the Tooele Army Depot power plant in Tooele, Utah, began producing electricity in July 2013 with a similar capacity and also employed the Stirling cycle. Unfortunately, both plants ceased operations before writing this book, reflecting the limited progress in scaling PDC technology for commercial electricity generation. These developments indicate that while PDC is a promising technology in theory, its practical application in commercial power production remains underdeveloped.

1.3.5 Optical Performance of Concentrating Solar Collectors

As seen in the preceding sections, there are different designs of concentrating solar collectors. In view of this, their optical performance varies with the geometric and surface properties of the system. Nonetheless, it is possible to use a general mathematical model in describing their optical performance. In this regard, the amount of solar radiation absorbed per unit area (G_a) of the unshielded aperture can be given by (Duffie and Beckman 2013):

$$G_a = G_{ef}\rho(\alpha\tau\psi)_n K_{\alpha\tau\gamma} \quad (1.1)$$

where G_{ef} is the effective solar radiation measured on the aperture plane (which includes only beam radiation except for concentrators with low concentration ratio), ρ is generally the specular reflectance except when the reflector is diffuse (ρ is replaced by transmittance when the concentrator is a refractor), α is absorptance of the absorber on the receiver, τ is the transmittance of the transparent cover on the receiver, and ψ the proportion of the reflected solar radiation which is incident on the absorbing surface of the receiver (values of ψ may be greater than 0.9), and $K_{\alpha\tau\gamma}$ is the incidence angle modifier (established from empirical data). All the three factors α, τ and ψ depend on the angle of incidence. Apparently, computation of α and τ may be considerably difficult due to the uncertainty that exists in establishing the angle of incidence of solar radiation from the reflector onto the cover.

1.3.6 Classification of Concentrating Solar Collectors

Concentrating solar collectors fall into imaging and non-imaging categories. In imaging collectors, parallel rays from the sun are focused on one point or line. Circular concentrators focus solar radiation on a point (small area). The parabolic dish concentrator is an example of an imaging collector. On the other hand, non-imaging collectors are unable to form the image of the sun. For instance, the compound parabolic concentrator (CPC) is a common non-imaging type of collector (Fig. 1.12). It is able to utilize beam as well as diffuse radiation and can be operated without sun-tracking.

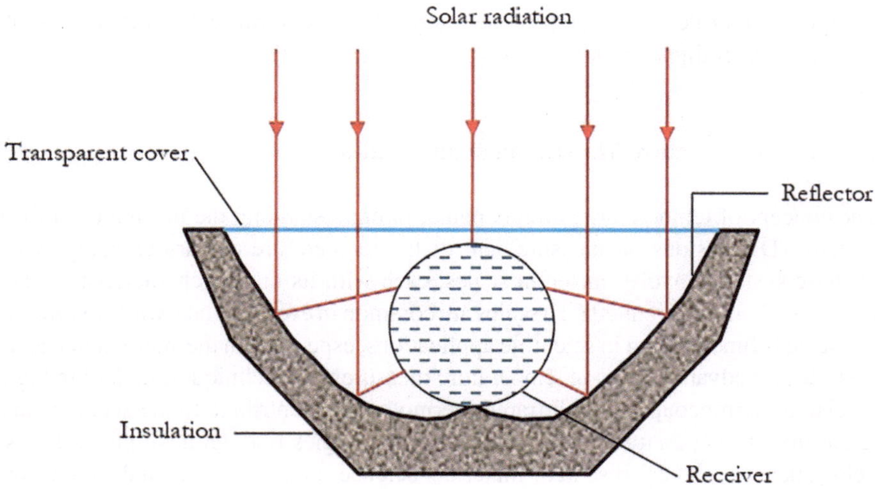

Fig. 1.12 Cross section of a non-imaging collector with a reflector type of concentrator. The reflector has two sections of a parabola

1.4 Thermodynamics of Solar Gas Turbines

Thermodynamics holds a pivotal position in the development and understanding of heat engines, including solar gas turbines. This scientific discipline involves the study of energy, its transformation from one form to another, and its correlation with the various states of matter (American Society of Heating, Refrigerating and Air-Conditioning Engineers 2001). In the realm of heat engines, thermodynamics provides the foundational principles that govern their design and operation. In this context, the study of thermodynamics is framed by four fundamental laws. These laws, as elucidated by researchers such as Ozdil et al. (2015) and Szőllősi and Márkus (2015), are the: Zeroth Law of Thermodynamics, First Law of Thermodynamics, Second Law of Thermodynamics and Third Law of Thermodynamics.

1.4.1 Zeroth Law of Thermodynamics: The Foundation of Temperature and Thermal Equilibrium

The Zeroth Law of Thermodynamics lays the groundwork for the concept of temperature and thermal equilibrium. It asserts that if two thermodynamic systems are each in thermal equilibrium with a third system, they are in thermal equilibrium with each other. This seemingly simple principle holds profound implications, serving as the basis for the definition and measurement of temperature. The Zeroth Law, though one of the foundational principles of thermodynamics, continues to influence and shape contemporary research and technological advancements across a wide spectrum of disciplines. Its role in temperature measurement and control remains indispensable in modern scientific and industrial pursuits.

1.4.1.1 Temperature Measurement and Scales

The concept of temperature is fundamental in understanding the behaviour of SGT systems. Discussions on measurement of the temperature of various components of these systems involve different scales, each with its unique characteristics. The Celsius (°C) and Fahrenheit (°F) scales are common in everyday use, while the Kelvin (K) scale is fundamental in scientific applications, especially in thermodynamic analysis. Recent advancements in temperature measurement techniques, including high-precision thermocouples and infrared thermometers, contribute to the accuracy and reliability of temperature readings. These technologies find applications in diverse fields such as medical research, materials science, and environmental monitoring (Machin et al. 2013; Rogalski 2011).

1.4.1.2 Practical Implications and Applications

The Zeroth Law has practical implications in the design and operation of SGT systems where temperature control is critical. For instance, the thermal efficiency of the SGT increases with turbine inlet temperature (TIT) that is directly correlated with the outlet temperature of the turbine. Harnessing concentrated solar energy for high-temperature applications is attainable but the main challenge is developing solar receivers that can efficiently perform this thermal conversion (Rednic et al. 2024). The gas turbine cycle requires heat at high temperature, and the third-generation CSP technology is expected to be based on this cycle. In view of this, the development of the next-generation CSP systems is aiming at designing high-temperature solar receivers with outlet temperatures above 973 K (He et al. 2020). Moreover, materials exploited in the SGT and other thermal systems have maximum service temperature. The Zeroth Law provides the foundation for developing thermal systems that operate effectively at different temperatures.

1.4.2 First Law of Thermodynamics (Conservation of Energy)

The First Law of Thermodynamics, often referred to as the Law of Conservation of Energy, is a fundamental principle that states that energy cannot be created or destroyed; it can only change forms. In the context of heat engines, this law emphasizes the conservation of energy, accounting for the various ways in which energy is transformed within a system. The First Law of Thermodynamics continues to be a cornerstone in the development of SGT systems, guiding researchers towards more efficient and sustainable solutions. Its principles are embedded in the ongoing efforts to optimize energy conversion by these systems.

1.4.2.1 Energy Transformation and Efficiency

Modern discussions around the First Law delve into the intricacies of energy transformation processes and the quest for improved efficiency. Researchers in SGT systems continually seek innovative ways to convert solar radiation to electricity with diminished losses. This is particularly relevant in the design of energy-efficient SGT systems for electricity generation. For example, the design power cycle efficiencies of the first and second generations of the CSP technology are respectively 28–38% and 38–44% while exploitation of gas turbine in the third generation of CSP is expected to yield cycle efficiencies that exceed 50% (Khan et al. 2022). Similarly, the annual solar-to-electricity efficiencies of the first and second generations of the CSP system are respectively 9–16% and 10–25% while exploitation of gas turbine in the third generation of CSP is expected to yield corresponding efficiencies that fall in the range of 25–30%. These improvements in system efficiency have been achieved through the application of the First Law of Thermodynamics.

1.4.2.2 Sustainable Energy Systems

In the context of sustainable energy, the First Law guides the development of efficient technologies for production and consumption of energy. Efficient conversion of solar energy into electricity by SGTs exemplifies the application of the First Law of Thermodynamics. Some of the important benefits of efficient production of energy are reduction in resource use and cost of energy. Power plants require different resources (including minerals, energy, water, and land). Exploitation of these natural resources entails environmental degradation (such as pollution). Moreover, some of them occur in limited quantities and so they can be depleted faster if they are inefficiently exploited. Fast depletion of energy resources can significantly impact energy security and economic progress (Pang et al. 2024). Thus, reduction in resource utilization does not only contribute to pollution reduction but also to the promotion of natural resource security. Similarly, the aim of energy efficiency is to produce more goods and services while utilizing fewer resources and minimizing the levels of waste and environmental pollution (Muthu 2020).

1.4.3 Second Law of Thermodynamics: Entropy and Irreversibility in Energy Transformations

The Second Law of Thermodynamics introduces the concept of entropy, a measure of the disorder or randomness in a system. It emphasizes that in any energy transfer or transformation, if no energy enters or leaves the system, the potential energy of the new state will always be less than that of the initial state. This law is concerned with the quality of energy, and the maximum useful work in a process is called exergy or irreversibility (Szőllősi and Márkus 2015).

The concept of exergy is especially relevant in the design of SGT systems and other devices where minimizing energy losses is a key consideration. The SGT system converts solar radiation to heat, heat to mechanical power and then mechanical power to electricity. During the conversion processes, some energy is lost by different components. It is therefore important to analyse the exergy of SGT systems. Exergy analysis aims at finding the amount of irreversibility that occurs in each component of a thermodynamic cycle and aims at effective parameters (Behbahaninia et al. 2017). This kind of analysis helps in: (a) establishing the type, location, and magnitude of energy losses, and (b) finding ways of reducing losses in order to maximize the thermal performance of the system. Compared with conventional gas turbines, the SGT system exhibits higher exergy loss (Rajaei et al. 2016). This is attributed to the variation in exergy losses in various components of the two systems. Most of the exergy destruction occurs in the solar section of the SGT system, especially in the solar receiver because of the low conversion of solar radiation to thermal energy that ultimately gets transferred to the working fluid (Giostri 2017). Consequently, solar receiver design is now aiming at developing receivers with a combination of high efficiency and outlet temperature.

1.4.4 Third Law of Thermodynamics: Exploring Systems at Absolute Zero

The Third Law of Thermodynamics is concerned with the behaviour of systems as they approach absolute zero temperature. It posits that as the temperature of a system approaches absolute zero, the entropy of the system approaches a minimum value. This law is particularly relevant in understanding the properties of matter at extremely low temperatures and provides insights into the behaviour of systems in the limit of zero thermal energy. The Third Law of Thermodynamics continues to shape our understanding of the behaviour of matter at extreme temperatures. From the exploration of quantum systems to the development of quantum technologies, the principles of the Third Law play a crucial role in advancing our knowledge of the fundamental nature of materials and energy states. SGTs use different materials ranging from structural to heat transfer fluids. In general, characteristics of materials are affected by thermal changes.

Recent discussions around the Third Law have focused on the exploration of quantum systems and the challenges associated with reaching temperatures close to absolute zero. Achieving ultra-low temperatures allows researchers to observe unique quantum phenomena and study the behaviour of matter in extreme conditions. Advancements in cryogenics and cooling technologies have enabled the creation of environments with temperatures close to absolute zero. Intercooling the working fluid after compression is one of the strategies for increasing the efficiency of the SGT system (see Chap. 6). The cooling systems may be chillers that operate based on the principles of cryogenics.

1.5 Heat Transfer in Solar Gas Turbines

Heat transfer, defined as the movement of thermal energy from regions of higher temperature to those of lower temperature, constitutes a fundamental aspect of energy systems. In the context of SGT systems, a thorough understanding of heat transfer mechanisms is essential for the efficient design, construction, testing, and operation of these systems. Heat transfer influences the efficiency and overall performance of gas turbines. As SGTs continue to evolve, a comprehensive understanding of heat transfer mechanisms, coupled with advancements in materials and design strategies, will play a key role in shaping the future of solar energy conversion technologies. Heat can be transferred through conduction, convection and radiation modes.

1.5.1 Heat Conduction

Heat conduction, a fundamental process in the transfer of energy within a medium, involves the exchange of kinetic energy among particles or groups of particles (American Society of Heating, Refrigerating and Air-Conditioning Engineers 2001). An illustration of this mode of heat transfer in one dimension is shown in (Fig. 1.13). Heat conduction occurs across the gaseous, liquid, and solid phases of a substance, dictating the flow of energy in the direction of decreasing temperature. Whether temperatures vary transiently or remain constant in a steady state, the principles governing heat conduction are elucidated through a mathematical lens.

In the realm of transient heat conduction in three dimensions, the renowned Fourier's theory (Gruber and Lesne 2005) provides a comprehensive understanding. The governing equation is expressed as:

$$\nabla^2 T = \frac{1}{\acute{\alpha}}\frac{\partial T}{\partial t} - \frac{q}{k} \tag{1.2}$$

where k denotes the material's thermal conductivity, q represents the rate of volumetric heat addition by thermal conduction per unit mass,. T is the temperature at a given point, t is time, ∇^2 represents the Laplacian operator, $\acute{\alpha} = k/(\phi C_p)$ is

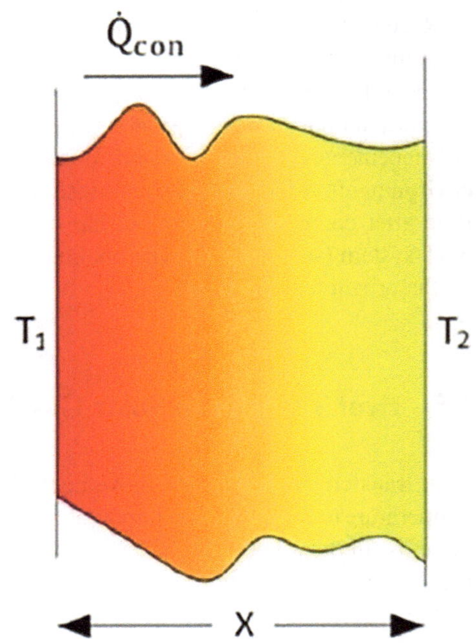

Fig. 1.13 Illustration of one-dimensional heat conduction ($\dot{Q}_{\text{con}}$) across a slab of thickness x, with two surfaces at temperatures T_1 and T_2, where $T_1 > T_2$

the thermal diffusivity, ϕ is density and C_p is the specific heat capacity at constant pressure.

In Cartesian coordinates, this equation takes the form:

$$\frac{\partial^2 T}{\partial x^2} + \frac{\partial^2 T}{\partial y^2} + \frac{\partial^2 T}{\partial z^2} = \frac{1}{\acute{\alpha}}\frac{\partial T}{\partial t} - \frac{q}{k} \tag{1.3}$$

This equation adapts to cylindrical or spherical coordinates, aligning with the geometry of the conductor under scrutiny. Further simplifying for one-dimensional scenarios, Eq. (1.3) transforms into:

$$\frac{\partial^2 T}{\partial x^2} = \frac{1}{\acute{\alpha}}\frac{\partial T}{\partial t} - \frac{q}{k} \tag{1.4}$$

Diving into the derivations leading to these equations involves intricate considerations of thermal conductive processes in diverse dimensions. Fourier's theory, a cornerstone in heat conduction studies, provides a robust foundation for comprehending the dynamic interplay of temperature, time, and heat transfer rates. These equations stand as mathematical expressions capturing the essence of heat conduction phenomena and serve as invaluable tools in unravelling the complexities of energy transfer in solar gas turbines.

The heat flux in three dimensions can be given by:

$$\vec{q}_{\text{con}} = -k\nabla T \tag{1.5}$$

where ∇T is the temperature gradient (a vector quantity).

In one dimension, conductive heat flux ($\dot{q}_{\mathrm{con}}$) can be calculated from:

$$\dot{q}_{\mathrm{con}} = -k\frac{\mathrm{d}T}{\mathrm{d}x} \tag{1.6}$$

It is noted from Eq. (1.6) that the rate of heat transfer increases with the coefficient of heat conduction (k) and temperature gradient. Thus, materials with relatively high values of k are suitable for the fabrication of solar absorbers and heat exchangers (that must exchange heat through conduction) while those with low values of k are appropriate for insulation to reduce heat loss from a given system to the environment. In addition, k may vary with the direction of heat flow and temperature of the conductor. The rate of heat conduction $\left(\dot{Q}_{\mathrm{con}}\right)$ across the slab without heat source, shown in (Fig. 1.13), can be given by (American Society of Heating, Refrigerating and Air-Conditioning Engineers 2001):

$$\dot{Q}_{\mathrm{con}} = \frac{kA(\Delta T)}{x} \tag{1.7}$$

$$\Delta T = T_1 - T_2 \tag{1.8}$$

1.5.2 Convection

Heat transfer through the movement of fluids characterizes convection. In a solar thermal system, convection is evident as heated fluid, whether air or a heat transfer fluid, carries thermal energy from the absorber to other components within the system. Heat convection, a fundamental aspect of energy transfer in fluids, manifests through natural and forced mechanisms, shaping the movement and exchange of thermal energy. In the realm of natural convection, fluid motion results from temperature-induced density gradients, creating a flow driven by temperature differences. Conversely, forced convection takes the stage when a fluid, set in motion by external forces such as fans or pumps, absorbs and carries away heat (Fig. 1.14). At the fluid–solid boundary, heat finds its path through conduction, transferring from hot surfaces to cooler fluids or vice versa.

The rate of heat transfer in natural convection $\dot{q}_{\mathrm{c}}$ can be expressed as:

$$\dot{q}_{\mathrm{c}} = h_{\mathrm{c}}(T_2 - T_1) \tag{1.9}$$

where $h_{\mathrm{c}} = \frac{\mathrm{Nu}\cdot k}{S}$, Nu is the Nusselt number, k is the thermal conductivity, and S represents the characteristic length or equivalent spacing. The Nusselt number, in turn, relies on the Grashof number (Gr) and Prandtl number (Pr) that are determined by fluid properties, temperature differentials, and surface geometry. The expressions

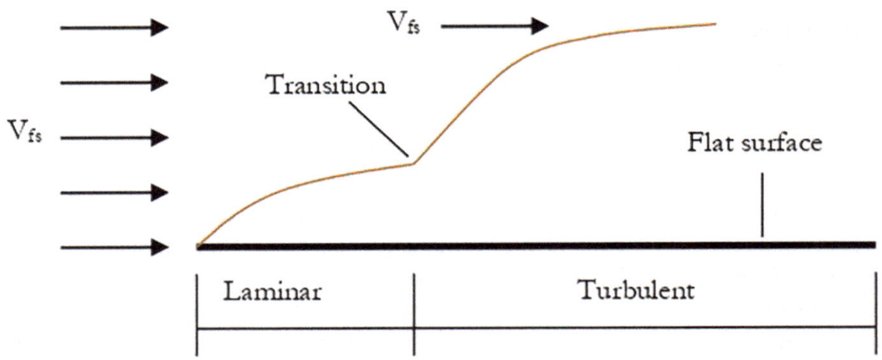

Fig. 1.14 Formation of laminar and turbulent boundary layers when a fluid flows over a flat surface. The velocity of the fluid increases with the distance from the surface and the edge. V_{fs} is free stream flow velocity

for Gr and Pr are given by:

$$\mathrm{Pr} = \frac{\mu C_p}{k} \tag{1.10}$$

where μ is the dynamic viscosity.

$$\mathrm{Gr} = \frac{g S^3 \beta \phi^2 \Delta T}{\mu^2} \tag{1.11}$$

where g is acceleration due to gravity, S is the specific length, and ΔT is the temperature difference.

The product (Gr · Pr) influences the convective heat transfer coefficient h_c, reflecting the intricate interplay of fluid characteristics and temperature dynamics.

Natural convection is categorized into turbulent, laminar, and transitional regions, each governed by the dimensionless parameter d. The coefficient of natural convection h_c in natural convection is calculated using the equation:

$$h_c = b\frac{k}{S}\left(\frac{S^3 \beta \phi^2 \Delta T}{\mu^2}\right)^d \tag{1.12}$$

In forced convection, where fluid flow is influenced by an external force, h_c is determined by:

$$h_c = b\frac{k}{S}\left(\frac{S\dot{m}'\mu}{k}\right)^e\left(\frac{\mu C_p}{k}\right)^d \tag{1.13}$$

This equation reveals the independence of h_c from cavity geometry when $e = 1$, emphasizing the influence of mass flux on forced convection.

For flow through a tube of diameter (D), the mean velocity (u) of the fluid through the tube can be calculated from the mass flow rate ($\dot{m}$) and density (ϕ) of the fluid:

$$u = \frac{\dot{m}}{\phi\left(\pi \frac{D^2}{4}\right)} \quad (1.14)$$

The fluid flow conditions, described by mean velocity u and Reynolds number Re, transition from laminar to turbulent. For instance, for circular tubes, laminar flow occurs when Re < 2300, calculated using $\vartheta =$ friction factor:

$$\mathrm{Re} = \frac{64}{\vartheta} \quad (1.15)$$

In turbulent flow, Re is determined by:

$$\mathrm{Re} = \frac{4\dot{m}'}{\pi \mu D} \quad (1.16)$$

The Nusselt number Nu for turbulent flow inside smooth pipes is expressed by:

$$\mathrm{Nu} = h\frac{D}{k} \quad (1.17)$$

A vital consideration is convective heat loss from components of an SGT system, such as the solar receiver. The rate of convective heat loss ($\dot{Q}_c$) from the central receiver is given by:

$$\dot{Q}_{\mathrm{c}} = A_{\mathrm{rc}} h_{\mathrm{c}} (T_{\mathrm{rc}} - T_{\mathrm{a}}) \quad (1.18)$$

For a solar tower, the coefficient of convective heat loss h_{c} can be computed using the Bejan correlation:

$$h_{\mathrm{c}} = 5.57 \times 10^{-7} \left(\frac{T_{\mathrm{rc}} - T_{\mathrm{a}}}{H_{\mathrm{to}}}\right)^{0.25} \quad (1.19)$$

where H_{to} is the tower height.

1.5.3 Radiation

Heat transfer occurring through electromagnetic waves is termed radiation. In SGT systems, components like absorbers and reflectors absorb and emit radiation. Briefly discussed in this section are heat radiation, emissive power, real surfaces, and radiative heat transfer between surfaces.

Heat radiation does not require a medium for propagation to take place, unlike heat conduction and convection. Consequently, the use of a vacuum to reduce heat loss only eliminates convective and conductive heat transfers. In fact, the presence of a medium between a radiator and receiver provides impedance to radiative heat transfer. The amount of energy emitted by a radiator depends on the nature of the material, microscopic structure, and temperature of the radiator and its surroundings.

A blackbody, for instance, absorbs all radiation incident on it. Its emissive power $(\dot{E}_{\text{bb}})$ to a hemispherical region above it is given by (Crepeau, 2007):

$$\dot{E}_{\text{bb}} = \sigma T^4 \tag{1.20}$$

$$\dot{E}_{\text{bb},\lambda} = \frac{c_1\lambda^{-5}}{e^{C_2/(\lambda T)}-1} \tag{1.21}$$

where T is the temperature of the blackbody, λ is the wavelength of radiation, σ is the Stefan-Boltzmann constant ($= 5.6697 \times 10^{-8}$ W/m^2·K^4), and $C_1 = 3.742 \times 10^{-16}$W m^2 and $C_2 = 0.014388$ m K are respectively the first and second Planck's constants.

A real surface absorbs part of the radiant energy which it receives. So, its emissive power to a hemispherical surface above it is given by:

$$\dot{E} = \varepsilon\sigma\left(T_i^4 - T_j^4\right) \tag{1.22}$$

$$\varepsilon\dot{E}_{\lambda} = \varepsilon\dot{E}_{\text{bb},\lambda} \tag{1.23}$$

$$\varepsilon = \frac{1}{\sigma T}\int_0^{\infty} \varepsilon_{\lambda}\dot{E}_{\text{bb},\lambda}\,\mathrm{d}\lambda \tag{1.24}$$

If ε_{λ} is independent of λ, then $\varepsilon = \varepsilon_{\lambda}$, and a surface with such a characteristic is known as a grey body. In practical calculations, surfaces are usually assumed to be grey because of the unavailability of information about the relationship between $\varepsilon = \varepsilon_{\lambda}$ and λ. It should also be mentioned that the energy from a non-black surface comprises the radiant and reflected components, and this energy may leave the surface specularly or diffusely.

To calculate the amount of radiation energy transferred between two surfaces, the following assumptions are often made (American Society of Heating, Refrigerating and Air-Conditioning Engineers·2001): (a) surfaces are grey or black, (b) radiation and reflection are diffuse, (c) $\alpha = \varepsilon$, and α does not depend on the temperature of the source of the incident radiation, and (d) surfaces are separated by a non-absorbing medium. For two given surfaces, the net radiative heat transfer $(\dot{Q}_r)$ can be given by (Hewitt et al. 1994):

$$\dot{Q}_{\text{r}} = \frac{\sigma\left(T_i^4 - T_j^4\right)}{\frac{1-\varepsilon_i}{A_i\varepsilon_i} + \frac{1-\varepsilon_j}{A_j\varepsilon_j} + \frac{1}{A_iF_{i,j}}} \tag{1.25}$$

Eq. (1.25) can also be expressed in a linear form by defining a coefficient $(h_{\text{r},i-j})$ of the radiative heat transfer:

$$\dot{Q}_{\text{r}} = A_i h_{\text{r},i-j}(T_i - T_j) \tag{1.26}$$

$$h_{r,i-j} = \frac{\sigma\left(T_i^2+T_j^2\right)(T_i+T_j)}{\frac{1-\varepsilon_i}{\varepsilon_i}+\frac{A_i\left(1-\varepsilon_j\right)}{A_j\varepsilon_j}+\frac{1}{F_{i,j}}}. \tag{1.27}$$

If the two surfaces are rectangular and parallel to each other, then $F_{i-j} \approx 1$ and $A_i = A_j$. In this case, Eq. (1.27) reduces to:

$$h_{\mathrm{r},i-j} = \sigma\varepsilon_{i,j}\left(T_i^2 + T_j^2\right)(T_i + T_j) \tag{1.28}$$

where $\varepsilon_{i,j}$ is given by:

$$\varepsilon_{i,j} = \frac{1}{\varepsilon_i} + \frac{1}{\varepsilon_j} - 1 \tag{1.29}$$

The view factor influences the exchange of radiation between two given surfaces, and it depends on the geometries of the surfaces in question. For isothermal and diffuse surfaces, the view factor F_{i-j} is defined as the proportion of energy leaving surface (i) that is incident on surface (j) (Mishra et al. 2008). Moreover, the energy leaving surface (i) may reach other surfaces surrounding it. So, using the law of conservation of energy, this yields:

$$F_{i-i} + F_{i-2} + F_{i-3} + \cdots + F_{i-j} = 1 \tag{1.30}$$

where (2), (3), (4)…(j) are surfaces that surround surface (i). If a surface views itself, then $F_{i-i} > 0$, as shown in Fig. 1.15.

A curved surface exchanges radiation with itself and other surfaces while a flat surface exchanges radiation with other surfaces only. For example, parabolic trough and parabolic dish concentrators in SGT systems are curved surfaces. The following

Fig. 1.15 **a** Curved surface exchanges radiation with itself and other surfaces. **b** Flat surface exchanges radiation with other surfaces only

relationships are also useful for the calculation of radiation exchange among surfaces (Bergman et al. 2011):

$$F_{1-(2,3)} = F_{1-2} + F_{1-3} \tag{1.31}$$

$$F_{(2,3)-1} = \frac{A_2 F_{2-1} + A_3 F_{3-1}}{A_2 + A_3} \tag{1.32}$$

$$A_1 F_{1-2} = A_2 F_{2-1} \tag{1.33}$$

It should be noted that surface (1) views a combination of surfaces (2) and (3) in Eq. (1.31) while surfaces (2) and (3) jointly view surface (1) in Eq. (1.32). Equation (1.33) expresses the reciprocity of view factors, which is particularly important for computation of radiation exchange between surfaces with finite and infinite areas. This equation is required in the calculation of radiation exchange between a solar collector (with finite area) and the sky (with infinite area). For example, the rate of radiative heat loss ($\dot{Q}_r$) from a solar tower receiver can be calculated from (Sheu and Mitsos 2013; Al-Sulaiman and Atif 2015):

$$\dot{Q}_{\mathrm{r}} = F_{\mathrm{rc}} A_{\mathrm{rc}} \varepsilon_{\mathrm{rc}} T_{\mathrm{rc}}^4 \tag{1.34}$$

where F_{rc} is the view factor of the receiver, A_{rc} radiative area of the receiver, $\varepsilon_{\mathrm{rc}}$ is the emissivity of the receiver, and T_{rc} is the temperature of the receiver.

Eq. (1.33) is part of the broader understanding of radiative heat transfer between surfaces, and the view factor plays a crucial role in determining how much of the emitted radiation from the receiver is incident on the sky. As can be seen, heat loss from the receiver increases with the emissivity. Considering the high operational temperatures of solar tower receivers, radiative heat transfer becomes a predominant mode of heat loss. Researchers often focus on developing receivers with enhanced properties (such as low emissivity or selective coatings) to minimize heat loss and improve overall efficiency. If one has specific values for F_{rc}, A_{rc}, $\varepsilon_{\mathrm{rc}}$, and T_{rc}, you can substitute these values into Eq. (1.33) to calculate the rate of radiative heat loss from the solar tower receiver. If not, these values would need to be obtained through experimentation, simulations, or detailed analysis based on the characteristics of the specific solar tower receiver in question.

1.5.4 Heat Losses from Solar Collectors

While heat transmission is crucial for energy conversion in solar collectors, it can also lead to undesired heat losses, impacting the overall thermal efficiency of the system. Efficient heat transmission (Q_{t}) in solar collectors is governed by the equation:

$$Q_t = \eta_t Q_{solar} \tag{1.35}$$

where η_t represents the transmission efficiency and Q_{solar} is the solar energy incident on the collector. The overall thermal efficiency ($\eta_{thermal}$) of the solar collector system is given by:

$$\eta_{thermal} = \frac{\dot{Q}_u}{\dot{Q}_{solar}} \tag{1.36}$$

where $\dot{Q}_u$ is the useful thermal energy output.

Efficient heat transmission is crucial for maximizing useful energy output and, in turn, optimizing the thermal efficiency. To balance heat transmission and minimize losses, engineers and scientists often consider the collector efficiency (η_{co}):

$$\eta_{co} = \frac{\dot{Q}_u}{\dot{Q}_{ap}} \tag{1.37}$$

where $\dot{Q}_u$ is the useful energy and $\dot{Q}_{ap}$ is the amount of solar energy intercepted by the collector aperture. Eq. (1.37) emphasizes the need to enhance the collector's ability to convert absorbed solar energy into useful output.

Solar receivers, operating at elevated temperatures, experience heat transfer to other components and the surrounding environment at lower temperatures. This phenomenon can result in thermal energy losses, thereby reducing the thermodynamic performance of the SGT. The heat transfer rate $\left(\dot{Q}_{transfer}\right)$ from the receiver to the surroundings can be modelled using Newton's Law of Cooling:

$$\dot{Q}_{transfer} = h_c A_{rc} (T_{rc} - T_{sur}) \tag{1.38}$$

where h_c is the convective heat transfer coefficient, A_{rc} is the receiver surface area, and T_{rc} and T_{sur} are the temperatures of the receiver and surroundings, respectively.

Thermal energy losses $\left(\dot{Q}_{loss}\right)$ due to radiation can be expressed as:

$$\dot{Q}_{loss} = \sigma \varepsilon A_{rc} \left(T_{rc}^4 - T_{sur}^4\right) \tag{1.39}$$

where ε is the emissivity of the solar receiver.

Heat is lost through convection and radiation from solar collectors. Thus, strategies for addressing heat losses in solar collectors involve minimization of heat convection and radiation from the collector system to the ambient environment. This can be achieved through optimization of various parameters of the system (see Chap. 8).

1.5.5 Significance of Heat Transfer in Solar Gas Turbines

From the solar receiver, heat is distributed to various components of the SGT system. Understanding these flows is critical for optimizing the design and operation of the system. Thermal energy (Q_{flow}) flows through the components. The thermal energy flow equation can be expressed as:

$$Q_{flow} = \sum_{k=1}^{k=n} \dot{Q}_k \tag{1.40}$$

Here, $\dot{Q}_k$ represents the heat transfer rate in each component and n is the number components within the SGT.

The first law of thermodynamics describes the balance of energy flows within the SGT system:

$$\dot{Q}_{in} + \dot{W}_{in} = \dot{Q}_{out} \tag{1.41}$$

where $\dot{Q}_{in}$ is the input heat, $\dot{Q}_{out}$ is the output heat, and $\dot{W}_{in}$ is the input work.

Eq. (1.41) states that the sum of heat input and work input must be equal to the heat output. Managing and optimizing these energy flows are critical for the overall performance of the SGT.

(a) Impact on performance

Heat transfer mechanisms directly influence the performance of key components such as the receiver, combustor, and turbine. Managing these mechanisms is vital for maintaining high thermodynamic efficiency. The heat transfer rate between components is a critical parameter. It directly affects the thermal efficiency ($\eta_{thermal}$) of an SGT system. This efficiency can be expressed as:

$$\eta_{thermal} = \frac{\dot{W}_{net}}{\dot{Q}_{in}} \tag{1.42}$$

where $\dot{W}_{net}$ is the net work and $\dot{Q}_{in}$ is the input power.

The compressor requires input work to operate, and this input work is often provided by the turbine. Thus, the net output work is also affected by the compressor efficiency. Details about computation of ideal and real cycle efficiencies are reported in Chap. 5. Understanding and applying the various equations are essential for engineers and researchers involved in the design and operation of SGTs, ensuring the effective performance of individual components and the entire system.

1.5.6 *Implications for Design and Operation*

Three main study fields are involved in SGT design: thermodynamics, aerodynamics, and mechanics. Thermodynamic design point is the first key stage in the SGT design process. Moreover, thermodynamics and heat transfer are complementary fields. Knowledge about design and operation is required in all these disciplines. Research is going on to enhance the efficiency of heat transfer in the SGT system.

1.5.6.1 Knowledge Requirement

Heat transfer in SGTs is a multifaceted process, encompassing radiation, conduction, and convection. Thus, a successful approach to the design and operation of SGTs requires in-depth knowledge about heat transfer mechanisms. Experts involved in SGT development must consider the implications of heat transfer on the system performance. It is imperative for these experts to possess in-depth knowledge about the intricate processes governing heat transfer within the system. SGT designers must be well-versed in the fundamental principles of thermodynamics, aerodynamics, and fluid mechanics to navigate the complexities of energy conversion within the system.

Knowledge about heat transfer is crucial for integrating various components seamlessly. For instance, understanding how heat is transferred from the receiver to the combustor and subsequently to the turbine allows for the effective coordination of these components in the overall energy conversion process. An in-depth knowledge of heat transfer mechanisms is vital for effective thermal management and control. System designers must be able to predict, analyse, and optimize heat flows to prevent overheating, material degradation, and thermal losses, ultimately ensuring the longevity and reliability of the SGT.

1.5.6.2 Efficiency Enhancement

Ongoing research focuses on enhancing the efficiency of heat transfer in SGTs. This involves advances in materials, coatings, and overall system design to minimize losses and maximize energy conversion. Continuous multidisciplinary efforts are underway to enhance the efficiency of heat transfer in SGTs. Research is exploring novel materials with improved thermal conductivity and resistance to high temperatures. For instance, the use of metal liquids as heat transfer fluids significantly improves the heat transfer characteristics of solar receivers (Abdelsalam et al. 2023). These materials aim to minimize optical and thermal losses during energy conversion processes, contributing to higher overall system efficiency.

Advances in coatings play a crucial role in enhancing heat resistance and overall component performance. Thermal barrier coatings, for example, are designed to mitigate heat transfer to external components and the environment, reducing thermal losses (Narita et al. 2023). Efficiency enhancement extends to the optimization of the overall SGT design. This includes the geometric arrangement of components, fluid dynamics considerations, and the incorporation of innovative technologies to maximize energy conversion efficiency. System designers adopt a holistic approach, considering the interplay of various factors influencing heat transfer. Computational modelling, simulations, and experimental studies contribute to a comprehensive understanding, enabling targeted interventions for efficiency enhancement.

The implications for design and operation underscore the critical role of knowledge and ongoing research in shaping the efficiency and effectiveness of SGTs. System designers must stay abreast of advancements in the field to contribute to the development of sustainable and high-performance energy systems.

1.6 Heat Exchangers in Solar Gas Turbines

A heat exchanger facilitates the transfer of heat between hot and cold media at different temperatures. It is a crucial component in thermal systems and consists of two main elements: heat transfer and fluid distribution components. The heat transfer core, often referred to as the matrix, provides the surface for heat exchange, while fluid distribution components include headers, manifolds, tanks, and pipes.

Generally, heat exchangers are deployed in cooling and heating processes within an SGT system (see Chap. 5). For instance, intercoolers are essentially heat exchangers exploited on the cold side of the turbine to increase the mass flow rate and the efficiency of a SGT. In a closed cycle solar gas turbine, a heat exchanger is used on the exit side of the turbine to extract excess heat from the exhaust and reject it to the ambient environment, thereby cooling the working fluid before it re-enters the compressor to start another cycle.

1.6.1 Classification of Heat Exchangers

Heat exchangers are classified based on five main criteria, as outlined by Kakaç and Liu (1998): recuperators and regenerators, heat transfer processes, geometry of construction, heat transfer mechanisms, and flow arrangements.

1.6.1.1 Recuperators and Regenerators

Recuperators have a solid surface separating hot and cold fluids, allowing continuous heat conduction while preventing fluid mixing. In regenerators, heat exchange occurs intermittently through the storage and release of thermal energy. Hot and cold fluids flow through the same passage alternately. A packed bed for thermal energy storage is an example of regenerators. During the charging process, a hot fluid (say, from the solar receiver) is circulated through the storage medium. The stored heat is extracted by passing a cold fluid during the discharging process.

1.6.1.2 Transfer Processes

Heat transfer processes can be direct or indirect. Heat is directly transferred between cold and hot fluids, with no solid wall between them. This includes interfaces between immiscible liquids, gas–liquid pairs, or solid particle–fluid pairs. Similar to recuperators, indirect heat exchangers involve a solid wall separating hot and cold fluids.

1.6.1.3 Geometry of Construction

Common geometries of construction are tubular, plate, and extended surface. Tubular heat exchangers are made of circular tubes with one fluid flowing inside and the other outside the tube (shell-and-tube). Specific configurations of the shell-and-tube heat exchanger depend on the number of shell-and-tube passes (Bergman et al. 2011). The simplest configuration involves single tube and shell passes. Plate heat exchanges are constructed from thin plates forming channels that separate hot and cold streams. In extended-surface heat exchangers, fins or appendages are included on the primary heat transfer surface to enhance the heat transfer area.

1.6.1.4 Heat Transfer Mechanisms

Three major heat transfer mechanisms are single-phase convection, double-phase convection, and single-phase and double-phase convection. Single-phase convection involves single-phase fluid flow on both sides of the exchanger, applicable in compressor intercoolers, oil coolers, etc. There is boiling on one side and condensing on the other side in double-phase convection. One side experiences single-phase convection, while the other side involves double-phase convection in single-phase and double-phase convection. This kind of heat transfer mechanism is found in condensers and boilers in power plants.

1.6.1.5 Flow Arrangements

In an SGT system, fluids flow in specific circuits. Configurations of the fluid paths fall into: parallel, counter, and cross-flows. Hot and cold fluids enter the exchanger on one end, flow in the same direction, and exit together on the other end in a parallel-flow arrangement. In a counter flow, hot and cold fluids flow in opposite directions. Counter-flow heat exchangers are the most widely exploited heat exchangers because of their higher performance compared with the other two types (Ting 2022). Hot and cold fluid paths are mutually perpendicular in the cross-flow arrangement.

1.7 Other Applications of Solar Gas Turbines

As of now, the major application of solar gas turbines is the generation of electricity. Nevertheless, this technology can also be exploited in: (a) combined power and desalination and (b) cogeneration. These developments can raise the competitiveness of the SGT technology on the market.

1.7.1 Combined Power and Desalination

Water is an important resource for economic and social development, and a large proportion of the earth surface is covered by seawater which is salty. However, there is limited access to fresh water because most of the available water does not meet the standards for consumption. One way of improving the quality of water is through desalination, which is one of the conventional methods of treating water. Desalination is gaining popularity as a critical method for providing fresh water globally but most of the desalination technologies are energy-intensive and fuelled by fossil fuels. It is possible to produce both power and water optimally by exploiting a gas turbine integrated with a desalination subsystem (Eveloy et al. 2017).

Desalination technologies can generally be classified into thermal and membrane categories. Thermal-based desalination processes involve a phase change, in which the feed water is heated to produce water vapour that is condensed into freshwater (or the feed water is converted to ice which is then melted to obtain fresh water). Thermal techniques of desalination include multi-stage flash (MSF), multi-effect distillation (MED), and vapour compression distillation (VCD). Membrane-based techniques use a membrane through which water diffuses with a high proportion of the salts being retained, and this category comprises reverse osmosis (RO), nanofiltration (NF), and electrodialysis (ED).

The temperature of the exhaust gas from a gas turbine is generally high enough to supply heat for thermal desalination. For instance, Elminshawy et al. (2015) reported an exhaust temperature of 817 K for a Siemens industrial gas turbine (model SGT-800). The exhaust gas was circulated through a heat recovery steam generator (HRSG) in a combined cycle configuration to reduce the temperature to 473 K which is more suitable for desalination. Thus, a thermal desalination system can be coupled to the exit side of a solar gas turbine (Fig. 1.16). González-Almenara et al. (2024) developed an innovative RO desalination unit driven by the exhaust gases (at 523–573 K) of a solar micro gas turbine for off-grid application. The brine from the RO unit had high salt concentration and so it underwent further partial treatment in a zero liquid discharge unit to dehydrate and concentrate the effluent by using the

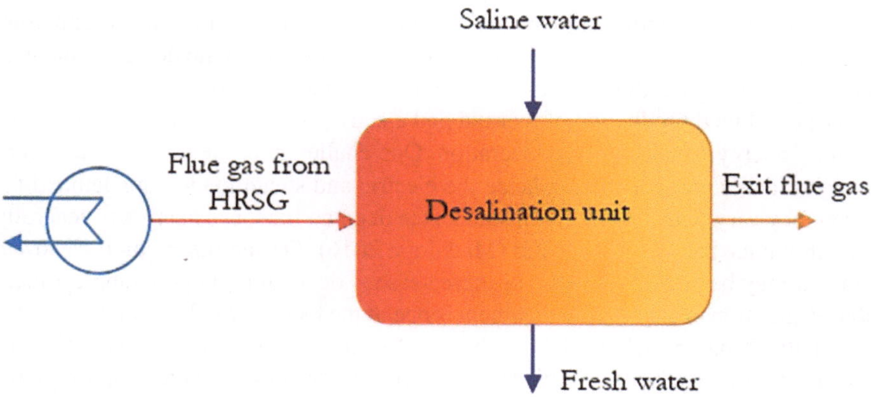

Fig. 1.16 Recovery of heat from flue gas (that has passed through a heat recovery steam generator (HRSG) or any other heat exchanger) by a thermal desalination system

exhaust gases. Results showed that the solar-to-electric efficiency increased by 5% at low loads. Zero liquid discharge was experimentally achieved by fully saturating the concentrate from the RO unit. Heat can also be recuperated from the exhaust gas (see Chap. 5).

The minimum permissible temperature for recovery of waste heat is influenced by the dew point of the flue gas. In turn, the dew point of the flue gas is affected by its chemical composition. For example, the combustion flue gas contains SO_x and NO_x that may react with water (H_2O) to form sulphuric acid (H_2SO_4) and nitric acid (HNO_3), respectively. The temperature at which the acid gas in flue gas starts to condense is known as acid due point (Zuo et al. 2020). This implies that when the flue gas is cooled to a temperature below or equal to the acid dew point, formation of liquid acid starts. These liquid acids enhance corrosion of the surfaces of the passage of the flue gas. Approximately, the dew point of sulphuric acid of flue gases from the combustion of fuels in thermal power plants falls in the range of 393–423 K. This means that the design of a desalination device for recovery of heat from flue gas needs to ensure that the temperature of the flue gas at the exit from the desalination system is above the acid dew point. In contrast, SGTs have low to zero emissions from the exhaust gas. For instance, the solar-only gas turbine does not emit pollutants associated with the combustion processes. However, acid dew point needs to be taken into account in flue gas from hybrid solar gas turbines, especially when fuels with a high concentration of the sulphur element are used.

1.7.2 Cogeneration

A gas turbine gives out a high proportion of waste heat through the exhaust gases. The waste heat can be recovered to generate steam for different purposes such as heat source for absorption chillers, and heating and process requirements. End users of the steam may have different requirements of steam, and so the design of the heat recovery steam generator should be tailored to specific needs of the user.

The combined production of thermal and electric power from the same source of primary energy is known as cogeneration. One challenge in the design of a cogeneration plant is about how to balance the electric and steam loads. The demand for steam may vary with season. For instance, the demand for space heating is generally high in winter but low in summer (Madhlopa 2016). On the other hand, electrical demand may be at peak in either of these seasons depending on geographical location. Ventilation in the built environment is one of the factors which affect demand for electricity. It may be achieved through natural or forced mechanisms. Forced mechanisms require electricity and contribute to electricity consumption. In hot regions, the cooling load is generally higher than the heating load with a reversed trend in cold regions. Space cooling/heating accounts for 60–70% of the energy consumption in the non-industrial buildings (Omer 2008), which is a significant proportion.

For instance, Mokheimer et al. (2017) simulated the technical and economic viability of integrating CSP technologies with cogeneration gas turbine systems for the production of electricity and steam in Saudi Arabia. They considered integration of the solar tower, parabolic trough collector, and linear Fresnel reflector systems with the conventional gas turbine cogeneration configuration. The solar tower was integrated to the gas side while the other two CSP technologies were integrated to the steam side. Thus, only the gas turbine cycle driven by a solar tower constituted a solar gas turbine. They found that the most optimal engine configuration was obtained when the linear Fresnel reflector was integrated with the steam side of a gas turbine cogeneration plant of 50 MWe. This configuration yielded a levelized cost of electricity of 0.051 US$/kWh and a reduction of 119k tonne in the annual CO_2 emission.

Dabwan and Mokheimer (2017) theoretically studied an integrated solar gas turbine cogeneration power plant (ISGTCPP) by using Thermoflow software. In their investigation, the LFR concentrating solar power technology was integrated with a conventional gas turbine for cogeneration of steam and electricity. Industrial process steam was produced at a flow rate of 81.44 kg/s, pressure of 4.588 MN/m^2, and temperature of 667 K. They found that ISGTCPP could reduce the levelized cost of electricity by 76–85% compared to the fully solar-powered LFR power plant (such as the Puerto Errado 2 Thermosolar Power Plant in Spain). At optimal design, the levelized cost of electricity was 0.05 US$/kW h. The Puerto Errado 2 Thermosolar power plant has a capacity of 30 MW (for electricity production only) and is based on a steam cycle driven by a LFR solar field with a single-tank thermocline storage system (for storage of up to 0.5 h) (National Renewable Energy Laboratory, 2024). The solar field of this plant covers an area of 302,000 m^2, and the outlet temperature of steam from the solar field is 546 K at a pressure of 5.5 MN/m^2.

1.8 Worked Examples

Example 1.1 A solar receiver is operating at a temperature of 870 K when ambient temperature is 288 K. If the receiver has a view factor of 0.45 and an emissivity value of 0.88, calculate the radiative heat loss per unit area of the surface of the receiver.

Solution

Radiative heat loss can be estimated from, $\dot{Q}_r = F_{rc}A_{rc}\varepsilon_{rc}T_{rc}^4$.

This implies that heat loss per unit area $= \frac{\dot{Q}_r}{A_{rc}} = F_{rc}\varepsilon_{rc}T_{rc}^4 =$ 0.45 x 0.88 x 5.6704 x $10^{-8}(870)^4$

$= 12.864$ kW/m^2

Example 1.2 Suppose we have a solid rod of length L, cross-sectional area A, and thermal conductivity k. The rod is initially at a higher temperature T_1 at one end and a lower temperature T_2 at the other end. We want to calculate the rate of one-dimensional heat conduction through the rod, given $L = 2$ m, $A = 0.01$ m^2, $k = 50$ W/m.K, $T_1 = 373$ K, and $T_2 = 323$ K.

Solution

Temperature difference $\Delta T = T_1 - T_2$.

From Eq. (1.6), heat conduction rate,

$$\begin{aligned}
\dot{Q}_c &= -kA\frac{\Delta T}{L} \\
&= -50 \times 0.01 \times \frac{373 - 323}{2} \\
&= -50 \times 0.01 \times \frac{373 - 323}{2} = -15\,\text{W} \\
&= -15\,\text{W}
\end{aligned}$$

Example 1.3 How do parabolic dish concentrators (PDC) differ from linear Fresnel reflectors (LFR) in terms of operating temperature and efficiency?

Solution

Parabolic dish concentrators (PDC) focus solar radiation onto a single point, achieving higher operating temperatures (up to 1773 K) compared to linear Fresnel reflectors (LFR) that focus radiation along a line. As a result, PDCs generally have higher thermodynamic efficiency than LFRs, making them more suitable for applications requiring high temperatures, such as solar gas turbines.

Example 1.4 Describe the principle of exergy analysis and its relevance in optimizing solar gas turbines.

Solution

Exergy analysis is a thermodynamic approach that evaluates the quality and usability of energy within a system, considering the irreversibilities associated with real processes. It provides a more accurate measure of system efficiency and performance than energy analysis. In solar gas turbines, exergy analysis helps identify areas where energy losses occur and where design improvements can be made to optimize system performance and reduce inefficiencies.

Example 1.5 What is the role of computational fluid dynamics (CFD) in the design and operation of solar gas turbines?

Solution

CFD plays a critical role in the design and operation of solar gas turbines by simulating complex interactions between turbulence, heat transfer, and combustion processes. It helps predict flame characteristics, pollutant formation, and thermal profiles, allowing for optimization of the combustor design and overall system performance.

Nomenclatures

A	Area (m^2)
C_p	Specific heat capacity at constant pressure (J/kg·K)
D	Diameter (m)
e	Cavity geometry (dimensionless)
$\dot{E}$	Emissive power (W/m^2)
$F_{i\text{-}j}$	View factor of surface i relative to surface j (dimensionless)
g	Acceleration due to gravity (9.8 m/s^2)
h	Coefficient of heat transfer (W/m^2.K)
G_{ef}	Effective solar radiation measured on the aperture plane (W/m^2)
Gr	Grashof number (dimensionless)
k	Thermal conductivity (W/m.K)
K	Incidence angle modifier (dimensionless)
$\dot{m}$	Mass flow rate (kg/s)
Nu	Nusselt number (dimensionless)
Pr	Prandtl number (dimensionless)
q	Volumetric heat addition by thermal conduction per unit mass (J/kg.K)
Q_{solar}	Solar energy incident on the collector (W/m^2)
$\dot{Q}$	Rate of heat flow (J/m^2.s)
Re	Reynolds number (dimensionless)
S	Specific length (m)
t	Time (s)
T	Temperature (K)
x	Distance (m)

u	Mean velocity (m/s)
$\dot{W}$	Rate of work (W)

Greek Symbols

α	Absorptance of the absorber on the receiver (dimensionless)
$\acute{\alpha}$	Thermal diffusivity (m^2/s)
ε	Emissivity of a surface (dimensionless)
ϕ	Density (kg/m^3)
ρ	Specular reflectance (dimensionless)
σ	Stefan-Boltzmann constant ($W/m^2 \cdot K^4$)
ϑ	Friction factor (dimensionless)
λ	Wavelength of radiation (m)
μ	Dynamic viscosity (N s/m^2)
τ	Transmittance (dimensionless)
ψ	Proportion of the reflected solar radiation (dimensionless)

Subscripts

a	Air/ambient
bb	Blackbody
c	Convention
co	Collector
con	Conduction
p	Pressure
r	Radiative
rc	Receiver

References

Abdelsalam TI, Tian Z, Robinson A (2023) Directly irradiated liquid metal film in an ultra-high temperature solar cavity receiver. Part 1: concepts and a quasi-steady-state analysis. Sol Energy 255:355–368

Alami AH, Olabi AG, Mdallal A, Rezk A, Radwan A, Rahman SMA, Shah SK, Abdelkareem MA (2023) Concentrating solar power (CSP) technologies: status and analysis. Int J Thermofluids 18:100340

Al-Sulaiman FA, Atif M (2015) Performance comparison of different supercritical carbon dioxide Brayton cycles integrated with a solar power tower. Energy 82:61–71

American Society of Heating, Refrigerating and Air-Conditioning Engineers (2001) Fundamentals handbook. American Society of Heating, Refrigerating and Air-Conditioning Engineers (ASHRAE), Atlanta

Behbahaninia A, Ramezani S, Hejrandoost ML (2017) A loss method for exergy auditing of steam boilers. Energy 140:253–260

Bergman TL, Lavine AS, Incropera F, DeWitt DP (2011) Fundamentals of heat and mass transfer. Wiley, New Jersey

Bhattacharyya SC (2011) Energy economics: concepts, issues, markets and governance. Springer, London

Blanco MJ, Santigos LR (eds) (2017) Advances in concentrating solar thermal research and technology. Amsterdam, Elsevier

Çengel YA, Boles MA (2006) Thermodynamics: an engineering approach, 5th edn. McGraw-Hill

Crepeau J (2007) Joseph Stefan: his life and legacy in the thermal sciences. Exp Therm Fluid Sci 31:795–803

Dabwan Y, Mokheimer E (2017) Optimal integration of linear Fresnel reflector with gas turbine cogeneration power plant. Energy Convers Manage 148:830–843

Dähler F, Ambrosetti G, Steinfeld A(2017) Optimal solar dish field layouts for maximum collection and shading efficiencies.Sol Energy 144:286–294

Duffie JA, Beckman WA (2013) Solar engineering of thermal processes, 4th edn. Wiley, New Jersey

Eastop T D, McConkey A (2009) Applied thermodynamic for engineers and technologists, 5th edn. Pearson Education Ltd.

Elminshawy N, Siddiqui F, Sultan G (2015) Development of a desalination system driven by solar energy and low-grade waste heat. Energy Convers Manage 103:28–35

Eveloy V, Rodgers P, Alili A (2017) Multi-objective optimization of a pressurized solid oxide fuel-cell gas turbine hybrid system integrated with seawater reverse osmosis. Energy 123:594–614

Giampaolo T (2003) The gas turbine handbook: principles and practices, 2nd edn. Lilburn, Fairmon

Giostri A (2017) Preliminary analysis of solarized micro gas turbine application to CSP parabolic dish plants. Energy Procedia 142:768–773

González-Almenara R, García-Rodríguez L, Muñoz A, Sánchez T, Sánchez D (2024) Innovative desalination system driven by a solar micro gas turbine for off-grid applications. Appl Therm Eng 236:121488

Gruber C, Lesne A (2005) Hamiltonian model of heat conductivity and Fourier law. Physica A 351:358–372

He YL, Qiu Q, Wang K, Yuan F, Wang WQ, Li M-J, Guo J-Q (2020) Perspective of concentrating solar power. Energy 198:117373

Hewitt GF, Shires GL, Bott TR (1994) Process heat transfer. CRC Press, Boca Raton

Jacobson MZ, Delucchi MA (2011) Providing all global energy with wind, water, and solar power, Part I: technologies, energy resources, quantities and areas of infrastructure, and materials. Energy Policy 39(3):1154–1169

Kakaç S, Liu H (1998) Heat exchangers: selection, rating and thermal design. CRC Press, Boca Raton

Khan MI, Asfand F, Al-Ghamdi SG (2022) Progress in technology advancements for next generation concentrated solar power using solid particle receivers. Sustain Energy Technol Assess 54:102813

Machin G, Anhalt K, Edler F, Pearce JV, Sadli M, Strnad R, Vuelban E M (2013). HiTeMS: a project to solve high temperature measurement problems in industry. In: AIP Conference Proceedings, vol 1552, No. 1. American Institute of Physics, pp. 958–963

Madhlopa A (2016) Effect of controlling airflow in a solar chimney on thermal load in a built environment. J Eng des Technol 14(2):286–309

Md Islam T, Huda N, Abdullah AB, Saidur R (2018) A comprehensive review of state-of-the-art concentrating solar power (CSP) technologies: current status and research trends. Renew Sustain Energy Rev 91:987–1018

Mishra SC, Shukla A, Yadav V (2008) View factor calculation in the 2-D geometries using the collapsed dimension method. Int Commun Heat Mass Transf 35:630–636

Mokheimer E, Dabwan Y, Habib M (2017) Optimal integration of solar energy with fossil fuel gas turbine cogeneration plants using three different CSP technologies in Saudi Arabia. Appl Energy 185:1268–1280

Montes MJ, Guédez R, D'Souza D, Linares R, González-Aguilar J, Romero M (2023) Proposal of a new design of central solar receiver for pressurised gases and supercritical fluids. Int J Therm Sci 194:108586

Muthu SS (2020) Assessing the Environmental Impact of Textiles and the Clothing Supply Chain. Elsevie

Narita T, Kato Y, Ara M (2023) Advances in diffusion barrier coatings for high-temperature applications. High Temp Corros Mater 100:399–411

National Renewable Energy Laboratory (2024) Concentrating solar power projects: dish projects. National Renewable Energy Laboratory(NREL). https://solarpaces.nrel.gov/by-technology/dish. Accessed 4 Dec 2024

Okoroigwe E, Madhlopa A(2016) An integrated combined cycle system driven by a solar tower: a review. Renew Sustain Energy Rev 57:337–350

Omer AM (2008) Renewable building energy systems and passive human comfort solutions. Renew Sustain Energy Rev 12(1):1562–1587

Olumayegun O, Wang M, Kelsall G (2016) Closed-cycle gas turbine for power generation: a state-of-the-art review. Fuel 180:694–717

Opiyo N (2016) A survey informed PV-based cost-effective electrification options for rural sub-Saharan Africa. Energy Policy 91:1–11

Ozdil N, Segmen M, Tantekin A (2015) Thermodynamic analysis of an Organic Rankine Cycle (ORC) based on industrial data. Appl Therm Eng 91:43–52

Pang L, Liu L, Zhou X, Hafeez M, Ullah S, Sohail MT (2024) How does natural resource depletion affect energy security risk? New insights from major energy-consuming countries. Energy Strat Rev 54:101460

Pierce W, Gauché P, von Backström T, Brent AC, Tadros A (2013) A comparison of solar aided power generation (SAPG) and stand-alone concentrating solar power (CSP): a South African case study. Appl Therm Eng 61:657–666

Poullikkas A (2005) An overview of current and future sustainable gas turbine technologies. Renew Sustain Energy Rev 9:409–443

Rajaei A, Barzegar A, Eslami E (2016) Exergy and environmental impact assessment between solar powered gas turbine and conventional gas turbine power plant. Int J Chem Eng 2016:8323859

Rashid M (2016) Electric Renewable Energy Systems. Elsevier

Rednic V, Gutt R, Bruj E, Belean B, Murariu T, Raita O, Turcu F (2024) Optimization and performance evaluation of a high-temperature solar receiver for concentrated solar power applications. Appl Therm Eng 250:123407

Rogalski A (2011) Infrared detectors, 2nd edn. Taylor and Francis Group and CRC Press, Boca Raton

Selwynraj AI, Iniyan S, Polonsky G, Suganthi L, Kribus A (2015) An economic analysis of solar hybrid steam injected gas turbine (STIG) plant for Indian conditions. Appl Therm Eng 75:1055–1064

Sheu EJ, Mitsos A (2013) Optimization of a hybrid solar-fossil fuel plant: solar steam reforming of methane in a combined cycle. Energy 51:193–202

Stein WH, Buck R (2017) Advanced power cycles for concentrated solar power. Sol Energy 152:91–105

Szőllősi T, Márkus F (2015) Searching the laws of thermodynamics in the Lorentz-invariant thermal energy propagation equation. Phys Lett A 379:1960–1965

Vahidinasab V, Habibi M (2021) Electric energy storage systems integration in energy markets and balancing services. In: Mohammadi-Ivatloo B, Shotorbani A, Anvari-Moghaddam A (eds) Energy Storage in Energy Markets. Elsevier, pp 287–316

Zhang HL, Baeyens J, Degreve J, Cacéres G (2013) Concentrated solar power plants: review and design methodology. Renew Sustain Energy Rev 22:466–481

Zuo W, Zhang X, Li Y (2020) Review of flue gas acid dew-point and related low temperature corrosion. J Energy Inst 93:1666–1677

Chapter 2
Gas Turbine Fuels and Fuel Systems

2.1 Introduction

Gas turbines can utilize a wide variety of fuels for power generation, including gaseous, liquid, and solid forms. Traditionally, fossil fuels dominate the landscape, with coal being used for closed and semi-closed cycle gas turbines, and oil and natural gas applicable to all types of gas turbines. However, the exploration of biofuels as a viable alternative is gathering momentum, driven by concerns over environmental degradation and energy security (see Chaps. 1 and 11).

Combustible fuels remain the primary energy source for gas turbines. These fuels are also pivotal in the development of hybrid solar gas turbine (SGT) systems aimed at achieving 100% renewable energy operation (see Sects. 5.4.2 and 5.4.4). The quality of fuel directly affects the thermodynamic, environmental, and economic performance of gas turbines. Key considerations when selecting a fuel include heating value, cleanliness, corrosiveness, deposition and fouling, and availability (Boyce 2002).

Recent advancements have seen an increased focus on hydrogen as a potential fuel for gas turbines (Nazari et al. 2022; Cecere et al. 2023; Langston 2019). Hydrogen, when used as a fuel, produces only water vapour, thus eliminating carbon emissions and offering a sustainable alternative to traditional fossil fuels. The development of hydrogen infrastructure and technology is critical to realizing its potential in gas turbine applications.

Nuclear power is also a feasible option for fuelling closed cycle gas turbines, although it is unsuitable for open and semi-closed cycles that require clean combustible fuels such as natural gas. Additionally, fuels are essential for hybridization of SGTs, ensuring a consistent base-load supply. This chapter explores the potential fuels that can provide backup heating for solar gas turbines, highlighting the ongoing innovations and the future landscape of gas turbine fuels.

A. Madhlopa and R. Nkhoma, *Principles of Solar Gas Turbines for Generating Electricity*, Green Energy and Technology,
https://doi.org/10.1007/978-3-031-84096-8_2

2.2 Fuel Specifications

2.2.1 Heating Value

The heating value of a fuel is a measure of the heat released during complete combustion of a unit mass of the fuel. It can be empirically determined using calorimetric methods, where a known quantity of fuel is burned under controlled conditions in a calorimeter, and the heat output is calculated from temperature changes (Ozyuguran et al. 2018). Alternatively, the heating value can be estimated from the chemical analysis of the fuel and the associated heats of reactions of its chemical elements (Ulbig and Hoburg 2002).

There are two types of heating values: higher (total) and lower (net) heating values. The higher heating value (HHV) includes the latent heat of vaporization of water, determined when the vapour in the combustion by-products is condensed (American Society of Heating, Refrigerating and Air-Conditioning Engineers 2001). The lower heating value (LHV) excludes the latent heat of vaporization. As a result, the LHV is often used in energy efficiency calculations. The heating value of a fuel influences the design of the fuel system, such as the size of storage tanks and piping components. A fuel with a high heating value requires a smaller fuel system to deliver the same amount of energy as a fuel with a lower heating value. Hydrogen has relatively high heating values, and so it is a promising candidate for reducing carbon emissions in gas turbine applications. Nevertheless, accurate measurement of the heating value is necessary to optimize its use.

2.2.2 Cleanliness of Fuel

Fuel cleanliness is crucial for the efficient operation of gas turbines. Fuels can be naturally dirty or contaminated with particulate matter during transportation or storage, which can damage components such as nozzles and turbine blades. Therefore, contaminants must be removed from the fuel before use. To ensure fuel cleanliness, various standards have been established. For instance, ISO Standard 4261 specifies the requirements for petroleum fuels used in gas turbines for electricity generation, industrial, and marine applications (International Organization of Standardization 2013).

Advances in fuel processing and filtration technologies continue to improve fuel cleanliness, reducing maintenance costs and enhancing the reliability of gas turbine systems. The ongoing development of biofuels and hydrogen as cleaner alternatives to traditional fossil fuels further underscores the importance of maintaining stringent cleanliness standards.

2.2.3 Corrosion and Particulate Deposition

Corrosion in gas turbine systems predominantly occurs in the fuel system and the hot sections of the engine, such as the combustion chamber and turbine blades. The presence of heavy metals like lead and vanadium accelerates corrosion, leading to the degradation of engine components and reducing the engine's lifespan (see Sect. 5.7.8 on degradation of gas turbine). The rate of corrosion increases with temperature, and preheating the fuel to enhance gas turbine efficiency, particularly for high-viscosity fuels, can further elevate corrosion rates (Seljak et al. 2016). Common fossil fuels (including natural gas, propane, and jet fuel) contain sulphur, sodium, potassium, vanadium, lead, and molybdenum that contribute to corrosion. The composition of the inlet air, influenced by the atmospheric environment, also plays a role. Moisture and oxygen in the air promote oxidation, leading to hot corrosion that occurs when metals are exposed to high-temperature oxidizing gases such as air. Hot corrosion is categorized into low-temperature (873–1023 K) and high-temperature (1073–1223 K) types (Bradshaw et al. 2013). Silicon-aluminide coatings can help mitigate this kind of corrosion.

Particulate deposition and fouling within the fuel system and the hot side of the turbine are influenced by the concentration of certain chemical compounds in the fuel. These compounds can be removed through fuel treatment, which is essential to prevent damage and maintain the efficiency of the turbine. Particle deposition can cause damage to turbines, such as erosion, corrosion, elevation of flow resistance, unbalanced vibrations, and fatigue damage of the blades (Jianjun et al. 2019). So, recent advances in materials science have introduced high-temperature resistant alloys and advanced coatings that further mitigate corrosion and deposition issues. These innovations are necessary for extending the life of gas turbine components and maintaining high efficiency.

2.2.4 Fuel Availability

Fuel availability is a critical factor in choosing a fuel for gas turbine operation. Understanding the available reserves of fuels aids in decision-making regarding fuel selection. Additionally, the market supply level of the fuel can impact the economic performance of the engine, as fuel costs significantly contribute to the total operational costs of the plant. Estimates of global fossil fuel reserves indicate that production of these fuels may peak around 2025, followed by a plateau lasting about 50 years before a decline (Mohr et al. 2015). This underscores the concern that fossil fuels will eventually be depleted. Consequently, the development of renewable energy technologies is crucial for achieving sustainable energy production and consumption.

Hydrogen, biofuels, and other renewable energy sources are being actively researched and developed as alternatives to traditional fossil fuels. These renewable options cannot only address concerns about fuel availability but they can also contribute to reducing environmental impacts. Governments and industries are investing in infrastructure and technology to support the transition to renewable fuels, ensuring a more stable and sustainable energy future (see Chap. 11).

2.3 Fossil Fuels

Fossil fuels are energy resources formed from the remains of dead animals and plants that have been buried and subjected to heat and pressure in the earth's crust over millions of years (Taipabu et al. 2024). There are three main types of fossil fuels: natural gas, oil, and coal. For a significant part of human history, these fuels have been the major source of energy, providing heat and power for domestic, commercial and industrial applications.

2.3.1 Gas and Oil

Natural gas is a gaseous fossil fuel used predominantly for cooking, heating, and electricity generation. This fuel is considered as the cleanest of fossil fuels because it generally produces lower greenhouse gas emissions (see Chap. 11). Consequently, natural gas is more preferred in fuelling power plants. Oil is a liquid fuel exploited primarily in transportation and heating applications. One of the advantages of oil is its high energy density that makes it to be an efficient and convenient fuel for transportation.

2.3.1.1 Formation

Oil and gas originate from dead marine organisms such as zooplankton, shellfish, and algae that become buried in sediments of sand and mud (Ngô and Natowitz 2009). The formation of oil and gas is governed by two primary processes: subsidence and compaction.

Subsidence refers to the downward movement of the earth's surface, which occurs as sediments continuously accumulate in estuaries and deltas, forming sedimentary basins. As the mass of sediments increases at the top of the basin, the bottom part subsides, leading to increased pressure and compaction of the lower sediment layers. In oxygen-free conditions, the organic matter in these sediments transforms into kerogen that exists in three types: I, II, and III (Lai et al. 2017). Type-I kerogen, formed mainly from algal and amorphous constituents, has a high potential to transform into liquid hydrocarbons. Type-II kerogen, consisting of mixed terrestrial and marine source materials, has a lower oil-producing potential but contains more aromatic and heteroatomic ketone and carboxylic acid groups. Type-III kerogen, derived from woody terrestrial material, is characteristic of coal and coaly shale.

A temperature above 373 K is generally sufficient to transform kerogen into hydrocarbons over a geological time scale, which is a system of sequential dating relating geological strata to time (Ngô and Natowitz 2009). Higher temperatures, typically found at deeper levels, increase the probability of gas formation. As pressure increases, hydrocarbons are expelled from the source rock. Oil and gas mix with saltwater in the porous surrounding medium, accumulating in layers due to density differences: gas at the top, oil in the middle, and saltwater at the bottom. This transformation process spans millions of years, underscoring the finite nature of fossil fuels. Reservoirs may contain a combination of gas, oil, and water, requiring extraction for use.

2.3.1.2 Oil and Gas Recovery

The best reservoirs for oil or gas recovery possess high permeability and porosity. Permeability measures how well a liquid or gas diffuses through a porous material, rated on a scale from 1 (lowest) to over 13,000 (Ngô and Natowitz 2009). Porosity refers to the volume of spaces within a solid material relative to the total volume, with reservoir porosity typically ranging between 7 and 40%.

Natural reservoir pressure allows for the recovery of a small fraction of the deposit (20–40%). However, the recovery factor can be increased by injecting water or gas to augment reservoir pressure or by using chemicals to reduce flow resistance. Enhanced recovery techniques can elevate the recovery factor to 30–60% or even higher. Advances in recovery technologies, such as hydraulic fracturing and horizontal drilling (Golembiewski et al. 2023), have significantly improved the efficiency and yield of oil and gas extraction. Nevertheless, hydraulic fracturing (fracking) can cause land degradation (see Chap. 11).

Additionally, the development of carbon capture and storage (CCS) techniques in oil and gas recovery is gaining traction. CCS not only enhances recovery by maintaining reservoir pressure but also mitigates environmental impacts by capturing and storing carbon dioxide produced during extraction. This dual benefit aligns with global efforts to reduce greenhouse gas emissions and combat climate change.

2.3.1.3 Peak Oil

As discussed in Sect. 2.3.1, the formation of oil takes millions of years. At any given time, only a portion of the existing oil can be recovered at an acceptable cost using the prevailing techniques. This recoverable portion constitutes the "known reserves". Currently, the annual consumption of conventional oil surpasses the amount discovered each year. It is believed that approximately 90% of crude oil reserves have already been discovered, indicating that production will eventually begin to decline (Ngô and Natowitz 2009).

Forecasting the ultimate recovery of a finite resource like oil is crucial for long-term energy planning. The Hubbert model, developed by King Hubbert in 1956, is

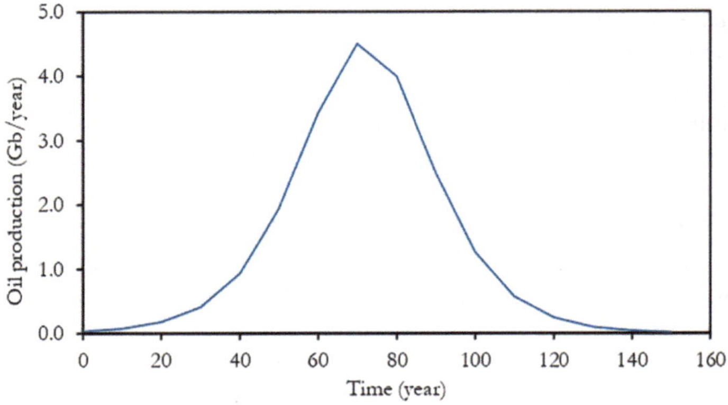

Fig. 2.1 Variation of oil production with time for a single cycle

a widely used method for predicting the ultimate recovery of fossil-fuel resources (Berka and Ediger 2016). Originally designed for the oil industry, the model is also applicable to coal and gas resources. According to the Hubbert model, production increases with time until it reaches a peak, after which it begins to decline (Fig. 2.1). This model is expressed as:

$$P = \frac{2P_{\mathrm{m}}}{1 + k\cosh[b(t - t_{\mathrm{m}})]} \tag{2.1}$$

where P is annual production in billion barrels for year t, P_{m} is peak production (maximum point of the curve), t_{m} is the year at which production peaks, and b and k are constants. The parameter b is the slope of the curve which can be calculated from:

$$U = \frac{4P_{\mathrm{m}}}{b} \tag{2.2}$$

where U is the ultimate estimated recovery.

The Hubbert model predicts that after reaching peak production, oil production will enter a decline phase. This decline is a natural consequence of depleting reserves and increasing extraction difficulty. The model highlights the finite nature of fossil-fuel resources and the need for alternative energy solutions.

In addition to conventional oil reserves, unconventional oil sources such as tar sands, oil shale, and deep-water oil are being developed to offset the decline in conventional oil production. However, these sources are typically more expensive and environmentally challenging to extract, highlighting the need for transitioning to sustainable energy sources. Production from other reserves may begin at different times, leading to multiple production cycles (Fig. 2.2). This phenomenon can be observed in the rise of shale oil production in the United States of America, which created a new peak in the global oil production. Advances in extraction technologies

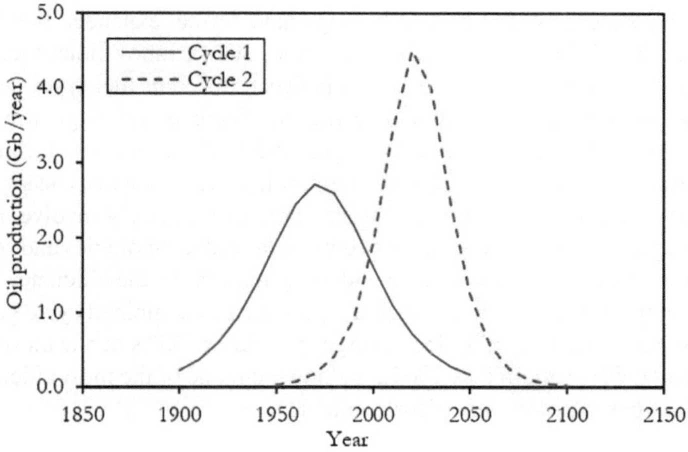

Fig. 2.2 Variation of oil production with time for two cycles

demonstrate the potential for multiple production cycles within the overall depletion trend.

In the Hubbert curve, four major assumptions are made (Berka and Ediger 2016): (i) annual production is modelled as the first derivative of the logistic function, (ii) the production profile is symmetric about the peak time, (iii) production trails behind discovery with a uniform time delay, and (iv) production rises and then decreases in a single cycle with one peak (bell-shaped curve). These assumptions were formulated to permit mathematical analysis without necessarily reflecting the reality of exploiting fossil-fuel resources (Brandt 2010). Despite these and other criticisms, the Hubbert model yields acceptable estimates.

Customarily, combustible fuels for gas turbines encompass natural gas, process gas, low-Btu coal gas, and vaporized fuel oil gas (Boyce 2002). Natural gas is the most preferred conventional fuel for the propulsion of gas turbines and is often used as a reference fuel for comparing the performance of gas turbine engines. Process gas is a general term that refers to gas (such as refinery gas, producer gas, coke-oven gas, and blast-furnace gas) produced through industrial processes. The properties of vaporized fuel oil gas are close to those of natural gas. Worldwide, about 40% of operational gas turbines are driven by liquid fossil fuels. These fuels range from light volatile naphtha through kerosene to heavy, viscous residuals. Light distillates and natural gas have similar fuel properties.

Fuel type, source, and composition are critical factors to consider when selecting fuels, despite the fact that aero-derivative and heavy-duty gas turbines feature wide fuel flexibility. Conventionally, gas turbines operate on high-calorific fuels such as natural gas (HV = 39–46 MJ/kg) and Diesel no. 2 fuel oil (HV = 42 MJ/kg), but modern gas turbines are being developed to operate on a variety of fuels (El-Hossaini 2013). The first closed cycle gas turbine (model AK36) used light oil as a source of heat (Olumayegun et al. 2016).

Recent advances in gas turbine technology have further expanded fuel flexibility (Feng et al. 2022; Ćosić et al. 2022; Webb et al. 2024). Innovations such as dual-fuel systems allow gas turbines to switch between different fuel types seamlessly, enhancing operational efficiency and resilience. There is also a growing interest in integrating renewable fuels like hydrogen and biofuels into gas turbine operations. Hydrogen, in particular, offers a clean-burning alternative that can significantly reduce carbon emissions. The future of gas turbines will likely involve a blend of traditional and alternative fuels, supported by advanced technologies and robust fuel management systems. Efforts are also underway to improve the efficiency and environmental impact of gas turbines. Technologies such as combined cycle gas turbine (CCGT) systems are becoming increasingly popular. CCGTs can achieve efficiencies exceeding 60% (Poullikkas 2005), making them one of the most efficient forms of fossil-fuel-based power generation available.

2.3.2 *Coal*

Coal is a solid fuel primarily utilized to generate electricity in power plants. It is commonly burned to produce steam that drives turbines to generate electricity. Nevertheless, coal is the most carbon-intensive and a major contributor to greenhouse gas emissions. There has been significant development in the exploitation of coal to propel closed cycle gas turbine engines. Coal is a sedimentary rock comprising more than 50% carbon, with both organic and inorganic constituents (Ngô and Natowitz 2009). The organic component predominantly includes carbon (C) but also smaller amounts of hydrogen (H), nitrogen (N), and oxygen (O). The inorganic content constitutes 9–30% of coal by weight and may include sulphur and hazardous chemical elements such as arsenic, beryllium, cadmium, mercury, uranium, and thorium.

The degree of coalification influences the value of coal and its suitability as an energy source. This degree of maturity, specified by rank, is determined by measuring the moisture content, gross heating value, and optical properties. Coal quality is usually divided into five ranks (in order of decreasing grade): anthracite, semi-anthracite, bituminous, sub-bituminous, and lignite (Mohr et al. 2015). Anthracite is a top-grade coal with the highest quality and an approximate lower heating value of 30–36 MJ/kg. Semi-anthracite and Bituminous (also known as black coals) have lower heating values compared to anthracite but are still high-grade. Sub-bituminous and lignite are brown coals that have lower grades with approximate lower heating values of 15–16 MJ/kg. Low-grade coals are often used in electricity generation. For example, in 1956, a 2.3 MW closed cycle power plant driven by pulverized bituminous coal was built in Ravensburg. It was observed that fossil-fired closed cycle plants using air as the working fluid could attain temperatures of up to 973 K, which was lower than those achieved by open cycle gas turbines. Generally, the thermodynamic performance of an open cycle gas turbine is superior to that of a closed cycle.

Research attention has also been given to the improvement of efficiency and environmental performance of coal-fired gas turbines. One of the developments is the integrated gasification combined cycle (IGCC) technology that converts coal into a cleaner-burning gas (Toporov 2014). An IGCC system comprises gasification of coal, syngas stream cleaning systems for hydrogen sulphide (H_2S), chlorides and particulate matter, and subsequent combustion in a gas turbine. The combustion of syngas drives gas turbines with diminished emissions. Nevertheless, the IGCC system entails high costs that are related to the required oxygen plant and various other subsystems (such as the air-separation unit, clean-up unit, syngas cooler, and gas turbine). So, the future of IGCC systems depends on the possibility of increasing the cycle efficiency and reducing its investment costs. A recent perspective of the IGCC technology is that it provides suitable conditions for carbon dioxide sequestration and the possibility of raising efficiencies to more than 50% through the production of hydrogen and combination of IGCC with fuel cell technology. Another notable technology for addressing carbon emissions from coal-fired power plants is carbon capture and storage (CCS). CCS was first considered as a mitigation technology in 1977, and it usually refers to the capture of carbon dioxide (CO_2) at large point sources of CO_2 (Määttä and de Gooyert 2025). The captured gas is compressed, transported, and injected deep into geological formations for permanent storage that requires constant checking. Barriers to this technology include social opposition to CCS projects, high investment costs, and uncertainty about the associated risks. As of now, CCS is still under active research and development.

2.4 Biofuels

Biofuels are renewable energy sources produced from biomass raw materials through processes such as anaerobic digestion and gasification. These fuels can support the supply of a base load with low carbon emissions and replenish themselves through natural mechanisms, thereby providing energy security for the future generations. Consequently, hybridizing biofuels with solar energy can enhance the reliability of SGT power plants.

2.4.1 Classification of Biofuels

Biofuels can be classified based on their generation number (first, second, third, and fourth generations), depending on the type of feedstock and the period of development. At the beginning of biofuel production (early twenty-first century) (Naik et al. 2010), bioethanol was produced from sugar plants such as corn, sugarcane, sugarcane molasses, cassava, wheat, sweet potato, and sweet sorghum. Biodiesel was produced from oil-producing biomass such as rapeseed oil, soybean oil, palm oil, groundnut oil, canola oil, and animal fats. Anaerobic digestion was used to produce biogas from manure, biodegradable biomass, and wastewater. However, the use of edible plants for biofuel production competes with food production for human and animal nutrition.

The second-generation biofuels were developed in response to the food-energy nexus. These biofuels are based on non-edible plant feedstock (Koizumi 2015). This includes agricultural residues, wood waste, municipal solid waste, energy crops (jatropha, castor, sunflower, aspen), water hyacinth, and lignocellulosic biomass/ grass such as switchgrass.

Produced from algae, third-generation biofuels utilize oil-rich algae, a group of photosynthetic organisms that are polyphyletic in nature. A polyphyletic group of organisms possesses two or more separate groups, each with a distinct common ancestor. Photosynthetic organisms range from microscopic species (microalgae) to large species like seaweeds (macroalgae). Microalgae are a promising alternative feedstock for biodiesel production due to their high lipid content and rapid growth rate. Studies have shown that strains such as Chlorella vulgaris are highly productive for biodiesel (Francisco et al. 2010). Chlorella vulgaris can even grow well in environments contaminated with crude oil, augmenting biodiesel productivity (Kalhor et al. 2016). However, commercial biodiesel production from microalgae has not yet been achieved (Huang et al. 2010).

The fourth-generation biofuels are produced from genetically modified crops that absorb more CO_2 from the atmosphere than the amount produced during their combustion (Demirbas 2011; Cuellar-Bermudez et al. 2015). This type of biofuel is produced using technology pathways such as pyrolysis, gasification, upgrading, solar-to-fuel, and genetic manipulation of organisms to secrete hydrocarbons (Demirbas 2011).

2.4.2 Conversion Pathways for Producing Energy Carriers from Biomass Raw Materials

Biomass can be converted to energy carriers via biochemical, chemical, and thermochemical processes (Faaij 2006; Chum et al. 2011). Biochemical processes include digestion and fermentation. The common energy carrier produced through digestion is biogas, while fermentation yields alcohols such as ethanol. Transesterification and hydrogenation are chemical processes for biofuel production. In the transesterification process, a lipid reacts with an alcohol to form esters and glycerol as by-products (Anastopoulos et al. 2009). This reaction proceeds in the presence of a catalyst, typically a strong acid or base. For instance, sodium hydroxide can be used as a catalyst. Vegetable oil, extracted from seeds using mechanical crushing or chemical solvents, is often used in this process. The protein-rich residue, known as cake, is typically sold as animal feed or fertilizer but can also be used as a raw material for producing high-value chemicals (Demirbas 2009).

$$RCOOR' + R''OH \rightarrow RCOOR'' + R'OH \tag{2.3}$$

where RCOOR′ is a lipid, R″ OH is an alcohol, RCOOR″ is an ester, and R′ and R″ are radicals. The reaction is reversible, but using a strong catalyst can significantly adjust the reaction equilibrium. The yield of the desired ester (RCOOR″) increases with an increase in the concentration of alcohol (R″OH).

Thermochemical processes include combustion, pyrolysis, and gasification (Chum et al. 2011). During combustion, carbon and hydrogen in the fuel react with excess oxygen to form carbon dioxide and water, thereby generating heat. The heat released can drive a steam or gas cycle to produce electricity. Pyrolysis is the thermal decomposition of biomass in the absence of oxygen, producing solid (charcoal), liquid (pyrolysis oil or bio-oil), and gas products. The relative quantities of these co-products are influenced by operating temperature and residence time. Elevated heating rates at temperatures of 723–823 K yield oxygenated oils as the main products (70–80%), with the remainder shared between biochar and gases. Gasification involves partial oxidation of biomass during heating, yielding a combustible gas mixture (fuel gas) rich in carbon monoxide (CO) and hydrogen (H_2) with a heating value of 5–20 MJ/kg. Fuel gas can be upgraded to biomass synthesis gas or syngas (Faaij 2006), which can replace diesel in appropriately designed internal combustion engines.

2.4.3 Exploitation of Biofuels in Gas Turbine Engines

Biofuels play a critical role in the power, transport, and other sectors. These fuels are utilized in various forms (solid, liquid, and gas) and for different purposes, such as engine fuelling and lubrication. The prime mover that converts the energy content of a fuel into a desirable form influences fuel consumption. For instance, a wide spectrum of fuel forms can be used to drive closed cycle gas turbines, while open cycle gas turbines can only accept clean liquid and gaseous fuels.

2.4.3.1 Solid Biofuels

Solid biomass, including wood chips, briquettes/pellets, forestry and agricultural residues, and biodegradable municipal solid waste, has been utilized in gas turbines. This supplies the thermal energy required to raise the temperature of the working fluid to the expected turbine inlet temperature (Al-Attab and Zainal 2015). However, solid biofuels are typically used with high-temperature heat exchangers, making them suitable for closed cycle gas turbine engines.

2.4.3.2 Liquid Biofuels

Liquid biofuels such as biodiesel, bioethanol, biomethanol, and bio-oil are essential and can be classified into biogasoline and biodiesel. Biogasoline includes bioethanol, biomethanol, bio-ETBE (ethyl-tertio-butyl-ether), and bio-MTBE (methyl-tertio-butyl-ether) (International Energy Agency 2016).

(a) Biodiesel

Biodiesel is produced through the transesterification process from renewable biological sources (Gupta et al. 2010a). It has become a crucial alternative to fossil fuels in the transportation and power sectors, recognized globally as the first public biofuel alternative (Ahmad et al. 2011). Biodiesel's attractive qualities include biodegradability, low toxicity, and low carbon emissions. It also exhibits good engine performance, low combustion emissions due to its high oxygen content, good blending qualities with petroleum fuel, a closed carbon cycle, and adaptability in existing diesel engines with or without modification (Demirbas 2002).

Palm methyl ester (PME), derived from palm oil, is a promising biofuel alternative to diesel fuel due to its properties that are similar to conventional diesel fuel. Research on PME's fundamental combustion characteristics for gas turbines has shown similar adiabatic flame temperatures to diesel no. 2, with flow rates increasing with atomizing pressure and no difference in ignition characteristics (Hashimoto et al. 2008). PME's complete combustion, owing to the absence of aromatic rings and presence of oxygen, confirms its suitability for gas turbine combustion.

Other biodiesel varieties have also been tested in gas turbines, including soybean oil biodiesel (Gupta et al. 2010b), Jatropha oil (Rehman et al. 2011), and canola oil biodiesel (Habib et al. 2010). Blending biodiesel with conventional diesel has been explored as well. Habib et al. (2010) studied the engine performance of Jet A and biodiesels from soy, canola, recycled rapeseed, and hog fat, and their 50% blends with Jet A fuel using a 30 kW gas turbine. Jet A fuel is a common aviation fuel for commercial aircraft powered by gas turbines. They observed that the static thrust developed with pure biofuels and their blends was comparable to Jet A, with pure biofuels exhibiting higher thermal efficiencies due to lower equivalence ratios and extra oxygen in the biofuel molecules, resulting in complete combustion. There was also an insignificant difference in the turbine inlet and exhaust gas temperatures of all tested fuels.

Blakey et al. (2011) reviewed biodiesel's potential in the aviation industry, comparing the properties of fuels from ethyl and methyl transesterification of vegetable oils with commercial Jet A-1 fuel. They found that fatty acid esters (FAEs) are unsuitable for aviation due to metal contaminants from the raw feedstock. Tan and Liou (2011) characterized a laboratory-sized turbojet engine using mixtures of conventional aviation fuel (Jet A-1) and biodiesel from waste vegetable oil, varying the blends from 0 to 100% biodiesel at 5% intervals. Their tests showed a proportional increase in engine thrust with increased biodiesel blend, with B100 producing about 200% more thrust than conventional aviation fuel at a 30 mL/min fuel flow rate. Lower fuel consumption rates were also observed in the same proportion, with

Table 2.1 Comparison of properties of Jet A-1 and biodiesel (B100) (Tan and Liou 2011)

Fuel property	Fuel	
	Jet A-1	B100
Density (kg/m^3)	810	880
Distillation range (K)	450–573	473
Kinematic viscosity at 313 K (mm^2/s)	1–2	2.97
Heating value (MJ/L)	> 35.1	32.08
Flashing point (K)	> 311	341
Freezing point (K)	226	> 273

B100 resulting in the lowest thrust-specific fuel consumption due to its high cetane index, indicating biodiesel's potential to replace conventional diesel.

As seen from Table 2.1, the kinematic viscosity of biodiesel is higher than that of Jet A-1. This level of viscosity has been used to discredit biodiesel, but the oxygen content and the high viscous nature of the biofuel have been found to improve fuel–air mixing ratio, thereby promoting complete combustion (Ali et al. 1995; Tan and Liou 2011).

(b) Bioethanol and butanol

Bioethanol can be produced from sugar and starch-rich materials which are sources of nutrition for people and livestock. Consequently, alternative lignocellulosic materials such as forestry residue, agricultural residue, energy crops, yard waste, wood products/shavings, animal and human wastes are being proposed, promoted, and used. Exploitation of lignocellulosic biomass for ethanol production involves two major processes that comprise: (a) hydrolysis of the cellulose in the lignocellulosic materials to fermentable reducing sugars, and (b) fermentation of the sugars to ethanol (Ye and Jiayang 2002). Hydrolysis (saccharification) is the chemical decomposition of a substance by the use of water, and it can entail the use of enzymes, acidic and alkaline substances to facilitate the cleavage of bonds in the presence of water. The use of enzymes for lignocellulose hydrolysis is considered to be the most viable strategy, offering more advantages compared to the chemical conversion pathways for its higher yields, minimal by-product formation, low energy requirements, mild operating conditions, and environmentally friendly processing technique (Saha 2000; Wingren et al. 2005). Often, enzymatic hydrolysis is stimulated by cellulase enzymes, and the fermentation is carried out by yeasts or bacteria. Pre-treatment of lignocellulosic biomass is required to eliminate or alter the lignin and cellulose matrices that surround the biomass, before enzymatic hydrolysis of cellulose and hemicelluloses. This pre-treatment process is necessary because of the structural and chemical complexity of plant biomass material. There are three common pre-treatment methods: (a) physical pre-treatment (uncatalysed steam-explosion, liquid hot water pre-treatment, mechanical comminution), (b) chemical pre-treatment (catalysed steam-explosion, acid pre-treatment, alkaline pre-treatment, etc.), and (c) biological pre-treatment (use of microorganisms).

Fermentation employs microorganisms to convert the hydrolysates produced during hydrolysis, to ethanol. As the hydrolysates contain glucose and various monosaccharides, such as xylose, mannose, galactose, and arabinose, and oligosaccharides, microorganisms are required for effective fermentation process. Due to lack of suitable microorganisms capable of efficient fermentation (with high yield and rate) of all sugars released during the pre-treatment and hydrolysis, commercial/industrial ethanol production is prevented (Talebnia et al. 2010). However, the *Sacchamyces cerevisiae* yeast and the *Zymomonas mobilis* bacterium are the best-known microorganisms for high-yield ethanol production from hexoses (90–97% of the theoretical value) (Claassen et al. 1999).

The chemical processes of biomass fermentation are complex, but the overall chemistry of the process can be summarized as the conversion of glucose sugar ($C_6H_{12}O_6$) to alcohol (C_2H_5OH) and carbon dioxide gas (CO_2), (Balat and Balat 2009):

$$C_6H_{12}O_6 \rightarrow 2C_2H_5OH + 2CO_2 \tag{2.4}$$

Theoretically, 1 kg of glucose yields about 0.51 kg of bioethanol and 0.49 kg of carbon dioxide for 100% conversion of the biomass raw materials but the yield is practically less than 100% because part of the glucose is used to sustain the microorganisms (Demirbas 2005; Balat and Balat 2009).

Bioethanol finds application in combustion processes especially in internal combustion engines. This fuel is considered as a suitable alternative to fossil fuels as a liquid transportation fuel with powerful economic, environmental, and strategic attributes especially when used in dedicated engines as the only fuel or in blends of up to 30% with petroleum fuels without modification to the engine (Talebnia et al. 2010; Brethauer et al. 2010). Bioethanol has a high-octane number of about 96–113 (Galbe and Zacchi 2002). The octane number of a fuel gives a measure of the fuel to resist knocking. Consequently, blending petroleum fuels with bioethanol increases the octane number of the resulting mixture, thereby eliminating or reducing the use of octane number improvers that are toxic and corrosive. Another advantage is the level of oxygen content (34.78%) in bioethanol that makes its combustion environmentally suitable because it helps reduce emission of NOx. The presence of oxygen in bioethanol also promotes complete combustion of the fuel, thereby resulting in reduction in or complete elimination of unburned hydrocarbons (UHC) produced during the combustion of pure hydrocarbon fuels from petroleum (Malça and Freire 2006). Ethanol application in combustion engines has the same overall transport efficiency as diesel in compression-ignition engines and about 15% more efficient than gasoline in optimized spark-ignition engines (Bailey 1996). This makes its combustion in gas turbines suitable because diesel is extensively used to fuel this variety of heat engines. In addition, ethanol burns with broader flammability limits, higher flame speeds, higher heats of vaporization, and shorter burn time although it has lower energy density than gasoline and shows some corrosiveness, low flame luminosity, lower vapour pressure, miscibility with water, toxicity to ecosystems, acetaldehyde emission from exhausts, and increase in vapour pressure when blended with gasoline (MacLean and Lave 2003).

Other researchers have also examined the possibility of using bioethanol in the gas turbine engine. Alfaro-Ayala et al. (2013) carried out numerical and thermodynamic analyses of the combustion of bioethanol in a 50 MW gas turbine that rotated at 3600 rpm. They compared the effect of bioethanol, diesel, and natural gas combustion on turbine inlet temperature and emission intensity. Based on equal quantity of fuel, bioethanol and diesel combustions exhibited lower power output of the plant than natural gas even though natural gas has a higher level of NOx. This observation is due to the influence of the heating value of fuel. Bioethanol has lower heating value than natural gas, which implies that the former fuel releases less heat in the combustion chamber, resulting in diminished turbine inlet temperature and output power. Due to high heating value, natural gas is capable of attaining high turbine inlet temperatures. Unfortunately, NOx formation is enhanced by elevated combustion temperatures. The analysis based on equal amount of energy (increased bioethanol and diesel) yielded an increase in CO_2 emission more than natural gas, possibly due to incomplete combustion of the fuels. The combustion of bioethanol is promising for low NOx emission. A blend of bioethanol (5 and 10%) and kerosene was investigated by Patra et al. (2015) to determine its combustion performance. Results showed that flame luminosity and brightness reduced progressively with an increase in the percentage of bioethanol, thereby reducing the peak wall temperature of the combustor in that order.

Biobutanol has a higher heating value (29.2 MJ/L) than bioethanol (19.6 MJ/L) and biomethanol (16 MJ/L), (Kamiński et al. 2011). Other properties of biobutanol are presented in Table 2.2. This table demonstrates that the evaporation heat of biobutanol is more attractive than that of bioethanol and biomethanol even though it has lower Research Octane Number (RON) and Motor Octane Number (MON) than the other fuels. Depending on the engine design, a low octane number can still be appropriate, in other words, some engines operate on low octane numbers.

Biobutanol has not been widely considered in gas turbine application in spite of its potential. Nevertheless, Mendez et al. (2014) studied the performance of biobutanol blend with Jet-A fuel as a possible alternative to complete dependence on fossil fuel in aviation gas turbine engines. As has been established, hydrocarbon fuels develop more power than oxygenated fuels (biofuels) due to higher heating value

Table 2.2 Properties of biobutanol and other fuels

Fuel	Combustion energy (MJ/L)	Evaporation heat (MJ/kg)	Research Octane Number (RON)	Motor Octane Number (MON)
Petrol	32	0.36	91–99	81–89
Biobutanol	29.2	0.43	96	78
Bioethanol	19.6	0.92	130	96
Biomethanol	16	1.2	136	104

Source of data Kamiński et al. (2011)

of the former, but their emission levels are a cause for concern. So, biobutanol and its blends, although they yield less thrust than the Jet-A fuel, have lower emission indices for CO and NO_x.

(c) Bio-oil

Plant cells are predominantly composed of three organic compounds: cellulose, hemicellulose, and lignin in varying proportions (Sannigrahi et al. 2008; Chen 2014). In addition, natural lignocellulosic materials contain small quantities of pectin, nitrogenous compounds, and other constituents. Depending on structural composition, it is difficult to convert some types of biomass materials to energy by using microbial or chemical processes. Such types of biomass can be transformed by using thermal decomposition processes such as pyrolysis (the thermal decomposition of woody substances in the absence of oxygen for the purposes of producing bio-oil, biochar, and gaseous fuel). The rate of pyrolysis depends on the temperature and particle size of the feedstock, resulting in fast and slow pyrolytic processes. In fast pyrolysis, woody biomass is thermally decomposed at a temperature range of 673–723 K and particle size of 250 μm to 1 mm for optimal bio-oil production. For slow pyrolysis, the temperature range is 303–623 K and involves large particle size resulting in large production of biochar as the major product. Fast pyrolysis is more preferred than slow pyrolysis because of its low production cost, high terminal efficiency, and low fossil-fuel input leading to CO_2 neutral fuel (Oasmaa et al. 2003). Similarly, fast pyrolysis is the most valuable bio-conversion process in that it accounts for no product loss. All the products of fast pyrolysis can be directly used in energy production.

Bio-oils are highly viscous, which creates challenges for their injection into the combustor of a gas turbine. In view of this, modifications to the injection and ignition systems, combustor geometry, and starting conditions are required. One way of reducing the viscosity of bio-oil is through blending with other fuels. A study conducted by Juste and Monfort (2000) showed that a blend of 20% bio-oil and 80% ethanol resulted in combustion characteristics that were similar to those of JP-4 (50–50 kerosene–gasoline jet fuel blend). The viscosity of bio-oil can also be decreased through fuel preheating. For instance, the fuel can be preheated to temperatures of 343–363 K to obtain levels of viscosity below 10 cSt, and a review of several studies has shown that the combustion of bio-oil in gas turbines does not have a technological issue but economical (Chiaramonti et al. 2007). This shows that there is potential to exploit bio-oil in solar gas turbine engines.

2.4.3.3 Gaseous Biofuels

(a) Syngas

Syngas (or synthesis gas) is produced through gasification which is a thermochemical decomposition process of biomass in the presence of controlled air. In this process, solid biomass is converted to a gaseous mixture of syngas that consists of hydrogen (H_2), carbon monoxide (CO), methane (CH_4), and carbon dioxide (CO_2)

through partial oxidation (Wang et al. 2008). Oxygen, carbon dioxide, steam or their mixtures can be used as oxidants or gasifying agents in the gasification process. A gasifier operates between temperatures of 600 and 1273 K. The gasification process is temperature-dependent, and higher temperatures favour the production of synthesis gas with very low concentrations of char and tar. Nonetheless, complete elimination of char and tar is not guaranteed regardless of the kind of feedstock employed. Factors that affect the quality and quantity of gas yield or performance of the gasification process include: the feedstock, temperature, gasifier design, static bed height, fluidizing velocity, equivalence ratio, gasifying agent, and catalyst (Lim and Alimuddin 2008; Ghani et al. 2009). In view of these varying factors, synthesis gas needs to be cleaned prior to use in gas engines.

The combustion of synthetic gas has been applied in industrial gas turbines (Lee et al. 2010, 2012). These studies found that the combustion of syngas varied with hydrogen and carbon monoxide. The concentration of NOx was proportional to H_2 with no release of CO. It was also observed that the dilution of non-flammable components of syngas (N_2, CO_2, and steam) reduced NOx emission while the dilution of steam elevated the emission of CO at 30 kW.

(b) Biogas

Biogas is produced by anaerobic digestion (AD) of biomass. AD is the decomposition of organic matter in the absence of oxygen for the production of combustible gas rich in methane known as biogas (Okoroigwe et al. 2010; Oparaku et al. 2013). This process is often carried out in a biodigester by the action of different microorganisms on the substrate through complex biochemical reactions that yield methane and effluent rich in plant nutrients (Fagbohungbe et al. 2017). Microorganisms that are involved in the AD process are classified into two groups: (a) acid producers (acidogens and acetogens) and (b) methane producers (methanogens). These groups of microorganisms have different physiological features, growth rates, and sensitivities to operational conditions. Consequently, operational conditions (such as temperature and pH) in the digester influence the rate of biogas production. Biogas has found application in the gas turbine and other engines.

The exploitation of biogas in gas turbines has an issue on the commercial scale of megawatt capacity because of insufficient biogas production capacity (Kang et al. 2012). Consequently, micro gas turbines have received more attention. A micro gas turbine (MGT) operating in Reus municipality, Spain, is fuelled with biogas from the wastewater treatment plant (Bruno et al. 2009). This MGT system operates in a trigeneration mode in which waste heat is employed to power an absorption chiller. The chilled water is used to condense water in the biogas pre-treatment process and in turn cools the MGT inlet air. Findings showed that the economic benefits of trigeneration using only the available biogas (without additional natural gas) were insufficient for this case study which was considered as representative for a sewage treatment plant serving a community of around 100,000 inhabitants. Other studies on the economic performance of real plants have demonstrated that the combustion of biogas is most economical in a combined heat and power system compared to only power or heat generation studies (Kang et al. 2014). Increased interest in the performance of biogas combustion in MGTs has been shown by other researchers (Lee et al. 2010; Breuhaus 2013).

Biogas is a promising biofuel with potential to replace natural gas in fuelling gas turbines. Nevertheless, its heating value is lower than that of natural gas. Increasing the rate of fuel flow helps to compensate for this deficiency. It has also been found that blending with hydrogen can improve the combustion stability of biogas in gas turbine engines (Hosseini and Wahid 2014).

Gas turbines develop a steady flame which renders them flexible to run on biofuels (Gupta et al. 2010a). Whereas combustibility has been demonstrated at laboratory and pilot scales and a lot of research carried out on biofuels for gas turbines and their mixtures with fossil hydrocarbons, combustion difficulties such as atomization and emission of NOx are still prohibiting problems with the present gas turbine combustion features (Gupta et al. 2010b). Formation of oxides of nitrogen (NOx) is unavoidable due to the combustion of air–fuel mixture at high temperatures expected in gas turbine combustors (Bergthorson and Thomson 2015). In addition, other thermodynamic conditions in the combustor such as primary air temperature, equivalence ratio and lower heating value of fuel, fuel temperature, and fuel composition affect the NOx and other pollutant emissions in gas turbines (Seljak et al. 2016). Cleaning and refining of biofuels may be required alongside calibration of fuel injection and combustion systems for gas turbines to accept a wide range of biofuels.

2.5 Nuclear Energy

Nuclear energy is obtained from nuclear fission and fusion processes. In a fission process, energy is released during the disintegration of a heavy nucleus when bombarded by fast-moving neutrons. Uranium is commonly used as a fuel in a nuclear reactor. For example, a high level of energy is released when uranium nuclei are bombarded by neutrons:

$$^{235}_{92}\text{U} + {}^{1}_{0}n \rightarrow {}^{148}_{57}\text{La} + {}^{85}_{35}\text{Br} + 3{}^{1}_{0}n + \text{E} \tag{2.5}$$

where E is the amount of released energy.

Neutrons produced in a fission process can produce other fissions, resulting in a chain reaction and very high quantities of thermal energy. Graphite (or other material) is used to slow down neutrons produced in the fission process, thereby controlling the rate of power production. The fission process is a mature means of generating power.

In nuclear fusion, energy is released when light nuclei are fused together to form a heavier nucleus. For instance, the nuclei of deutrium (${}^{2}_{1}\text{H}$) can be fused together to form an isotope of helium (He). This type of nuclear energy is still under active research and development.

$${}^{2}_{1}\text{H} + {}^{2}_{1}\text{H} \rightarrow {}^{3}_{2}\text{He} + {}^{1}_{0}n \tag{2.6}$$

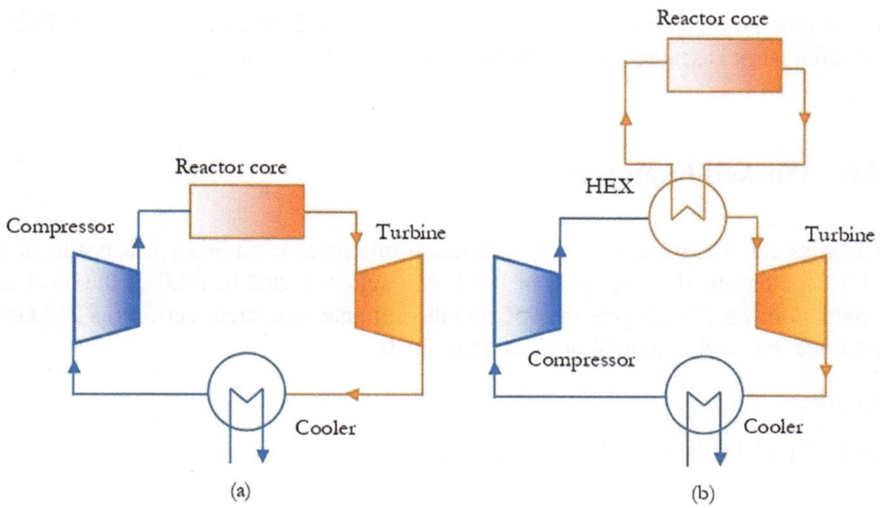

Fig. 2.3 Closed cycle gas turbine fuelled by nuclear reactor: **a** direct heating and **b** indirect heating. HEX = heat exchanger

Most of nuclear power plants operate on the principles of the steam Rankine cycle. Two common steam-based reactor systems are: (a) boiling water reactor and (b) pressurized water reactor. Nevertheless, the thermodynamic efficiency of the Rankine cycle is lower than that of the Brayton cycle. Consequently, nuclear resources have also been used to fuel the closed cycle gas turbine. The working fluid of the gas turbine can be directly heated in the reactor or indirectly heated via a heat exchanger (Fig. 2.3).

The thermal dynamic performance of a gas turbine increases with the temperature of the heat source. In case of the open cycle and semi-closed gas turbine systems, combustible clean fuels can be used directly, which tends to elevate the possible attainable temperature of the working fluid more compared to the closed cycle gas turbine. Use of a high-temperature heat source is one of the most important strategies of improving the thermodynamic and economic performances of a closed cycle gas turbine. Nonetheless, most of the high-temperature heat sources that are suggested for application in the closed cycle gas turbine system are still under active research and development (Olumayegun et al. 2016). Research work is going on in the area of high-temperature gas reactors with helium as a potential working fluid. For instance, the Generation-IV (Gen IV) consortium was established in 2000 to develop six categories of next-generation nuclear reactors which are expected to be ready for commercialization over the time horizon between 2020 and 2030 or later. Some research work is under way in the areas of reactor core design, design of pressure vessel, design of control rods, materials, internal heat exchangers, fuel and cladding design, and nuclear fuel reprocessing. These developments may assist in enhancing the thermal performance of the closed cycle gas turbine. It is expected that the Gen

IV reactor power plant will be operational between 2030 and 2050. As of 2025, research work is scarce on solar-nuclear hybrid gas turbines.

2.6 Worked Examples

Example 2.1 The forecast of oil production from a reservoir shows a peak production of 5 Giga barrels (Gb) per year in 1972. Production started in 1900. If the level of production was 2.8 Gb/year in 1956 and the ultimate estimated recovery is 230 Gb, estimate the level of annual production in 2020.

Solution

From Eq. (2.1), the annual oil production is:

$$P = \frac{2P_m}{1 + k\cosh\{b(t - t_m)\}}$$

The parameter b can be computed from:

$$b = \frac{4P_m}{U}$$

But $P_m = 5$ Gb/year and $U = 230$ Gb.
Therefore,

$$b = \frac{4 \times 5}{230} = \frac{2}{23}$$

when $t = 1956$, $P = 2.8$ Gb/year
So,

$$2.8 = \frac{2 \times 5}{1 + k\cosh\left\{\frac{2}{23}(1956 - 1972)\right\}}$$

This implies that,

$$1 + k\cosh\{2(1956 - 1972)/23\} = 2 \times 5/2.8 = 10/2.8$$
$$k = 1.2047.$$

This implies that when $t = 2020$,

$$P = \frac{2 \times 5}{1 + 1.2047 \times \cosh\left\{\frac{2}{23}(2020 - 1972)\right\}}$$
$$= 0.25\ \text{Gb/year}$$

Example 2.2 A sample of fuel weighing 0.5 kg was burned in a calorimeter, and the temperature of 2 kg of water increased from 25 to 75 °C. Calculate the higher heating value (HHV) of the fuel, given that the specific heat capacity of water is 4.18 kJ/kg·°C. Assume no heat losses.

Solution

Heat absorbed by the water (Q):

$$Q = mc\,\Delta T$$

where m is the mass of water, c is the specific heat capacity of water and ΔT is temperature change.

Therefore, $Q = 2 \text{ x } 4.18 \text{ x } 50 = 418$ kJ.

HHV of the fuel:

HHV = Q/m_f,

where m_f is mass of fuel burned.

Therefore,

$$\begin{aligned}\text{HHV} &= 418\ \text{kJ}/0.5\ \text{kg} \\ &= 836\ \text{kJ/kg}.\end{aligned}$$

Example 2.3 Using the Hubbert model, calculate the level of annual oil production in 2030, given a peak production P_m of 4.5 billion barrels per year in 2010, with $U = 250$ billion barrels, and constants $b = 0.07$ and $k = 1.2$.

Solution

Using the Hubbert model equation:

$$\begin{aligned}P &= \frac{2P_m}{1 + k\ \cosh\{b(t - t_m)\}} \\ &= \frac{2 \times 4.5}{1 + 1.2 \times \cosh\{0.07 \times (2030 - 2010)\}} \\ &= 2.5\ \text{billion barrels per year}.\end{aligned}$$

Example 2.4 If 1 kg of biogas with a specific heat capacity of 1.75 kJ/kg·K is heated from 298 to 573 K, calculate the energy required to achieve this temperature change.

Solution

$Q = \text{mC}\,\Delta T$.

But $m = 1$ kg, $C = 1.75$ kJ/kg · K, and $\Delta T = 573–298 = 275$ K.

Therefore,

$$Q = 1 \times 1.75 \times 275 = 581.25\ \text{kJ}.$$

Example 2.5 Calculate the amount of CO_2 produced during the combustion of 5 kg of bioethanol (C_2H_5OH), assuming complete combustion and the following chemical equation of combustion:

$$C_2H_5OH + 3O_2 \rightarrow 2CO_2 + 3H_2O$$

(Molecular weights: C = 12, H = 1, O = 16).

Solution

Molar mass of bioethanol C_2H_5OH:

$$M_{C_2H_5OH} = 2 \times (12) + 6 \times (1) + 1 \times (16) = 46 \text{ g/mol}$$

Number of moles of bioethanol in 5 kg:

$$n_{C_2H_5OH} = 5000/46 = 108.7 \text{ moles}$$

Moles of CO_2 produced (from the balanced equation, 1 mol of C_2H_5OH produces 2 mol of CO_2):

$$n_{CO_2} = 2 \times n_{C_2H_5OH} = 2 \times 108.7 = 217.4 \text{ moles}$$

Molar mass (M_{CO_2}) of bioethanol CO_2:

$$M_{CO_2} = 1 \times (12) + 2 \times (16) = 44 \text{ g/mol}$$

Mass of CO_2 (m_{CO2}) produced:

$$\begin{aligned} m_{CO_2} &= n_{CO_2} M_{CO_2} = 217.4 \times 44 \\ &= 9565.6 \text{ g} \\ &= 9.57 \text{ kg}. \end{aligned}$$

Therefore, the amount of CO_2 produced is 9.57 kg.

Nomenclatures

P	Annual production in billion barrels for the year (Gb/year)
P_m	Peak production (maximum point of the curve) (Gb/year)
HV	Heating value (J/kg)
E	Amount of released energy (J)
c	Heat capacity (kJ/kg.K)
m	Mass (kg)
T	Temperature (K)

References

Ahmad A, Yasin N, Derek C, Lim J (2011) Microalgae as a sustainable energy source for biodiesel production: a review. Renew Sustain Energy Rev 15:584–593

Al-Attab K, Zainal Z (2015) Externally fired gas turbine technology: a review. Appl Energy 138:474–487

Alfaro-Ayala J, Gallegos-Muñoz A, Uribe-Ramírez A, Belman-Flores J (2013) Use of bioethanol in a gas turbine combustor. Appl Therm Eng 61:481–490

Ali Y, Hanna M, Leviticus L (1995) Emissions and power characteristics of diesel engines on methyl soyate and diesel fuel blends. Bioresour Technol 52:185–195

American Society of Heating, Refrigerating and Air-Conditioning Engineers (2001) Fundamentals handbook. American Society of Heating, Refrigerating and Air-Conditioning Engineers (ASHRAE), Atlanta

Anastopoulos G, Zannikou Y, Stournas S, Kalligeros S (2009) Transesterification of vegetable oils with ethanol and characterization of the key fuel properties of ethyl esters. Energies 2:362–376

Bailey B (1996) Performance of ethanol as a transportation fuel. In: Wyman C (ed) Handbook on bioethanol: production and utilization. Taylor & Francis, Bristol, pp 37–60

Balat M, Balat H (2009) Recent trends in global production and utilization of bio-ethanol fuel. Appl Energy 86:2273–2282

Bergthorson J, Thomson MJ (2015) A review of the combustion and emission properties of advanced transportation biofuels and their impact on existing and future engines. Renew Sustain Energy Rev 42:1393–1417

Berka I, Edigerb VŞ (2016) Forecasting the coal production: Hubbert curve application on Turkey's lignite fields. Resour Policy 50:193–203

Blakey S, Rye L, Wilson C (2011) Aviation gas turbine alternative fuels: a review. Combustion Institute

Boyce MP (2002) Gas turbine engineering handbook, 2nd edn. Gulf Professional Publishing, Boston

Bradshaw A, Simms N, Nicholls J (2013) Hot corrosion tests on corrosion resistant coatings developed for gas turbines burning biomass and waste derived fuel gases. Surf Coat Technol 228:248–257

Brandt A (2010) Review of mathematical models of future oil supply: historical overview and synthesizing critique. Energy 35:3958–3974

Brethauer S, Wyman CE (2010) Review: continuous hydrolysis and fermentation for cellulosic ethanol production. Bioresour Technol 101:4862–4874

Breuhaus P (2013) Biogas production technologies. International research institute of Stavanger, Stavanger, Project number 7882727

Bruno J, Ortega-López V, Coronas A (2009) Integration of absorption cooling systems into micro gas turbine trigeneration systems using biogas: case study of a sewage treatment plant. Appl Energy 86:837–847

Cecere D, Giacomazzi E, di Nardo A, Calchetti G (2023) Gas turbine combustion technologies for hydrogen blends. Energies 16(19):6829

Chen H (2014) Biotechnology of lignocellulose: theory and practice. Chemical Industry Press, Beijing

Chiaramonti D, Oasmaa A, Solantausta Y (2007) Power generation using fast pyrolysis liquids from biomass. Renew Sustain Energy Rev 11:1056–1086

Chum H, Faaij A, Moreira J, Berndes G, Dhamija P, Dong H, Gabrielle B, Goss EA, Lucht W, Mapako M, Masera CO, McIntyre T, Minowa T, Pingoud K (2011) Bioenergy. In: Edenhofer O, Pichs-Madruga R, Sokona Y, Seyboth K, Matschoss P, Kadner S, Zwickel T, Eickemeier P, Hansen G, Schlömer S, Von Stechow C (eds) IPCC special report on renewable energy sources and climate change mitigation. Cambridge University Press, New York, pp 209–332

Claassen PAM, Lopez Contreras AM, Sijtsma L, Weusthuis RA, Lier JB, Niel EWJ, Stams AJM, Vries SS (1999) Utilisation of biomass for the supply of energy carriers. Appl Microbiol Biotechnol 52:741–755

Ćosić B, Reiß F, Blümer M, Frekers C, Genin F, Pähr J, dWaßmer D (2022) Development and integration of the dual fuel combustion system for the MGT gas turbine family. J Eng Gas Turbine Power 144(2):021021

Cuellar-Bermudez S, Garcia-Perez J, Rittmann B, Parra-Saldivar R (2015) Photosynthetic bioenergy utilizing CO_2: an approach on flue gases utilization for third generation biofuels. J Clean Prod 98:53–65

Demirbas A (2002) Diesel fuel from vegetable oil via transesterification and soap pyrolysis. Energy Sources 24:835–841

Demirbas A (2005) Bioethanol from cellulosic materials: a renewable motor fuel from biomass. Energy Sources Part A 27:327–337

Demirbas A (2009) Progress and recent trends in biodiesel fuels. Energy Convers Manag 50:14–34

Demirbas M (2011) Biofuels from algae for sustainable development. Appl Energy 88:3473–3480

El-Hossaini MK (2013) Review of the New Combustion Technologies in Modern Gas Turbines. In: Benini E (ed). Progress in Gas Turbine Performance. InTech Open, pp 145–164

Faaij A (2006) Modern biomass conversion technologies. Mitig Adapt Strategies Glob Chang 11:343–375

Fagbohungbe OM, Herbert BMJ, Hurst L, Ibeto CN, Li H, Usmani SQ, Semple KT (2017) The challenges of anaerobic digestion and the role of biochar in optimizing anaerobic digestion. Waste Manag 61:236–249

Feng L, Qiang Y, Xiao L, Ming-jia LI, Jun-hui R, Ya-jun L (2022) Experimental study on fuel-switching of dual-fuel gas turbine combustor. Front Energy Res 9:796220

Francisco È, Neves D, Jacob-Lopes E, Franco T (2010) Microalgae as feedstock for biodiesel production: Carbon dioxide sequestration, lipid production and biofuel quality. J Chem Technol Biotechnol 85:395–403

Galbe M, Zacchi G (2002) A review of the production of ethanol from softwood. Appl Microbiol Biotechnol 59:618–628

Ghani W, Moghadam R, Salleh M, Alias A (2009) Air gasification of agricultural waste in a fluidized bed gasifier: hydrogen production performance. Energies 2:258–268

Golembiewski DA, Slothower BM, Bishop ER, Smith FS, Stumberg PM, Rideout-Hanzak S, Wester DB (2023) Fracking pond restoration in the western Rio Grande Plains, Texas: stockpiled topsoils may not be necessary, but seeding helps. J Environ Manag 346:118932

Gupta K, Rehman A, Sarviya R (2010a) Biofuels for the gas turbine: a review. Renew Sustain Energy Rev 14:2946–2955

Gupta K, Rehman A, Sarviya R (2010b) Evaluation of soya biodiesel as a gas turbine fuel. Iranica J Energy Environ 1(3):205–210

Habib Z, Parthasarathy R, Gollahalli S (2010) Performance and emission characteristics of biofuel in a small-scale gas turbine engine. Appl Energy 87:1701–1709

Hashimoto N, Ozawa Y, Mor N, Yuri Y, Hisamatsu T (2008) Fundamental combustion characteristics of palm methyl ester (PME) as alternative fuel for gas turbines. Fuel 87:3373–3378

Hosseini S, Wahid M (2014) Development of biogas combustion in combined heat and power generation. Renew Sustain Energy Rev 40:868–875

Huang GH, Chen F, Wei D, Zhang XW, Chen G (2010) Biodiesel production by microalgal biotechnology. Appl Energy 87:38–46

International Energy Agency (2016) IEA bioenergy countries' report: bioenergy policies and status of implementation. International Energy Agency (IEA), Paris

International Organization of Standardization (2013) Standard 4261: Petroleum Products-Fuels (class F) - Specifications of gas turbine fuels for industrial and marine applications. International Organization of Standardization (ISO), Geneva

Jianjun W, Yajun L, Weiwei X, Qiang L, Dandekar A (2019) Effects of blade roughness on particle deposition in flue gas turbines. Powder Technol 353:426–432

Juste G, Monfort J (2000) Preliminary test on combustion of wood derived fast pyrolysis oils in a gas turbine combustor. Biomass Bioenergy 19:119–128

Kalhor AX, Mohammadi-Nassab AD, Abedi E, Bahrami A, Movafeghi A (2016) Biodiesel production in crude oil contaminated environment using Chlorella vulgaris. Biores Technol 222:190–194

Kamiński W, Tomczak E, Górak A (2011) Biobutanol—production and purification methods. Ecol Chem Eng 18(1):31–37

Kang D, Kim T, Hur K, Park J (2012) The effect of firing biogas on the performance and operating characteristics of simple and recuperative cycle gas turbine combined heat and power systems. Appl Energy 93:215–228

Kang JY, Kang DW, Kim TS, Hur KB (2014) Economic evaluation of biogas and natural gas co-firing in gas turbine combined heat and power systems. Appl Therm Eng 70:723–731

Koizumi T (2015) Biofuels and food security. Renew Sustain Energy Rev 52:829–841

Lai D, Jin-Hui Zhan JH, Tian Y, Gao S, Xu G (2017) Mechanism of kerogen pyrolysis in terms of chemical structure transformation. Fuel 199:504–511

Langston LS (2019) Hydrogen fueled gas turbines. Mech Eng 141(03):52–54

Lee M, Seo SB, Chung JH, Kim SM, Joo YJ, Ahn DH (2010) Gas turbine combustion performance test of hydrogen and carbon monoxide synthetic gas. Fuel 89:1485–1491

Lee MC, Seo SB, Yoon J, Kim M, Yoon Y (2012) Experimental study on the effect of N_2, CO_2, and steam dilution on the combustion performance of H_2 and CO synthetic gas in an industrial gas turbine. Fuel 102:431–438

Lim M, Alimuddin Z (2008) Bubbling fluidized bed biomass gasification—performance, process findings and energy analysis. Renew Energy 33:2339–3234

MacLean H, Lave L (2003) Evaluating automobile fuel/propulsion system technologies. Prog Energy Combust Sci 29:1–69

Malça J, Freire F (2006) Renewability and life-cycle energy efficiency of bioethanol and bio-ethyl tertiary butyl ether (bioETBE): assessing the implications of allocation. Energy 31:3362–3380

Määttä S, de Gooyert V (2025) A critical review of social scientific research on carbon capture and storage. Renew Sustain Energy Rev 208:115063

Mendez C, Parthasarathy R, Gollahalli S (2014) Performance and emission characteristics of butanol/Jet A blends in a gas turbine engine. Appl Energy 118:135–140

Mohr SH, Wang J, Ellem G, Ward J, Giurco D (2015) Projection of world fossil fuels by country. Fuel 141:120–135

Naik S, Gou V, Rout P, Dalai A (2010) Production of first- and second-generation biofuels: a comprehensive review. Renew Sustain Energy Rev 14:578–597

Nazari MA, Alavi FM, Salem M, Assad MEH (2022) Utilization of hydrogen in gas turbines: a comprehensive review. Int J Low-Carbon Technol 17:513–519

Ngô C, Natowitz J (2009) Our energy future: resources, alternatives, and the environment. John Wiley and Sons, New Jersey

Oasmaa A, Kuoppala E, Solantausta Y (2003) Fast Pyrolysis of forestry residue 2: physicochemical composition of product liquid. Energy Fuels 17(2):433–443

Okoroigwe E, Ibeto C, Okpara C (2010) Comparative study of the potential of dog waste for biogas production. Trends Appl Sci Res 5(1):71–77

Olumayegun O, Wang M, Kelsall G (2016) Closed-cycle gas turbine for power generation: a state-of-the-art review. Fuel 180:694–717

Oparaku N, Ofomatah A, Okoroigwe E (2013) Biodigestion of cassava peels blended with pig dung for methane generation. Afr J Biotechnol 12(40):5956–5961

Ozyuguran A, Akturk A, Yaman S (2018) Optimal use of condensed parameters of ultimate analysis to predict the calorific value of biomass. Fuel 214:640–646

Patra J, Ghose P, Datta A, Das M, Ganguly R, Sen S, Chatterjee S (2015) Studies of combustion characteristics of kerosene ethanol blends in an axi-symmetric combustor. Fuel 144:205–213

Poullikkas A (2005) An overview of current and future sustainable gas turbine technologies. Renew Sustain Energy Rev 9(5):409–443

Rehman A, Phalke D, Pandey R (2011) Alternative fuel for gas turbine: esterified Jatropha oil-diesel blend. Renew Energy 36:2635–2640

Saha B (2000) Alpha-L-Arabinofuranosidases: biochemistry, molecular biology and application in biotechnology. Biotechnol Adv 18(5):403–423

Sannigrahi P, Ragauskas A, Miller S (2008) Effects of two-stage dilute acid pretreatment on the structure and composition of lignin and cellulose in Loblolly Pine. Bioenerg Res 1:205–214

Seljak T, Širok B, Katrašnik T (2016) Advanced fuels for gas turbines: fuel system corrosion, hot path deposit formation and emissions. Energy Convers Manag 125:40–50

Taipabu MI, Viswanathan K, Wu W, Aziz M, Kuo P-C, Madhankumar S (2024) Co-production of Bi-methanol from biomass. In: Reference module in chemistry, molecular sciences and chemical engineering. Elsevier. https://doi.org/10.1016/B978-0-443-15740-0.00036-7

Talebnia F, Karakashev D, Angelidaki I (2010) Production of bioethanol from wheat straw: an overview on pretreatment, hydrolysis and fermentation. Bioresour Technol 101:4744–4753

Tan E, Liou W (2011) Micro gas turbine engine characteristics using biofuel. The Hilltop Rev 5(1):40–50

Toporov D (2014) Combustion of pulverised coal in a mixture of oxygen and recycled flue gas. Elsevier

Ulbig P, Hoburg D (2002) Determination of the calorific value of natural gas by different methods. Thermochim Acta 382:27–35

Wang L, Weller C, Jones D, Hanna M (2008) Contemporary issues in thermal gasification of biomass and its application to electricity and fuel production. Biomass Bioenergy 32:573–581

Webb M, O'Connor J, Noble DR, Steele R (2024) Outlook for renewable liquid fuels in dual-fuel gas turbine combustors. Spring technical meeting of the eastern states section of the combustion institute. https://sites.psu.edu/rfdl/files/2024/03/ESSCI2024_GTLiquidFuel_240229b-f3a3f81fb185bebb.pdf

Wingren A, Galbe M, Roslander C, Rudolf A, Zacchi G (2005) Effect of reduction in yeast and enzyme concentrations in a simultaneous saccharification- and fermentation-based bioethanol process. Appl Biochem Biotechnol 122:485–499

Ye S, Jiayang C (2002) Hydrolysis of lignocellulosic materials for ethanol production: a review. Bioresour Technol 83:1–11

Chapter 3
Solar Radiation Resource

3.1 Introduction

A solar gas turbine (SGT) system converts concentrated solar radiation to heat that is transferred from the solar receiver to other components of the system. In this context, solar radiation is the most influential environmental factor of solar energy system performance (Nafey et al. 2000). Thus, knowledge about the physics of this resource is required in the development of the SGT system.

3.1.1 Solar Radiation Resource

Solar radiation is the electromagnetic radiation emitted by the sun, and it is concentrated around the short wavelength ($0.29\ \mu m \le \lambda \le 3.00\ \mu m$) region of the electromagnetic spectrum (Thekaekara 1976). This radiation is emitted at an equivalent black body temperature of about 6000 K with a constant intensity outside the earth's atmosphere (Kane 2005), and it propagates through the outer space and the earth's atmosphere. Part of the radiation reaches the surface of the earth located at a mean distance of 1.495×10^{11} m (one astronomical unit) from the sun (Duffie and Beckman 2013). The sun is almost spherical (diameter of about 1.39×10^{11} m) and subtends a maximum angle of $32'$ on the earth (diameter of 1.27×10^{9} m). This angle limits the concentration ratio (CR) of linear solar concentrators to 212 while the maximum value of CR for circular concentrators is 45,000. These concentrators focus solar radiation on point receivers and yield higher magnitudes of concentrated solar flux. Therefore, point receivers generally attain higher outlet temperatures. Theoretically, this type of receivers can achieve outlet temperatures of up to 1773 K that are suitable for driving advanced thermodynamic cycles such as the gas turbine.

A. Madhlopa and R. Nkhoma, *Principles of Solar Gas Turbines for Generating Electricity*, Green Energy and Technology,
https://doi.org/10.1007/978-3-031-84096-8_3

3.1.2 Conversion of Solar Radiation to Electricity

Applications of solar technologies are generally classified into photovoltaic (PV) and thermal categories. PV systems directly convert solar radiation into electricity through the photoelectric effect while solar thermal systems generate heat from solar radiation. The current major solar thermal technology for converting solar energy to electricity is concentrating solar power (CSP).

CSP technology converts solar radiation to heat that can propel thermodynamic cycles such as the gas turbine and steam engine cycles. It is possible to integrate CSP with a thermal energy storage (TES) component, which gives it an advantage over other energy systems that are confined to the deployment of batteries to store energy (Kuravi et al. 2013). Integration of CSP with TES provides the potential of full dispatchability of bulk power (Pfenninger et al. 2014). Moreover, TES enables stocking of thermal energy that can be used in power generation as well as heating and cooling applications. In view of this, recent CSP projects are integrated with TES, in different parts of the world. In an SGT system, the heat produced in the solar receiver is directly (or indirectly) transferred to the turbine. In this chapter, fundamentals of solar radiation are discussed and applied to the process of gas turbine solarisation. It is shown that the worldwide technical potential of CSP technology is much higher than the global consumption level of electricity. This magnitude of the available solar resource is very attractive for the advancement of the SGT technology.

3.2 Components of Solar Radiation

Irradiance on a surface which is perpendicular to the direction of propagation of solar radiation at an average distance between the earth and the sun outside the atmosphere is known as the solar constant (G_{sc}). This parameter is required in solar radiation analysis as shown in Sect. 3.4. Consequently, different authors have attempted to establish its accurate value for applications in solar engineering and science. A range of values of 1322–1395 W/m^2 has been reported in literature but the World Radiation Centre (WRC) adopted a value of 1367 W/m^2 (Gueymard 2004; Li et al. 2011), and that is used in this book.

3.2.1 Beam and Diffuse Solar Radiation

Outside the earth's atmosphere, there is only beam radiation. Nevertheless, as solar radiation propagates through the atmospheric matter, it undergoes scattering, absorption and transmission (Duffie and Beckman 2013). The scattering process produces the diffuse component of global solar radiation. So, the total solar radiation that reaches the earth's surface is made up of beam and diffuse components

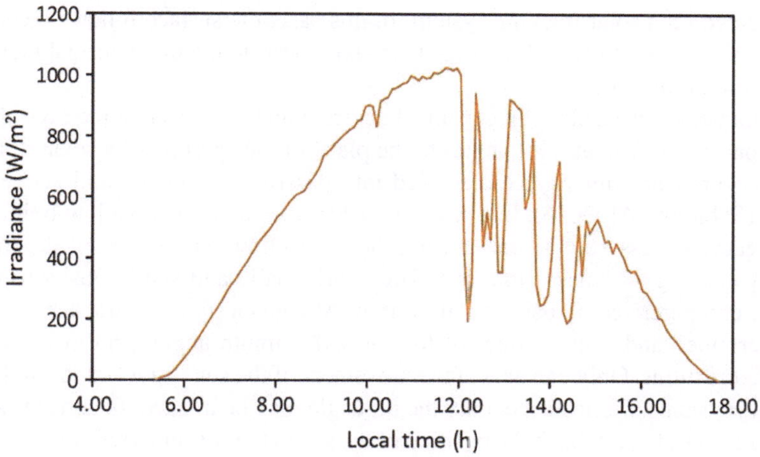

Fig. 3.1 Variation of global irradiance on a horizontal surface on a sample day (9th October 2008) at Malawi Polytechnic (15° 48′ S, 35° 02′ E)

(Yang et al. 2001; Lozano et al. 2023). These two components of solar radiation have different optical properties when incident on a surface. Beam radiation travels directly from the sun's disc to a receiver surface, and its rays can be traced from the sun's position and used in the computation of solar angles. The beam component of radiation can also be focussed to increase the amount of solar energy intercepted per unit area of an absorber surface. It is, therefore, not surprising that the CSP technology utilizes this component of solar radiation. On the other hand, diffuse radiation comes from the whole sky vault and it cannot be focussed. Moreover, its rays are not traceable from the sun's position, and the amount of diffuse solar energy directly received by a given surface depends on the proportion of the sky viewed by the surface (Duffie and Beckman 2013).

Due to atmospheric attenuation, the intensity of solar radiation on the earth's surface is much lower than the solar constant. An example of this variation is shown in Fig. 3.1. On this sample clear day, the highest recorded value of global irradiance on a horizontal surface was 1026 W/m^2, which is about 75% of the solar constant adopted by the WRC. It is also observed that the intermittency of irradiance was higher around 12:00–15:00 h due to passing clouds.

3.2.2 *Beam Normal Irradiance*

Beam or direct normal irradiance is the amount of solar radiation intercepted per unit area by a surface which is held at right angles to the rays that propagate in a straight line from the sun at its instantaneous position in the sky (Duffie and Beckman 2013). However, the position of the sun varies with time. In view of this, the surface has to be mounted on a solar tracking system. In this case, the surface follows the sun on daily (one-axis) or daily and seasonal (two-axis) basis to maintain normal incidence of the beam radiation.

A sun tracker is an electromechanical device which ensures that the rays of solar radiation are incident at right angles to the plane of the aperture of a solar collector. Sun trackers can primarily be classified into passive, time-based and active categories (Sabir and Ali 2016). In a passive tracker, the sun heats up a low-boiling gas that creates a pressure gradient causing the gas to flow on one end of the tracker, thereby causing the tracker to move. The hourly, daily and yearly positions of the sun are computerized by using an algorithm. At a given time, the tracker compares predetermined and actual values of the tilt and azimuth angles and makes appropriate corrections. Only one axis of the movement of the sun is tracked in single-axis tracking systems. For instance, only the tilt angle may be tracked. In active trackers, sensors are used to establish the position of the sun in the sky and a reference signal is generated to stimulate DC motor-based actuators to properly align the solar collector with the rays of the sun at a given time. Active trackers can be single axis or dual-axis. A single-axis solar tracker has a fixed tilt angle but the azimuth angle changes throughout the year. This category of active trackers is cheap to apply. In a dual-axis tracker, both the tilt and azimuth angles vary with time.

Global (G_{gh}) and diffuse (G_{dh}) components of solar radiation are measured on a horizontal surface at many locations in the world. In this case, beam normal irradiance (G_{bn}) can be calculated from (Kotti et al. 2014):

$$G_{bn} = \frac{G_{gh} - G_{dh}}{\cos\theta_z} = \frac{G_{bh}}{\cos\theta_z} \tag{3.1}$$

where G_{gh} is global irradiance on a horizontal surface, G_{bh} is beam irradiance on a horizontal surface, G_{dh} is diffuse irradiance on a horizontal surface, and θ_z is the solar zenith angle.

The level of direct normal irradiance is a good indicator of the suitability of a site for deployment of CSP technologies. The level of direct normal irradiance varies in space and time. In this vein, mean daily levels of direct irradiation and annual totals are shown in Fig. 3.2. Globally, a minimum annual direct normal irradiation (DNI) of 2,000 kWh/m^2 is recommended for CSP to be economically viable (Dawson and Schlyter 2012; Spyridonidou and Vagiona 2023), which indicates that many locations (in regions such as Australia, South Africa, North Africa, South America and Southern part of the United States of America) have great potential for the exploitation of CSP technology. About 50% of the global locations meet the minimum requirement of DNI. The worldwide technical potential of CSP is estimated at 3,000,000 TWh/year (Spelling 2013), which significantly exceeds the world electricity demands of 24,700 TWh in 2021 and projected world electricity demand of 62,159 TWh in 2050 (under the net zero emission scenario) (International Energy Agency 2022). This level of solar resource potential highlights the importance of developing SGTs that are driven by concentrated solar radiation.

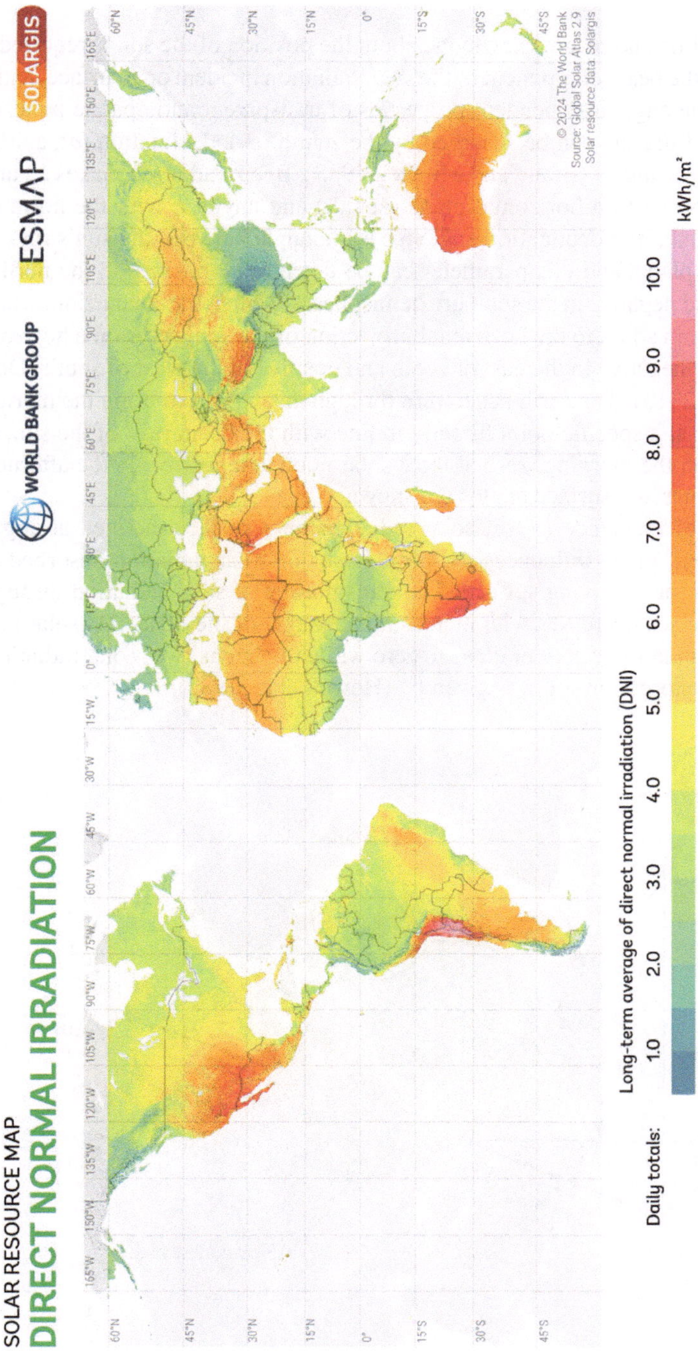

Fig. 3.2 Distribution of direct normal irradiation worldwide (https://solargis.com)

3.3 Sun Position and Direction of Beam Radiation

At any given time and place, knowledge about the position of the sun is required for calculation of the beam component of the solar radiation incident on a surface, and for determining the angular-dependent properties of transparent and opaque materials. The position of the sun can be specified by the zenith (θ_z), declination (δ), azimuth (γ_s) and hour (ω) angles (Braun and Mitchell 1983). Beam radiation makes an angle θ_z with the normal to a horizontal plane (zenith) and angle θ_i with the normal to an inclined plane. The declination angle (δ) is the angle between the sun's rays and the equatorial plane, and this parameter can be taken to be positive in the northern hemisphere and negative in the southern hemisphere. Further, the solar azimuth angle is measured from south to the horizontal projection of the sun's rays on a horizontal plane, and it is negative in the east of south and positive in the west of south (Duffie and Beckman 2013). The earth has to turn through an angle ω to bring the meridian of an observer at a specific point directly in line with the beam rays of the sun, and ω is negative in the morning, zero at local solar noon and positive in the afternoon. The various solar and surface angles are shown in Fig. 3.3.

The angle of incidence (θ_i) of solar radiation on a surface inclined at angle β to the horizontal plane influences the proportions of solar radiation absorbed and transmitted by the receiving surface. The transmission of solar radiation through a transparent surface decreases with increasing the angle of incidence. At solar noon, the angle of incidence is zero or close to zero while the intensity of solar radiation is optimum. An expression for θ_i is given by (Howell et al. 1982):

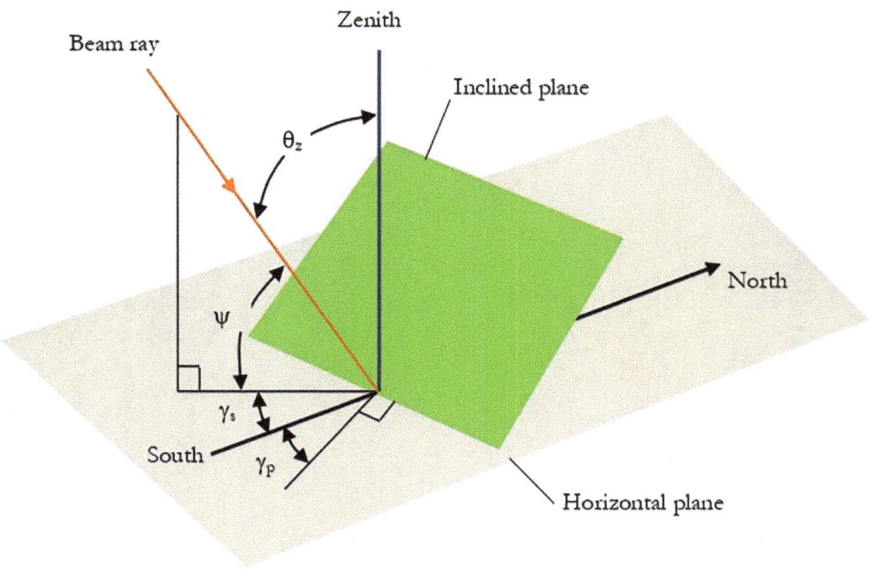

Fig. 3.3 Beam ray from the sun onto an inclined surface, and solar and surface angles (in the northern hemisphere)

$$\cos\theta_i = \cos\theta_z\cos\beta + \sin\theta_z\sin\beta\cos(\gamma_s - \gamma_p) \tag{3.2}$$

$$\cos\theta_z = \sin\delta\sin\phi + \cos\delta\cos\phi\cos\omega \tag{3.3}$$

$$\cos\gamma_s = \frac{\sin\phi\cos\delta\cos\omega - \cos\phi\sin\delta}{\cos\psi} \tag{3.4}$$

$$\omega = 15(t_s - 12) \tag{3.5}$$

$$t_s = t_c + t_e + \frac{\Theta_{sd} - \Theta_l}{15} \tag{3.6}$$

where t_s is the solar time, t_c is the civil time, t_e is the equation of time, Θ_l is the local longitude and Θ_{sd} is standard longitude.

$$t_e = 229.2\{0.000075 + 0.001868\cos\Gamma - 0.032077\sin\Gamma \\ -0.014615\cos(2\Gamma) - 0.04089\sin(2\Gamma)\}/60 \tag{3.7}$$

$$\psi = 90° - \theta \tag{3.8}$$

It is noted that Eqs. (3.3) and (3.4) also involve the declination angle (δ), which can be calculated from (Jain 1988):

$$\delta = 0.006918 - 0.399912\cos\Gamma + 0.07257\sin\Gamma - 0.006758\cos 2\Gamma + 0.000907\sin 2\Gamma \\ - 0.002697\cos 3\Gamma + 0.00148\sin 3\Gamma \tag{3.9}$$

$$\Gamma = \frac{360(N - 1)}{365} \tag{3.10}$$

where N is day of the year starting from 1 January.

When the sun is directly overhead on a horizontal surface ($\theta_z = 0$), solar radiation transmittance through the atmosphere is maximum at a given location on the earth's surface. The zenith angle is equal to zero during some time in the year at sites within the tropical zone, and it is always greater than zero at sites outside this region. The magnitude of the zenith angle is smallest and the potential for stable meteorological conditions is highest around solar noon (International Organization for Standardization 1994). So, irradiance on the earth's surface is highest around solar noon on a clear day. The values of solar angles and extraterrestrial radiation are required in the computation of solar radiation received by a surface on the earth.

3.4 Extraterrestrial Radiation and Solar Radiation on Inclined Surfaces

Extraterrestrial radiation is the theoretical amount of solar energy that would be available on a horizontal plane on the earth's surface if the earth were not surrounded by an atmosphere. In reality, solar radiation is attenuated as it propagates through the particulate matter surrounding the earth, resulting in a decrease of its intensity when measured on a plane on the earth's surface. At any specific time within the day, extraterrestrial irradiance (G_o) can be given by (Duffie and Beckman 2013):

$$G_o = G_{sc}\left\{1 + 0.033\cos\left(\frac{360N}{365}\right)\right\}\cos\theta_z \tag{3.11}$$

where G_{sc} is the solar constant.

In modelling and experimental studies of solar energy technologies, hourly (I) and daily (H) solar radiation data is sometimes required. So, integrated values of hourly (I_o) and daily (H_o) extraterrestrial radiation on a horizontal plane are used in such computations:

$$I_o = \frac{43{,}200}{\pi}G_{sc}\left\{1 + 0.033\cos\left(\frac{360N}{365}\right)\right\} \times \left\{\cos\phi\cos\delta(\sin\omega_2 - \sin\omega_1)\frac{\pi(\omega_2 - \omega_1)}{180}\sin\phi\sin\delta\right\} \tag{3.12}$$

$$H_o = \frac{86400}{\pi}G_{sc}\left\{1 + 0.033\cos\left(\frac{360\text{ N}}{365}\right)\right\}\left\{\cos\phi\cos\delta\sin\omega_{ss} + \frac{\pi\omega_{ss}}{180}\sin\phi\sin\delta\right\} \tag{3.13}$$

The hourly and daily extraterrestrial radiation is influenced by the day of the year and solar angles at a given location. Extraterrestrial radiation can be used to compute available solar radiation on a horizontal plane on the earth's surface. However, solar radiation on a tilted surface is needed in the design and performance studies of solar collectors. The total irradiance on a tilted plane consists of the beam, diffuse and ground-reflected components, and it can be calculated from:

$$G_{gp} = G_{bp} + G_{dp} + G_{gr} \tag{3.14}$$

where G_{bp} is beam irradiance on tilted plane, G_{dp} is diffuse irradiance on tilted plane and G_{gr} is ground-reflected irradiance.

In practice, global irradiance is commonly measured on a horizontal surface. Models are therefore employed to generate a data base of solar radiation received by inclined surfaces. For beam radiation, a geometric factor (R_b) is used to compute beam irradiance on a tilted surface from horizontal beam irradiance:

$$R_b = \frac{G_{bp}}{G_{bh}} = \frac{\cos\theta_i}{\cos\theta_z} \tag{3.15}$$

For CSP, only the beam radiation is considered because it can be focussed onto a smaller area. The diffuse and ground-reflected components cannot be focussed and are, therefore, ignored when computing the amount of solar radiation intercepted by the concentrator.

3.5 Available Solar Radiation on the Earth's Surface

Solar radiation that reaches the earth's surface is lower than the extraterrestrial radiation. In addition, commonly available solar radiation data is measured on a horizontal surface. Consequently, solar radiation data on an inclined plane is mostly computed from solar radiation data captured on a horizontal plane. For radiation analysis and other applications, it is necessary to measure solar radiation received by a horizontal surface at a given site.

Available global solar radiation on a horizontal surface can be directly measured by using a pyranometer (Thekaekara 1976), and accurate data of solar radiation is needed for simulation, testing and operation of solar energy technologies. In line with this requirement, it is important to calibrate pyranometers by using standard methods in order to capture accurate hemispherical solar radiation data (International Organization of Standardization 1993). The ratio of the solar radiant flux (including the beam solar radiant flux) received from the sky hemisphere by a given plane to the area of the plane is called hemispherical irradiance (International Organization of Standardization 1992). According to the ISO 9060 classification standard, there are three classes of pyranometers: secondary standard, first and second classes, with the secondary standard class being the best (Kipp & Zonen 2006). First class pyranometers are also recommended for measuring irradiance for performance evaluation of solar collectors (International Organization of Standardization 1994). Nevertheless, it is often desirable to know the individual levels of beam and diffuse radiation for purposes of solar energy system modelling, design, testing and deployment. Consequently, these components of global radiation are measured on a given plane.

Beam radiation can be measured by using direct or indirect methods. In the direct approach, a pyrheliometer is used to measure beam radiation at normal incidence but this instrument is expensive (Kudish and Evseev 2008). Consequently, only limited meteorological sites around the world have pyrheliometers. In the indirect method, beam radiation is determined by subtracting diffuse radiation from global radiation. In this case, concurrent measurements of global and diffuse radiation are taken in the same plane.

Diffuse radiation can be measured by using a pyranometer fitted with an occulting disc or shadow ring, with the occulting disc yielding more accurate data. However, the disc requires an expensive sun-tracking system. In view of this, diffuse radiation is commonly measured by using a pyranometer with a stationary ring in the east–west axis. The ring blocks off beam radiation and a small amount of diffuse radiation from reaching the sensor (Fig. 3.4), which results in slightly lower readings of diffuse irradiance. A correction factor is therefore applied to the data to take care of this error (Thekaekara 1976). The difference between global and diffuse irradiance yields beam irradiance which can then be used to compute direct normal irradiance required for CSP systems.

Fig. 3.4 A Kipp & Zonen pyranometer with a shadow ring mounted on roof top at Malawi Polytechnic: **a** pyranometer and **b** shadow ring

3.6 Interaction of Solar Radiation with Opaque and Transparent Surfaces

For opaque surfaces, part of the incoming solar radiation is reflected while the remainder is absorbed by the surface. The surface absorbs (α) and reflects (ρ) part of the radiation (Fig. 3.5). The absorbed proportion is converted to heat (unless the surface has photocells which convert part of solar radiation into electricity). For a smooth surface (such as a mirror), the angle of incidence (θ_i) is equal to the angle of reflection (θ_{re}). The absorptance and reflectance of solar radiation is influenced by the wavelength (λ) and direction (θ_i). This implies that the spectral distribution of the solar flux has to be known in order to enable computation of the solar weighted property. To enable comparison of the performance of different solar technologies, it is necessary to use a reference standard solar spectral irradiance distribution (International Organization of Standardization 1992). Part 1 of the ISO 9845 standard gives a suitable standard spectral irradiance distribution which can be employed to find the relative performance of solar thermal, photovoltaic and other systems, and components and materials where the direct and hemispherical irradiance is required. Based on energy balance, the sum of the proportions of reflected and absorbed radiation is equal to one:

$$\alpha(\lambda, \theta_i) + \rho(\lambda, \theta_i) = 1 \tag{3.16}$$

A surface can reflect solar radiation specularly or diffusely (Fig. 3.6). Radiation propagates in one direction from a specular plane reflector, and specular geometric (view) factor and reflectance are required in the computation of reflected beam radiation (Wijeysundera 1978). This property of radiation is useful in the advancement of CSP systems. Only beam radiation can be focussed, and therefore the performance of focussing solar technologies increases with the intensity of beam radiation. For a given set of surface properties, an opaque surface absorbs a maximum level of incoming radiation when the angle of incidence is zero. Under this condition, the surface is said to receive beam normal irradiance.

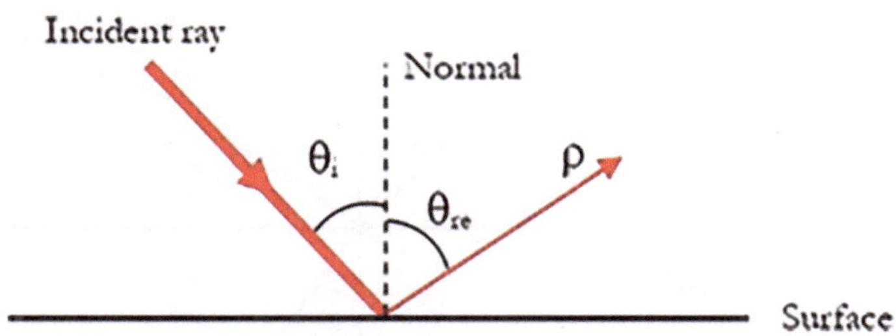

Fig. 3.5 Attenuation of solar radiation by an opaque surface

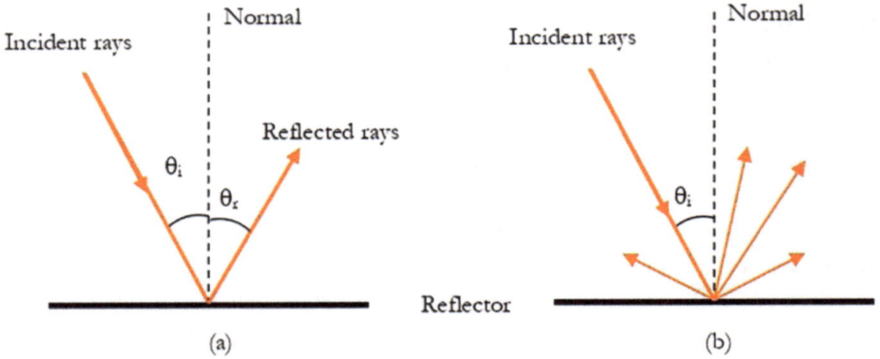

Fig. 3.6 Attenuation of solar radiation by an opaque surface **a** specular reflection and **b** diffuse reflection

On the other hand, reflected radiation goes in different directions in diffuse reflection, and diffuse view factor and reflectance are required for calculation of the magnitude of reflected diffuse solar radiation. Consequently, a lower proportion of diffuse radiation is desirable for satisfactory performance of CSP technologies.

In addition to reflection and absorption, a transparent surface transmits part of the intercepted solar radiation (Fig. 3.7). If the surface is surrounded by air, solar radiation travels through air before it reaches the transparent surface. The surface absorbs (α), reflects (ρ) and transmits (τ) part of the radiation. The sum of the proportions of reflected, absorbed and transmitted radiation is equal to one and these fractions are influenced by the optical properties of incoming solar radiation.

$$\alpha(\lambda, \theta_\mathrm{i}) + \rho(\lambda, \theta_\mathrm{i}) + \tau(\lambda, \theta_\mathrm{i}) = 1 \tag{3.17}$$

where θ_i is the angle of incidence at the boundary between two media and λ is the wavelength of the radiation.

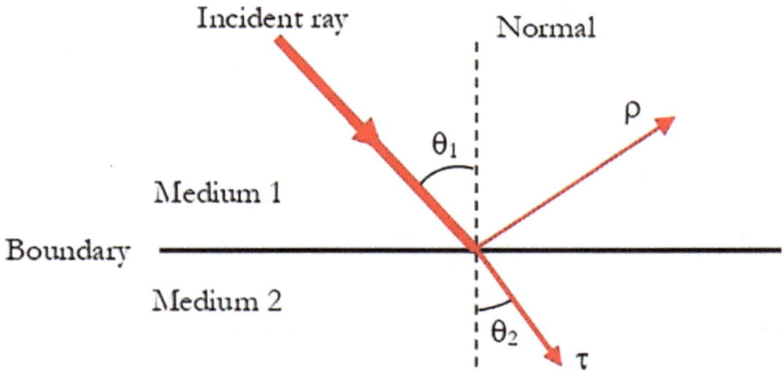

Fig. 3.7 Attenuation of solar radiation through a transparent surface

A ray of light changes direction when it traverses the boundary between Medium 1 and Medium 2 that have different values of optical density ($n_1 \neq n_2$). This phenomenon is defined as refraction. Snell's law states that:

$$n_1 \sin\theta_1 = n_2 \sin\theta_2 \tag{3.18}$$

where n_1 is the optical density of Medium 1, n_2 is the optical density of Medium 2, θ_1 is the angle of incidence in Medium 1 and θ_2 is the angle of refraction in Medium 2. This principle is useful in the design and operation of optical systems with transparent components (such as lenses).

Although most of the solar radiation received by the earth's surface falls within the range of short wavelength, many types of window glazing have a weak spectral selectivity in the solar spectrum (American Society of Heating Refrigerating and Air-Conditioning Engineers 2017). In view of this, only their angular dependence can be taken into consideration.

Solar radiation is polarized when it propagates through a transparent material. Perpendicular and parallel components of unpolarized radiation are polarized differently by the transparent material, which results in differences in the reflectance, absorptance and transmittance of the material for the two components. The transmittance of the material can be computed by including the angular dependence of solar attenuation. For the perpendicular component of unpolarized radiation, the absorptance (α_{n}), reflectance (ρ_{n}) and transmittance (τ_n) can be calculated as follows (Duffie and Beckman 2013):

$$\tau_{\mathrm{ab}} = e^{(-Lx/\cos\theta_2)} \tag{3.19}$$

$$\alpha_{\mathrm{n}} = (1-\tau_{\mathrm{ab}})\left(\frac{1-\rho_{\mathrm{n}}}{1-\rho_{\mathrm{n}}\tau_{\mathrm{ab}}}\right) \tag{3.20}$$

$$\rho_{\mathrm{n}} = \frac{\sin^2(\theta_2-\theta_1)}{\sin^2(\theta_2+\theta_1)} \tag{3.21}$$

$$\tau_{\mathrm{n}} = \frac{\tau_{\mathrm{ab}}(1-\rho_n)^2}{1-\rho_{\mathrm{n}}\tau_{\mathrm{ab}}} \tag{3.22}$$

where L is the coefficient of extinction and x is the thickness of the transparent material.

Similar equations are used to calculate the corresponding values of the absorptance, transmittance and reflectance for the parallel component of unpolarized solar radiation, with ρ_{n} replaced by ρ_{p}:

$$\rho_{\mathrm{p}} = \frac{\tan^2(\theta_2-\theta_1)}{\tan^2(\theta_2+\theta_1)} \tag{3.23}$$

The required angular-dependent optical properties of a single transparent cover are obtained from:

$$\alpha = \frac{(\alpha_n + \alpha_p)}{2} \tag{3.24}$$

$$\tau = \frac{(\tau_n + \tau_p)}{2} \tag{3.25}$$

$$\rho = \frac{(\rho_n + \rho_p)}{2} \tag{3.26}$$

The reflectance of a single transparent cover can be given by:

$$\rho \cong \tau_{ab} - \tau \tag{3.27}$$

The values of ρ and τ at normal incidence are commonly available for different transparent materials with specified thicknesses. So, Eqs. (3.14) and (3.22) can be used to compute the value of K (at normal incidence) for a transparent cover fitted over a solar collector. This parameter is then used to determine the values of α, τ and ρ at different angles of incidence.

3.7 Estimation of Beam and Diffuse Solar Radiation on the Earth's Surface

Knowledge about the individual levels of beam and diffuse radiation is required when developing solar technologies (Arias-Rosales and LeDuc 2020). Separation of the components is necessary for two reasons: (a) methods for computing the total radiation on tilted surfaces from data on a horizontal plane require distinct treatments of beam and diffuse radiation, and (b) prediction of the long-term performance of CSP technology is based on estimates of available beam radiation. Nevertheless, unlike global solar radiation, the beam and diffuse components of radiation are measured at limited sites around the world. Consequently, many mathematical models have been developed to estimate these radiation components on the earth's surface. About 250 models and correlations for estimating the diffuse radiation are found in the literature (de Miguel et al. 2001). In this book, selected models for estimating the diffuse radiation are presented. The difference between the global and diffuse radiation can yield the beam radiation.

3.7.1 Hourly Beam and Diffuse Radiation on a Horizontal Surface

Solar gas turbines use only beam radiation, which is measured at very limited sites. In contrast, global radiation on the horizontal surface is measured at many sites around the world. The beam radiation can be estimated from the difference between the global radiation and diffuse radiation. For cases where only data on global radiation is available, the diffuse component can be derived from correlations between the hourly diffuse fraction (I_{dh}/I_{gh}) and hourly clearness index (k). Clearness index is the ratio of global solar radiation measured at ground level to its counterpart estimated at the top of the atmosphere (Paulescu et al. 2021). The available solar radiation increases with the magnitude of clearness index. We can have instantaneous, hourly, daily or monthly clearness indices. The hourly clearness index is given by:

$$k = \frac{I_{gh}}{I_o} \tag{3.28}$$

where I_{gh} is the hourly irradiation on a horizontal surface at ground level and I_o is the hourly irradiation of the counterpart extraterrestrial solar radiation.

Piecewise polynomial models are commonly used to estimate the diffuse radiation. Generally, the conditions of the skies are classified into overcast (low cleanness index), partly cloudy (medium clearness index) and clear (high clearness index). For example, Orgill and Hollands (1977) used data from Canadian weather stations to obtain a piecewise polynomial correlation of the first order:

$$\frac{I_{dh}}{I_{gh}} = \begin{cases} 1.0 - 0.249k, & \text{for } 0 \le k \le 0.35 \\ 1.557 - 1.84k, & \text{for } 0.35 < k \le 0.75 \\ 0.117, & \text{for } k > 0.75 \end{cases} \tag{3.29}$$

Using data from four weather stations in the United States of America and one Australian station, Erbs et al. (1982) established the following correlation:

$$\frac{I_{dh}}{I_{gh}} = \begin{cases} 1.0 - 0.09k, & \text{for } k \le 0.22 \\ 0.9511 - 0.1604k + 4.388k^2 - 16.638k^3 + 12.336k^4, & \text{for } 0.22 < k \le 0.80 \\ 0.165, & \text{for } k > 0.80 \end{cases} \tag{3.30}$$

The correlation of Orgill and Hollands (1977) yields results that are comparable with those of Erbs et al. (1982), for practical purposes. Another important correlation was obtained by Reindl et al. (1990) using independent datasets from the United States of America and Europe. This correlation includes a constraint: $I_{dh}/I_{gh} \le 1$:

$$\frac{I_{dh}}{I_{gh}} = \begin{cases} 1.020 - 0.248k, & \text{for } 0 \le k \le 0.30 \\ 1.45 - 1.67k, & \text{for } 0.30 < k \le 0.78 \\ 0.147, & \text{for } k > 0.78 \end{cases} \tag{3.31}$$

These and many other models are available for estimating hourly diffuse radiation on a horizontal plane. The difference ($I_{gh} - I_{dh}$) yields the hourly diffuse radiation on this plane.

3.7.2 Daily Beam and Diffuse Radiation on a Horizontal Surface

In some cases, the daily radiation data is available and studies have shown that the daily diffuse fraction H_{dh}/H_{gh} is associated with the daily clearness index (K). For instance, by using data from four sites from the United States of America, Collares-Pereira and Rabl (1979) derived a fourth order polynomial correlation:

$$\frac{H_{dh}}{H_{gh}} = \begin{cases} 0.99, & \text{for } K \le 0.17 \\ 1.188 - 2.272K + 9.473K^2 - 21.865K^3 + 14.648K^4, & \text{for } 0.17 < K \le 0.75 \\ 0.632 - 0.54K, & \text{for } 0.75 < K < 0.80 \\ 0.20, & \text{for } K \ge 0.80 \end{cases} \tag{3.32}$$

Third-order piecewise polynomial models have also been applied by some authors. For instance, Frutos et al. (1985) used data from Madrid in Spain to derive the correlation:

$$\frac{I_{dh}}{I_{gh}} = \begin{cases} 0.920, & \text{for } K \le 0.17 \\ 0.770 + 1.930K - 6.860K^2 + 4.270K^3, & \text{for } 0.17 < K \le 0.80 \\ 0.190, & \text{for } K > 0.80 \end{cases} \tag{3.33}$$

This correlation is similar to that of de Miguel et al. (2001) which was based on a large volume of solar radiation data from various stations in France (three stations), Greece (one station), Portugal (five stations), and Spain (two stations):

$$\frac{I_{dh}}{I_{gh}} = \begin{cases} 0.952, & \text{for } K \le 0.13 \\ 0.868 + 1.335K - 5.782K^2 + 3.721K^3, & \text{for } 0.13 < K \le 0.80 \\ 0.141, & \text{for } K > 0.80 \end{cases} \tag{3.34}$$

Piecewise polynomial models proposed by Frutos et al. (1985), de Meguel et al. and other authors have sudden changes in the diffuse fraction at boundaries. Nevertheless, the change in the diffuse fraction is expected to be non-abrupt for a continuous change in the clearness index. Consequently, Madhlopa (2001) proposed a differential calculus method of smoothening the junctions to match reality more closely. This method was applied to the data from Chileka weather station (15° 40′ S, 34° 58′ E) in Malawi to obtain Eq. (3.30). An error analysis showed that curve smoothening improved the accuracy of estimating the diffuse fraction.

$$\frac{I_{dh}}{I_{gh}} = \begin{cases} 0.910, & \text{for } K \le 0.06 \\ 0.894 + 0.575K - 5.168K^2 + 4.351K^3, & \text{for } 0.06 < K \le 0.73 \\ 0.252, & \text{for } K > 0.73 \end{cases} \quad (3.35)$$

3.7.3 Beam Radiation for Clear Skies

For clear skies, it is possible to estimate beam radiation. Hottel (1976) proposed a model for estimating the beam radiation transmitted through clear atmospheres. This model takes into account the zenith angle and altitude for a standard atmosphere and it covers four types of climate. The transmittance (τ_b) of beam radiation through the atmosphere is G_{bn}/G_{on}, where G_{bn} is the beam radiation on a plane normal to the direction of the beam radiation, and G_{on} is the extraterrestrial radiation incident on a plane normal to the direction of the extraterrestrial radiation. Alternatively, transmittance is G_{bp}/G_{op}, where G_{bp} is beam radiation on tilted plane and G_{op} is extraterrestrial radiation on a tilted plane. This transmittance can be given by:

$$\tau_b = a + b\exp\left(\frac{-c}{\cos\theta_z}\right) \quad (3.36)$$

where *a*, *b*, and *c* are constants. For the standard atmosphere with 23 km visibility, these constants are found from a^*, b^* and c^* which (for altitudes below 2.5 km) are given by:

$$a^* = 0.4237 - 0.00821(6 - A)^2 \quad (3.37)$$

$$b^* = 0.5055 + 0.00595(6.5 - A)^2 \quad (3.38)$$

$$c^* = 0.2711 + 0.01858(2.5 - A)^2 \quad (3.39)$$

where *A* is the altitude of the site. The factors a^*, b^* and c^* are corrected to allow for variations in climate types. Applicable correction factors are $r_o = a/a^*$, $r_1 = b/b^*$ and $r_2 = c/c^*$ presented in Table 3.1.

Using the transmittance factor, the clear-sky normal beam radiation (G_{cnb}) is then given by:

Table 3.1 Correction factors r_o, r_1, and r_2 for climate type

Climate type	Correction factor		
	r_o	r_1	r_2
Tropical	0.95	0.98	1.02
Mid-latitude summer	0.97	0.99	1.02
Subarctic summer	0.99	0.99	1.01
Mid-latitude winter	1.03	1.01	1.00

$$G_{\text{cnb}} = \tau_{\text{b}} G_{\text{on}} \tag{3.40}$$

$$G_{\text{on}} = G_{\text{sc}}\left\{1 + 0.033\cos\left(\frac{360N}{365}\right)\right\} \tag{3.41}$$

Clear-sky horizontal beam radiation (G_{chb}) is expressed as:

$$G_{\text{chb}} = \tau_{\text{b}} G_{\text{on}} \cos\theta_{\text{z}} \tag{3.42}$$

Hourly clear-sky horizontal beam radiation (I_{chb}) is given by:

$$I_{\text{chb}} = \tau_{\text{b}} I_{\text{on}} \cos\theta_{\text{z}} \tag{3.43}$$

In this case, the value of G_{chb} at the mid-hour is computed and used to obtain the horizontal hourly beam irradiation. Hourly values of the beam irradiation can then be summed up to obtain the daily beam irradiation on a horizontal plane. So, the hourly beam irradiation on a horizontal plane can be expressed as:

$$I_{\text{chb}} = 3600 G_{\text{chb}} \tag{3.44}$$

where G_{chb} is the value of horizontal beam irradiance at the middle of the hour.

3.8 Worked Examples

Example 3.1 Calculate the solar time at 11:30 AM (civil time) on 9 October at the Malawi Polytechnic (15° 48′ S, 35° 02′ E).

Solution

Solar time,

$$t_{\text{s}} = t_{\text{c}} + t_{\text{e}} + (\Theta_{\text{sd}} - \Theta_{\text{l}})/15$$

$$t_{\text{e}} = 229.2\{0.000075 + 0.001868\ \cos\Gamma - 0.032077\ \sin\Gamma \\ -0.014615\ \cos(2\Gamma) - 0.04089\sin(2\Gamma)\}/60$$

But $\Gamma = 360(N-1)/365 = 277.1507$.
This implies that:

$$\begin{aligned} t_e &= \{0.000075 + 0.001868 \ \cos 277.1507 - 0.032077 \ \sin(277.1507) \\ &\quad -0.014615 \ \cos(2 \times 277.1507) - 0.04089 \ \sin(2 \times 277.1507)\}/60 \\ &= -0.2362 \text{ h} \end{aligned}$$

In Malawi, standard longitude is $\Theta_{sd} = 30°$ for all sites. Therefore,

$$\begin{aligned} t_s &= 11.5 - 0.2362 + (30 - 35.0333)/15 \\ &= 10.9283 \text{ h} \\ &= 10{:}56 \text{ AM} \end{aligned}$$

Example 3.2 In Example 3.1, the global irradiance on a horizontal surface at this site was 1010 W/m^2 at 11:30 AM (civil time) on 9th October 2008. Calculate the instantaneous clearness index over the site.

Solution

Instantaneous clearness index (I) is defined as:

$$I = \frac{G}{G_o}$$

From Eq. (3.11),

$$G_o = G_{sc}\left\{1 + 0.033\cos\left(\frac{360N}{365}\right)\right\}\cos\theta_z$$

But,

$$\cos\theta_z = \sin\delta\sin\phi + \cos\delta\cos\phi\cos\omega$$

From Eq. (3.10), the declination angle is:

$$\begin{aligned} \delta &= 0.006918 - 0.399912 \ \cos\Gamma + 0.07257 \ \sin\Gamma \\ &\quad - 0.006758\cos 2\Gamma + 0.000907\sin 2\Gamma \\ &\quad - 0.002697 \ \cos 3\Gamma + 0.00148\sin 3\Gamma \\ \Gamma &= 360(N-1)/365 \\ &= 360(282-1)/365 \\ &= 277.1507 \end{aligned}$$

So,

$$\begin{aligned}\delta &= 0.006918 - 0.399912\ \cos(277.1507) + 0.07257\sin(277.1507)\\ &\quad - 0.006758\ \cos(2\times 277.1507) + 0.000907\sin(2\times 277.1507)\\ &\quad - 0.002697\ \cos(3\times 277.1507) + 0.00148\ \sin(3\times 277.1507)\\ &= -0.2529 \text{ radians} = -0.2529\times 180/\pi = -14.4927^\circ\end{aligned}$$

The hour angle ω is:

$$\begin{aligned}\omega &= 15(t_s - 12) = 15\times(10.9283 - 12)\\ &= -16.0761^\circ\end{aligned}$$

Therefore,

$$\begin{aligned}\cos\theta_z &= \sin(-14.4927^\circ)\ \sin(15.8^\circ) + \cos(-14.4927^\circ)\\ &\quad\times\cos(15.8^\circ)\ \cos(-16.0761^\circ)\\ &= 0.9633\end{aligned}$$

$$G_o = 1367\left\{1 + 0.033\ \cos\left(\frac{360\times 282}{365}\right)\right\}\times 0.9633 = 1312.24\ \text{Wm}^{-2}$$

$$I = \frac{1010\ \text{Wm}^{-2}}{1312.24\ \text{Wm}^{-2}} = 0.77$$

Example 3.3 A north-facing surface is inclined at 35° from the horizontal at Bonfoi weather station (33° 56′ S, 18° 47′ E) in South Africa. Calculate the angle of incidence of beam radiation at 11:00 AM on 1st January at this site.

Solution

The angle of incidence on an inclined surface is computed from Eq. (3.2), which requires the equation of time (t_e), solar time (t_s), hour angle (ω), the declination angle(δ), zenith angle (θ_z) and solar (γ_s) and surface (γ_p) azimuth angles, and inclination angle (β).

Equation of time on 1st January,

$$\begin{aligned}t_e &= 229.2\{0.000075 + 0.001868\ \cos(0) - 0.032077\sin(0)\\ &\quad -0.014615\ \cos(2\times 0) - 0.04089\ \sin(2\times 0)\}/60\\ &= 229.2\times\{0.000075 + 0.001868 - 0.014615\}/60\\ &= -0.04841\end{aligned}$$

Solar time,

$$t_s = t_c + t_e + (\Theta_{sd} - \Theta_l)/15 = 11 - 0.04841 + (30 - 18.782)/15 = 11.6995 \text{ h}$$
$$\omega = 15(t_s - 12) = 15(11.6995 - 12) = -4.507^\circ$$
$$\cos\theta_i = \cos\theta_z \cos\beta + \sin\theta_z \sin\beta\cos(\lambda_s - \gamma_p)$$

But $\cos\theta_z = \sin\delta\sin\phi + \cos\delta\cos\phi\cos\omega$

$$\begin{aligned}\delta &= 0.006918 - 0.399912\cos\Gamma + 0.07257\sin\Gamma \\ &\quad - 0.006758\cos 2\Gamma + 0.000907\sin 2\Gamma \\ &\quad - 0.002697\cos 3\Gamma + 0.00148\sin 3\Gamma = 0.006918 - 0.399912\cos(0) \\ &\quad + 0.07257\sin(0) - 0.006758\cos(2 * 0) + 0.000907\sin(2\times 0) \\ &\quad - 0.002697\cos(3\times 0) + 0.00148\sin(3\times 0) \\ &= -23.059^\circ\end{aligned}$$

Zenith angle,

$$\begin{aligned}\theta_z &= \cos^{-1}(\sin\delta\sin\phi + \cos\delta\cos\phi\cos\omega) \\ &= \cos^{-1}\big[\sin(-23.059^\circ)\sin(-33.935^\circ) \\ &\quad + \cos(-23.059^\circ)\cos(-33.935^\circ)\cos(-4.507^\circ)\big] \\ &= 11.569^\circ\end{aligned}$$

Solar altitude,

$$\psi = 90^\circ - \theta_z = 90^\circ - 11.569^\circ = 78.431^\circ$$

Solar azimuth angle,

$$\begin{aligned}\gamma_s &= \cos^{-1}\{(\sin\phi\cos\delta\cos\omega - \cos\phi\sin\delta)/\cos\psi\} \\ &= \cos^{-1}\big\{\big[\sin(-33.935^\circ)\cos(-23.059^\circ)\cos(-4.507^\circ) - \cos(-33.935^\circ) \\ &\quad \times \sin(-23.059^\circ)\big]/\cos(78.431^\circ)\big\} \\ &= 158.900^\circ\end{aligned}$$

Angle of incidence,

$$\begin{aligned}\theta_i &= \cos^{-1}\{\cos\theta_z\cos\beta + \sin\theta_z\sin\beta\cos(\gamma_s - \gamma_p)\} \\ &= \cos^{-1}\{\cos(11.569^\circ)\cos(35^\circ) + \sin(11.569^\circ)\sin(35^\circ)\cos(158.900^\circ - 180^\circ)\} \\ &= 24.5^\circ\end{aligned}$$

Example 3.4 On 13 October 1989, the hourly global and diffuse radiation levels measured on a horizontal plane at Chileka weather station (15° 40′ S, 34° 58′ E, 760 m above sea level) in Malawi were respectively 3761 and 317 kJ/m^2 at 12:00 noon (civil time). Estimate the hourly beam radiation by using the Hottel (1976) clear-sky model and compare it with the empirical value.

Solution

On 13 October:

$$\Gamma = 360 \times (286 - 1)/365 = 281.0959,$$

Declination,

$$\begin{aligned}\delta &= 0.006918 - 0.399912\ \cos\Gamma + 0.07257\ \sin\Gamma - 0.006758\ \cos 2\Gamma \\ &\quad + 0.000907\ \sin 2\Gamma - 0.002697\ \cos 3\Gamma + 0.00148\ \sin 3\Gamma \\ &= 0.006918 - 0.399912\ \cos 281.0959 + 0.07257\ \sin 281.0959 \\ &\quad - 0.006758\ \cos(2 \times 281.0959) + 0.000907\ \sin(2 \times 281.0959) \\ &\quad - 0.002697\ \cos(3 \times 281.0959) + 0.00148\ \sin(3 \times 281.0959) \\ &= -1.5568^\circ\end{aligned}$$

The irradiation over the period 11: 00 AM to 12: 00 noon is recorded at 12: 00 noon. Therefore the mid-hour is at 11:30 AM.

Equation of time on 13th October,

$$\begin{aligned}t_e &= 229.2\{0.000075 + 0.001868\ \cos(281.0959) - 0.032077\ \sin(281.0959) \\ &\quad -0.014615\ \cos(2 \times 281.0959) - 0.04089\ \sin(2 \times 281.0959)\}/60 \\ &= -0.1532\ \text{h}\end{aligned}$$

Solar time,

$$t_s = t_c + t_e + (\Theta_{sd} - \Theta_l)/15 = 11 - 0.1532 + (30 - 34.9667)/15 = 11.3221\ \text{h}$$

Hour angle,

$$\omega = 15(t_s - 12) = 15(11.3221 - 12) = -10.1681^\circ$$

$$\begin{aligned}\cos\theta_z &= \sin(-1.5568^\circ)\ \sin(-15.6667^\circ) + \cos(-1.5568^\circ) \\ &\quad \times \cos(-15.6667^\circ)\ \cos(-10.1681^\circ) \\ &= 0.9547\end{aligned}$$

Using correction factors for tropical climate from Table 3.1,

$$a = r_0 a^* = 0.95 \times \left[0.4237 - 0.00821(6 - 0.76)^2\right] = 0.1884$$
$$b = r_1 b^{\times} = 0.98 \times \left[0.5055 + 0.00595(6.5 - 0.76)^2\right] = 0.6875$$
$$c = r_2 c^* = 1.02 \times \left[0.2711 + 0.01858(2.5 - 0.76)^2\right] = 0.3339$$

Therefore, from Eq. (3.36), the atmospheric transmittance is

$$\tau_b = 0.1884 + 0.6875 \exp\left(\frac{-0.3339}{0.9547}\right) = 0.673$$

From Eq. (3.41),

$$G_{on} = G_{sc}\left\{1 + 0.033\ \cos\left(\frac{360N}{365}\right)\right\} = 1367 \times \left\{1 + 0.033\ \cos\left(\frac{360 \times 286}{365}\right)\right\}$$
$$= 1402.61$$

$$G_{chb} = \tau_b G_{on}\ \cos\theta_z = 0.673 \times 1402.61 \times 0.9547 = 901.16\ \text{W/m}^2$$

$$I_{chb} = 3600 \times 901.16 = 3.24\ \text{MJ/m}^2$$

Empirical hourly beam irradiation on a horizontal plane, $I_{gh} - \text{I}_{dh} = 3.761 - 0.317 = 3.44\ \text{MJ/m}^2$

Thus, there is good agreement between theoretical and empirical values of the hourly beam irradiation on the horizontal plane.

Example 3.5 Using Example 3.3, the intensity of beam irradiance on a horizontal surface is 600 W/m^2. Calculate the level of irradiance on the north-facing surface inclined at 35° at the specified time.

Solution

Irradiance on the inclined surface (G_{bp}) can be computed by using Eq. (3.15):

$$G_{bp} = R_b G_{bh} = (\cos\theta_i/\cos\theta_z)\, G_{bh}$$
$$= (\cos 24.5°/\cos 11.6°) \times 600$$
$$\approx 557 W/m^2$$

Nomenclatures

G	Irradiance (W/m^2)
H	Daily irradiation (J/m^2)
I	Hourly irradiation (J/m^2)
k	Hourly clearness index (dimensionless)
K	Daily clearness index (dimensionless)
L	Coefficient of extinction (1/m)
N	Day of the year starting from 1 January (dimensionless)
n	Optical density (dimensionless)
R	Geometric factor (dimensionless)
t	Time (h)
x	Thickness (m)

Greek Symbols

α	Absorptance (dimensionless)
β	Tilt angle to the horizontal (degree)
δ	Declination angle (degree)
ϕ	Latitude (degree)
γ	Azimuth angle (degree)
λ	Wavelength of radiation (m)
θ	Angle (degree)
Θ	Longitude (degree)
ρ	Reflectance (dimensionless)
τ	Transmittance (dimensionless)
ω	Hour angle (degree)

Subscripts

1	First
2	Second
b	Beam
bh	Beam on a horizontal plane
bn	Beam at normal incidence
bp	Beam on tilted plane
c	Civil
chb	Clear-sky horizontal beam
cnb	Clear-sky normal beam
dh	Diffuse on a horizontal plane

dp Diffuse on a tilted plane
e Equation
gh Global on a horizontal plane
gp Global on tilted plane
gr Ground-reflected
i Incident
l Local
n Normal
o Extraterrestrial
on Extraterrestrial on normal plane
op Extraterrestrial on tilted plane
p Plane/parallel
re Reflection
s Solar
sc Solar constant
sd Standard
ss Sunset
z Solar zenith

References

American Society of Heating Refrigerating and Air-Conditioning Engineers (2017) ASHRAE Handbook Fundamentals. American Society of Heating, Refrigerating and Air-Conditioning Engineers (ASHRAE), Atlanta

Arias-Rosales A, LeDuc PR (2020) Modeling the transmittance of anisotropic diffuse radiation towards estimating energy losses in solar panel coverings. Appl Energy 268:114872

Braun J, Mitchell J (1983) Solar geometry for fixed and tracking surfaces. Sol Energy 31:439–444

Collares-Pereira M, Rabl A (1979) The average distribution of solar radiation-correlations between diffuse and hemispherical and between daily and hourly insolation values. Sol Energy 22(2):155–164

Dawson L, Schlyter P (2012) Less is more: strategic scale site suitability for concentrated solar thermal power in Western Australia. Energy Policy 47:91–101

Duffie JA, Beckman WA (2013) Solar Engineering of thermal processes, 4th edn. Wiley, New Jersey

Erbs DG, Klein SA, Duffie JA (1982) Estimation of the diffuse radiation fraction for hourly, daily, and monthly-average global radiation. Sol Energy 28:293–302

Frutos F, Ruiz V, Gutierrez J (1985) Correlations between the solar diffuse and global radiation for Madrid. Revue Internationale Heliotechnique I.7

Gueymard CA (2004) The sun's total and spectral irradiance for solar energy applications and solar radiation models. Sol Energy 76(4):423–453

Hottel HC (1976) A simple model for estimating the transmittance of direct solar radiation through clear atmospheres. Sol Energy 18(2):129–134

Howell J, Bannerot R, Liet G (1982) Solar-thermal Energy systems: analysis and design. McGraw_Hill Book Company, New York

International Energy Agency (2022) World Energy Outlook 2022. International Energy Agency, Paris

International Organization for Standardization (1994) 9806–1:1994(E): Test Methods for Solar Collectors –Part 1: Thermal performance of glazed liquid heating collectors including pressure drop. International Organization for Standardization, Geneva
International Organization of Standardization (1992) Standard 9845-1: Solar energy - Reference solar spectral irradiance at the ground at different receiving conditions—Part 1: Direct normal and hemispherical solar irradiance for air mass 1.5. International Organization of Standardization (ISO), Geneva
International Organization of Standardization (1993) Standard 9846: Solar energy—calibration of a pyranometer using a pyrheliometer. International Organization of Standardization, Geneva
Jain PC (1988) Monthly average daily extraterrestrial irradiation and the maximum possible sunshine duration. Sol Wind Energy Technol 5:41–53
Kane RP (2005) Sun-earth relation: historical development and present status—a brief review. Adv Space Res 35(5):866–881
Kipp & Zonen (2006) Manual: CMP series 11/CMA series pyranometers / albedometer. Kipp & Zonen, Delft
Kotti MC, Argiriou AA, Kazantzidis A (2014) Estimation of direct normal irradiance from measured global and corrected diffuse horizontal irradiance. Energy 70:382–392
Kudish AI, Evseev E (2008) Assessment of four different correction models applied to the diffuse radiation measured with a shadow ring using global and and normal beam radiation measurements for Beer Sheva, Israel. Sol Energy 82:144–156
Kuravi S, Trahan J, Goswami YD, MM. Rahma MM, Stefanakos EK, (2013) Thermal energy storage technologies and systems for concentrating solar power plants. Prog Energy Combust Sci 39(4):285–319
Li H, Lian Y, Wang X, Ma W, Zhao L (2011) Solar constant values for estimating solar radiation. Energy 36(3):1785–1789
Lozano IL, Alados I, Foyo-Moreno I (2023) Analysis of the solar radiation/atmosphere interaction at a Mediterranean site: the role of clouds. Atmos Res 296:107072
De Miguel A, Bilbao J, Aguiar R, Kambezidis H, Negro E (2001) Diffuse solar irradiation model evaluation in the North Mediterranean Belt area. Sol Energy 70(2):143–153
Madhlopa A (2001) Evaluation of piecewise polynomial models for computation daily diffuse solar radiation in Malawi. In Proceedings of the International Solar Energy Society (ISES) 2001 Solar World Congress, 25th November – 2nd December, 2001 Adelaide, Australia
Nafey AS, Abdelkader M, Abdelmotalip A, Mabrouk AA (2000) Parameters affecting solar still productivity. Energy Convers Manag 41(16):1797–1809
Orgill JF, Hollands KGT (1977) Correlation equation for hourly diffuse radiation on a horizontal surface. Sol Energy 19(4):357–359
Paulescu M, Paulescu BV (2021) Nowcasting solar irradiance for effective solar power plants operation and smart grid management. In: Deo R, Samui P, Roy SS (eds) Predictive modelling for energy management and power systems engineering. Elsevier, pp 249–270
Pfenninger S, Gauche P, Lilliestam J, Damerau K, Wagner F, Patt A (2014) Potential for concentrating solar power to provide baseload and dispatchable power. Nat Clim Change 4, 689–692
Reindl DT, Beckman WA, Duffie JA (1990) Diffuse fraction correlations. Sol Energy 45(1):1–7
Sabir MM, Ali T (2016) Optimal PID controller design through swarm intelligence algorithms for sun tracking system. Appl Math Comput 274:690–699
Spelling JD (2013) Hybrid Solar Gas-Turbine Power Plants: A Thermoeconomic Analysis. KTH Royal Institute of Technology
Spyridonidou S, Vagiona DG (2023) A systematic review of site-selection procedures of PV and CSP technologies. Energy Rep 9:2947–2979
Thekaekara M (1976) Solar radiation measurement: techniques and instrumentation. Sol Energy 18:309–325
Wijeysundera NE (1978) Geometric factors for plane specular reflectors. Sol Energy 20:81–85
Yang K, Huang GW, Tamai N (2001) A hybrid model for estimating global solar radiation. Sol Energy 70(1):13–22

Chapter 4
Main Components of Solar Gas Turbines

4.1 Introduction

A solar gas turbine (SGT) typically consists of a solar field (including a concentrator and a receiver), a compressor, a combustion chamber (combustor), a turbine, and a generator (see Fig. 4.1). The concentrator captures solar radiation, which is then converted into heat by the receiver. Some of this heat is transferred to a heat transfer fluid, while the remainder dissipates into the surrounding environment. The heated fluid expands as it passes through the turbine. This fluid expansion process rotates the turbine that, in turn, drives the generator to produce electricity.

Solar radiation is intermittent, and therefore, a solar gas turbine can be supplemented with combustion and energy storage components to enhance its capacity factor. The various elements of an SGT impact overall system performance and must be managed to ensure the generator produces the desired output power. These components are explored and discussed in this chapter.

4.2 Solar Field

The solar field consists of two main components: the concentrator and the solar receiver. The concentrator captures and focuses solar radiation, either on a point or along a line, using an array of reflective mirrors or lenses. Concentrators such as parabolic troughs and linear Fresnel reflectors focus solar radiation along a line, while parabolic dishes and solar towers focus the radiation on a point. The four primary concentrating solar power (CSP) technologies widely utilized are parabolic trough concentrator (PTC), linear Fresnel reflector (LFR), solar tower (ST) and parabolic dish concentrator (PDC), and these technologies are briefly introduced in Chap. 1. Reflecting concentrators use mirrors to concentrate solar radiation onto the solar receiver.

A. Madhlopa and R. Nkhoma, *Principles of Solar Gas Turbines for Generating Electricity*, Green Energy and Technology,
https://doi.org/10.1007/978-3-031-84096-8_4

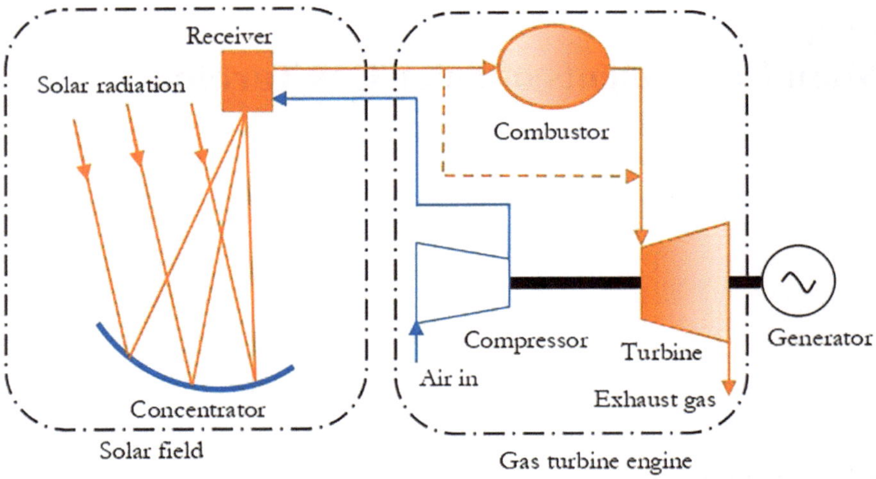

Fig. 4.1 Schematic representation of a solar gas turbine, showing the solar field, gas turbine cycle and generator

4.2.1 Mirrors for Solar Concentration

Mirrors are a critical component of CSP systems because they are responsible for reflecting incoming solar radiation onto the solar receiver. Given their role, a large surface area of mirrors is typically required, which significantly contributes to both the capital investment and the operational and maintenance costs of a solar field. Research into improving solar mirrors has gained significant traction because of their importance. Current research efforts focus on developing anti-soiling coatings, enhancing the reflectance of mirror materials, and designing high-temperature mirrors for secondary concentrators.

4.2.1.1 Anti-soiling Coatings

To maximize reflectance, mirrors are generally coated with silver, and ongoing research aims to push the reflectance of silvered reflectors to 95–96% by 2020–2030 (Blanco and Santigosa 2017). Additionally, advances in coating technology are expected to extend the lifespan of solar reflectors to 10–30 years, even under harsh outdoor conditions, contributing to lower maintenance costs and improved overall economic performance of CSP plants.

One of the biggest challenges in maintaining optimal mirror performance is the accumulation of dust on mirror surfaces, particularly in desert regions where direct normal irradiance is highest. Unfortunately, these are also areas where water scarcity limits the feasibility of wet cleaning, making dust buildup a serious issue. To combat this, researchers are focussing on developing anti-soiling coatings that prevent dust particles from adhering to mirror surfaces without affecting their optical performance. These coatings must also withstand extreme environmental conditions, such as high temperatures, sandstorms, and prolonged exposure to ultraviolet (UV), visible, and infrared (IR) radiation. Additionally, the coatings must endure routine cleaning procedures without degradation (Sarver et al. 2013).

The characteristics of effective low-soiling surfaces are outlined by Cuddihy (1980), and they include: hardness, smoothness, hydrophobicity, low surface energy, non-sticking properties and cleanliness. Hardness is the resistance of a solid material to indentation/abrasion. Therefore, hard surfaces have higher resistance to damage from dust particle embedding or impacts. The likelihood of trapping dust particles decreases with an increase in the smoothness of the surface. A hydrophobic surface repels water and hydrophobicity can range from hydrophobic to superhydrophobic. Superhydrophobicity produces significantly less water wetting than hydrophobicity (Dimitrakellis and Gogolides 2018). Superhydrophobic surfaces have a higher ability of self-cleaning, and so they help in minimizing soiling. Chemical reactions on the mirror surface increase with surface energy. Surface energy refers to the work required per unit area to slice an infinite crystal into two parts (Kiejna 1999). It can be classified into electrostatic, exchange–correlation, kinetic and structureless pseudopotential energy terms. Consequently, low surface energy assists in reducing soiling. Surfaces with non-sticking properties prevent the accumulation of sticky materials. Soiling of surfaces can also be reduced by cleaning to ensure that surfaces are free of water-soluble salts and other contaminants.

4.2.1.2 Mirror Materials with High Reflectance

Beyond anti-soiling coatings, enhancing the reflectance of mirrors is crucial for improving the overall efficiency of CSP systems. Higher reflectance reduces the size of the solar field needed for a given electricity output, thus lowering both the upfront investment in collectors or heliostats and the long-term operational costs.

One approach to achieving high reflectance involves using mirrors made of thin, transparent materials with a silver backing (Fig. 4.2). These mirrors are designed to minimize energy losses while maximizing the amount of solar radiation reflected onto the receiver. Reflectance improvements are also vital for secondary concentrators that focus the solar energy even further, and are critical in applications where high-temperature operations are required, such as in solar towers and parabolic dish systems.

When incoming rays of light strike a solar mirror, they first pass through a transparent front layer before being reflected by a silver layer at the back of the reflector. This transparent layer can be made of either glass or polymer. Silvered-polymer thin films, while capable of achieving reflectance values around 94%, generally perform slightly worse than 4 mm silvered-glass films that have a reflectance of 94.7% (Blanco and Santigosa 2017). The flexibility of polymer films allows for the construction of a wider variety of solar collector geometries, and researchers have investigated their potential in CSP systems (Almanza et al. 2009). However, glass-based mirrors have

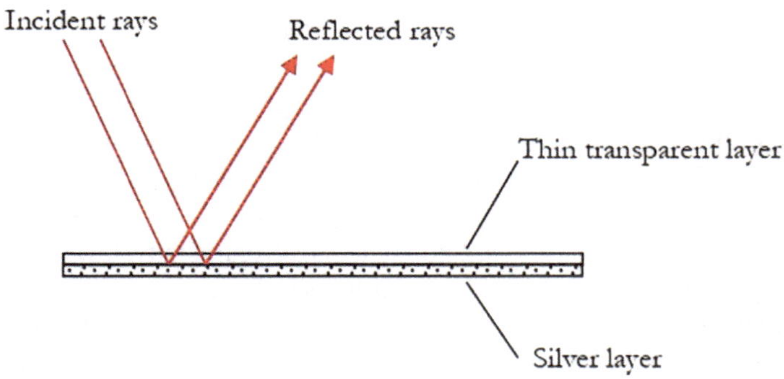

Fig. 4.2 Reflection of beam rays on a reflector made of thin transparent layer and silver layer

consistently shown superior performance, leading to a continued focus on improving glass reflectors in CSP applications.

4.2.1.3 High-temperature Mirrors for Secondary Concentrators

In concentrating solar power systems, solar radiation is concentrated onto a solar receiver that converts it into high-temperature heat. However, some of this heat is inevitably lost to the environment, and heat losses increase with the size of the receiver. To minimize these losses, secondary concentrators are employed to direct solar radiation from the primary concentrators onto a smaller focal area.

Compound parabolic concentrators (CPC) are commonly used as secondary reflectors, concentrating solar radiation either in two dimensions (2D) along a line or in three dimensions (3D) on a point. Typically, 2D-CPCs are used with linear Fresnel reflectors, while 3D-CPCs are applied in solar tower systems. Secondary concentrators in 2D-CPC configurations generally rely on natural convection cooling and must endure high temperatures of up to 623 K (Fernández-García et al. 2014). In contrast, 3D-CPC systems are exposed to even higher temperatures due to the intense concentration of solar radiation, compounded by convective and radiative heat exchange with the environment. Secondary mirrors with a reflectance of 94.7% absorb around 4.3% of the concentrated solar radiation, which can be significant, particularly given the high concentration ratios typical of CSP systems, ranging from 10 for linear Fresnel reflectors to 1500 for solar towers.

4.2.2 Solar Receivers

Solar receivers play a crucial role in CSP systems by converting concentrated solar radiation into high-temperature heat, which is then transferred to the working fluid.

This heat is essential for driving gas turbines and other power generation technologies, and the receiver serves as the critical link between the solar field and the power block. The efficiency of a solar receiver has a direct impact on the overall performance of solar gas turbines. Due to their exposure to high-intensity solar radiation and extreme temperatures, a variety of solar receivers have been developed to optimize performance.

Solar receivers can be categorized based on the type of heat transfer medium they use: gas receivers (e.g., air, carbon dioxide), liquid receivers (e.g., water, molten salt, liquid sodium), and solid particle receivers (e.g., ceramic particles) (Ho and Iverson 2014; Ho 2017). Another classification is based on geometric design, with tubular, volumetric, and microchannel receivers being the most prominent. Among these, tubular and volumetric receivers are considered to be the most mature technologies (Stein and Buck 2017).

(a) Gas receivers

Gas receivers typically employ tubular, volumetric (honeycomb and channels), or microchannel configurations to heat gases like air for gas turbine cycles or thermal storage. In a tubular receiver, concentrated solar radiation is absorbed by a series of tubes that convert the radiation into heat that is transferred to a pressurized gas flowing inside the tubes. Multiple tubes are assembled into panels with distribution and collector headers, and these panels can be connected in either series or parallel arrangements (Fig. 4.3). Tubular receivers operate in a closed-loop system where the gas is pumped at high pressure through the irradiated tubes (Ho 2017).

The design of tubular receivers can be categorized into external or cavity configurations. In external tubular receivers, panels are arranged in cylindrical or polygonal shapes, exposing a large absorber area to direct solar radiation. However, this also increases heat losses through convection and radiation. Cavity receivers, on the other hand, have panels arranged along the internal surfaces of an insulated cavity, which significantly reduces heat loss to the environment. Due to their superior thermal efficiency, cavity receivers are often preferred for the SGT technology, though they require taller towers to ensure that the receiver has an unobstructed view of the entire heliostat field, which is critical for maximizing power output (Ho 2017).

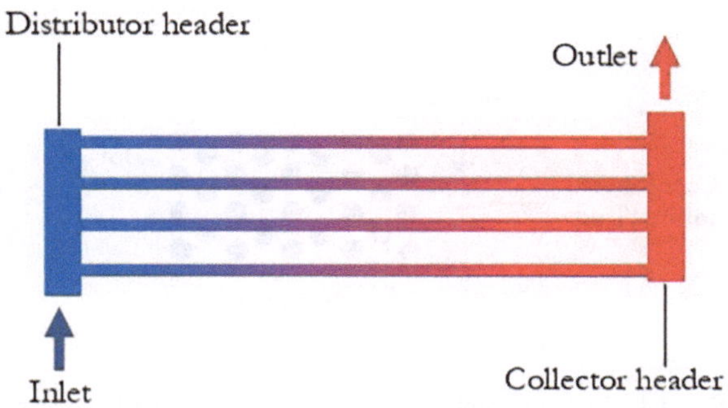

Fig. 4.3 Schematic illustration of a tubular panel

Direct heating of CO_2 in tubular receiver designs has emerged as a promising approach, especially with receivers capable of withstanding high internal fluid pressures and temperatures. These designs are showing potential for achieving thermal-to-electricity conversion efficiencies near or even above 50% (Ho 2017). For example, Le Roux et al. (2014) simulated the performance of an open-cavity tubular solar receiver for application in micro gas turbines with capacities ranging from 1 to 100 kW. Their findings emphasized the importance of tube diameter in optimizing the collector's efficiency, particularly for systems at the micro gas turbine capacity level. This observation is consistent with the principle that resistive pressure increases as the fluid passage diameter decreases, assuming a constant surface finish (American Society of Heating, Refrigerating and Air-Conditioning Engineers 2001).

In contrast, volumetric receivers utilize a porous honeycomb material or a block of channels to absorb concentrated solar radiation. Air, functioning as the heat-transfer fluid, is passed through this porous structure (Fig. 4.4). The intended design goal is for solar radiation to penetrate deeply into the receiver, creating the hottest region inside the porous material, away from the aperture (Ho 2017). However, many studies indicate that the aperture itself tends to exhibit the highest temperatures, leading to significant radiative heat losses to the environment. Despite this, air in these receivers can be heated to temperatures ranging from 1073 to 1273 K for metals, up to 1473 K for ceramics, and up to 1773 K for silicon carbide. This heated air can then be directly used to drive a gas turbine or to charge a thermal storage component.

Recent advancements in volumetric receivers include the integration of microchannels. These microchannels are designed to increase the surface area of contact between the heat transfer fluid and the channel walls, thereby enhancing heat transfer and improving the overall system efficiency. These innovations reflect ongoing efforts to optimize volumetric receivers for higher performance of CSP systems.

There are two fundamental principles for integrating volumetric receivers into power plant systems: (a) the open-loop receiver system with a Rankine cycle, and (b) the closed-loop receiver system with a Brayton cycle (Ávila-Marín 2011;

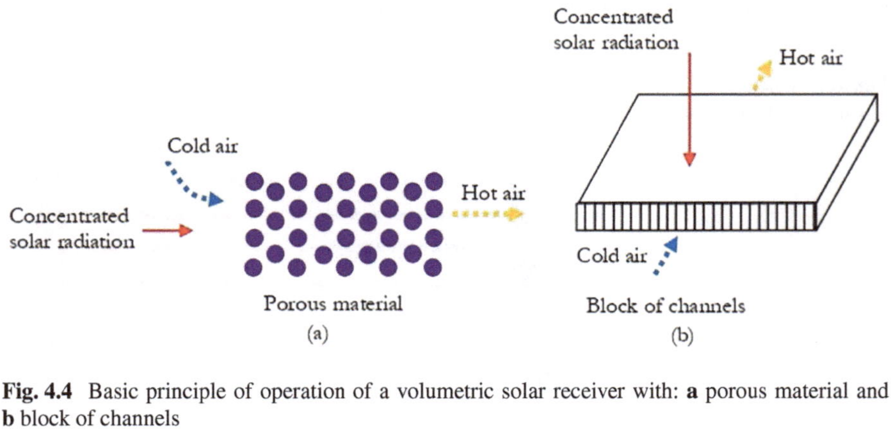

Fig. 4.4 Basic principle of operation of a volumetric solar receiver with: **a** porous material and **b** block of channels

Ho 2017). In the open-loop system, atmospheric air is heated by a metal or ceramic volumetric receiver. The heated air is then used to generate steam in a heat recovery steam generator, which includes a separate superheater. This steam is used to drive a steam turbine, producing electricity. Meanwhile, the closed-loop system, which operates with a Brayton cycle, introduces solar thermal energy via pressurized volumetric receivers. This configuration has potential applications in combined cycle solar gas turbine (CCSGT) systems, where the solar tower (ST) technology offers thermodynamic benefits in such hybridized systems.

Volumetric receivers are increasingly being recognized as superior alternatives to traditional tube receivers. Ávila-Marín (2011) conducted an extensive review of volumetric receivers, highlighting their diverse configurations, materials, and performance. The review also outlined the advantages and the basic plant configurations suitable for volumetric receivers. In a related development, windowed receivers that utilize transparent covers to protect the receiver, are gaining traction in solar power applications.

A more recent innovation involves microchannel receivers that consist of enclosed panels or plates with numerous tiny channels that enhance the contact surface area between the heat transfer fluid and the irradiated walls of the receiver. Theoretically, microchannel receivers have demonstrated the ability to heat supercritical carbon dioxide (at a flow rate of 1 kg/s) from 803 to 973 K under solar intensities of 500 kW/m^2 (Besarati et al. 2015).

(b) Liquid receivers

Liquid receivers are commonly used for direct steam generation in Rankine cycles, with water being the primary heat transfer fluid. For indirect heating, molten salt is widely utilized due to its ability to function as both a heat transfer fluid and a thermal storage medium. However, molten salt has certain limitations: (i) its temperature must be kept above 493 K to avoid solidification, (ii) it operates effectively within a narrow temperature range of 533 to 894 K (González-Roubaud et al. 2017), and (iii) it can only be used indirectly to drive a steam cycle (Ho 2017). In liquid receivers, concentrated solar radiation heats the liquid as it flows through panels of tubes that may be arranged in various configurations such as cavity, cylindrical, or cubical receivers. Cavity receivers, which reduce radiative and convective heat losses, are typically preferred for their efficiency. These systems are also used to preheat working fluids in SGT systems.

(c) Solid particle receivers

Solid particle receivers utilize small particles as the heat transfer medium, and they can achieve temperatures exceeding 1273 K (Ho 2017). These particles absorb solar energy directly, allowing for efficient heat capture and storage. One key advantage of solid particle receivers is that they avoid the heat transfer resistance and flux limitations encountered with tubular receivers. The working fluid in an SGT system must still be heated indirectly via a heat exchanger, which may reduce the overall solar-to-heat conversion efficiency. Nevertheless, solid particle receivers hold promise due to their improved material compatibility and reduced corrosion issues, though challenges remain in particle conveyance, attrition, and transportation.

4.3 Enhancing the Capability of Solar Gas Turbines

CSP technology represents a promising approach to reducing reliance on fossil-fuel power plants. With its high thermodynamic efficiency, CSP has great potential as a driving force for solar gas turbines. However, the intermittent nature of solar energy presents challenges to the capacity factor of solar-only gas turbine systems. To address this, solar gas turbines (SGTs) can be hybridized with combustion and energy storage technologies, ensuring consistent year-round performance and the ability to meet base-load power demands.

4.3.1 System Hybridization

In hybrid solar gas turbines, the solar receiver and other heat sources provide the thermal energy necessary to drive the engine cycle. Typically, a backup heat source, such as a combustor, is used to supplement solar energy. In a serial configuration, solar energy preheats the compressed air before it enters the combustor, reducing the amount of fuel required to reach the desired turbine inlet temperature (Fig. 4.5). If the solar-heated air reaches the necessary turbine inlet temperature, the combustor can be entirely bypassed. Parabolic trough concentrators (PTCs) or linear Fresnel reflectors (LFRs) are often utilized in these systems to preheat the working fluid before it enters the combustor, thereby increasing the overall system efficiency.

In the parallel hybrid configuration, compressed air from the compressor is divided into two streams: one is heated by the solar receiver while the other is heated by the combustor. The two streams are then merged before entering the turbine. This setup facilitates easier isolation of the solar and combustion subsystems for maintenance purposes. Additionally, the parallel hybridization configuration typically experiences lower pressure losses between the compressor and turbine outlets. However, due to the intermittent nature of solar radiation, the two air streams can possess different thermodynamic properties, such as temperature and pressure. Material limitations restrict the solar-heated air stream to a maximum temperature of approximately 1473 K, which is lower than the combustion-heated air stream that can exceed 1673 K in modern F-Class gas turbines (Spelling 2013). For these reasons, the series configuration, where solar energy preheats the air before combustion, is often preferred.

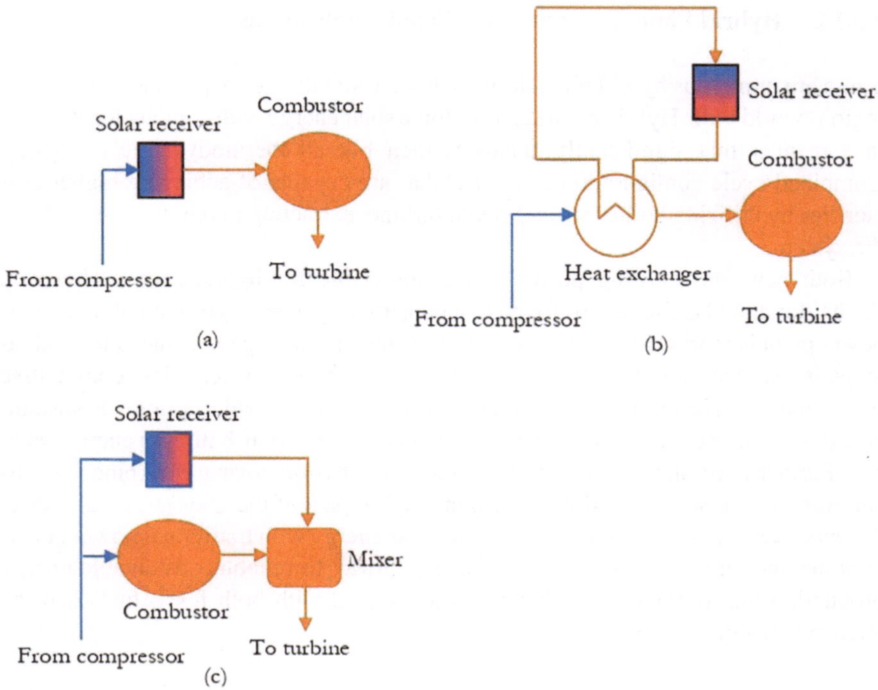

Fig. 4.5 Schematic representation of a hybrid solar gas turbine in: **a** serial configuration with direct heating, **b** serial configuration with indirect heating and **c** parallel configuration

4.3.1.1 Advantages of Solar Hybridization

Solar hybridization offers several distinct advantages over both renewable and conventional power plants (Schwarzbözl et al. 2006; Korzynietz et al. 2016), including:

- *Full dispatchability and grid stability*: Hybrid systems can operate independently of solar conditions by utilizing combustion for power generation during periods of low solar radiation.
- *Quick startup of the gas turbine*: Gas turbines can quickly begin generating power, which is a significant benefit in grid management.
- *Ease of operation*: Hybrid solar gas turbines are simpler to operate compared to other solar technologies due to their compatibility with existing infrastructure.
- *Low water consumption*: Unlike some other solar technologies, hybrid systems can operate with minimal to no water consumption, particularly beneficial in arid regions.
- *Air as a heat transfer fluid*: Air is inexpensive, non-toxic, and widely available, making it an attractive choice for solar hybrid systems.

4.3.1.2 Hybrid Plant Concepts and Global Applications

There are numerous hybrid plant designs that are suitable for deployment in various regions worldwide. Hybrid plants can combine solar energy with combustion systems in a manner that significantly enhances their overall thermodynamic efficiency. Combined cycle configurations, in particular, are capable of achieving higher efficiencies by utilizing both a gas and steam turbine, extracting maximum energy from the system.

Both new and existing power plants are viable for hybridization. However, Saghafifar and Gadalla (2016) found that constructing a new hybrid combined cycle power plant is more cost-effective than retrofitting an existing one. One notable challenge is that declining fossil fuel prices may make hybrid systems less competitive in the market. Nevertheless, hybridization remains a valuable option for sustainable development due to its environmental benefits and contribution to energy security. Furthermore, the integration of biofuels into hybrid solar gas turbine systems can further enhance the positive environmental impact of these systems and reduce dependence on fossil fuels. By combining solar energy with traditional power generation technologies, hybrid systems offer a pathway to reliable, sustainable energy production that mitigates the challenges associated with both fossil fuels and the intermittency of solar power.

4.3.2 Thermal Storage

Thermal storage involves temporarily storing heat for later use, particularly when there is excess thermal energy. The stored heat is released when the supply is insufficient, forming a thermal cycle. In gas turbines with thermal storage, excess thermal energy from the receiver is stored in a medium. There are three primary techniques: sensible heat, latent heat, and thermochemical heat storage (Dutta 2017).

4.3.2.1 Sensible Thermal Storage

Sensible heat storage (SHS) stores thermal energy by raising the temperature of a storage medium without inducing a phase change. When heat is added, the medium's temperature increases; when released, the temperature decreases. Key attributes of an ideal SHS material include low cost, high specific heat capacity, good thermal conductivity, long-term thermal stability, containment compatibility, recyclability, and a low carbon footprint.

Materials for SHS are categorized into liquids and solids. Common liquids include water and oil, where the stored energy ($\dot{Q}$) is related to the medium's temperature change. The energy equation for a fully mixed liquid medium is:

$$\dot{Q} = m_{\mathrm{sm}} \int_{T_i}^{T_f} C_{\mathrm{p,sm}}(T)\mathrm{d}T \tag{4.1}$$

where $C_{\mathrm{p,sm}}$ is the specific heat capacity of the storage material at a constant pressure, and m_{sm} is mass of storage medium. If the specific heat capacity ($C_{\mathrm{p,sm}}$) is constant, the equation, within the range T_i to T_{f}, then Eq. (4.1) simplifies to:

$$\dot{Q} = m_{\mathrm{sm}} C_{\mathrm{sm}} (T_{\mathrm{f}} - T_i) \tag{4.2}$$

In an un-stratified storage vessel, the energy balance is:

$$m_{\mathrm{st}} C_{\mathrm{p,sm}} \frac{\mathrm{d}T_{\mathrm{sm}}}{\mathrm{d}t} = \dot{Q}_{\mathrm{u}} - L_{\mathrm{sm}} - UA(T_{\mathrm{sm}} - T_{\mathrm{a}}) \tag{4.3}$$

where $\dot{Q}_{\mathrm{u}}$ is the charging rate, L_{sm} is the heat removal rate, U is the heat loss coefficient, A is the vessel surface area, and T_{a} is the ambient temperature. For low-temperature applications, water is a common storage medium. Various other liquid materials are listed in Table 4.1.

Solid materials, especially particulate solids, are also used for sensible heat storage. A packed-bed storage system uses loosely packed particulate materials, like rocks, bricks, and ceramics, to store heat. Heat is exchanged by circulating fluids through the bed, making ceramics (e.g., rocks, sand) suitable for high-temperature storage due to their high melting points. Selected solid materials for sensible heat storage are presented in Table 4.2.

4.3.2.2 Latent Heat Storage

Latent heat storage (LHS) relies on phase changes in materials to store and release energy. During these phase transitions, the storage medium shifts between solid, liquid, or gas states. Among the various transitions, solid–liquid phase changes are the most promising for CSP applications due to their manageable volume changes and high energy density (Blanco and Santigosa 2017).

The energy stored (Q) in a solid–liquid phase change can be calculated using:

$$Q = m_{\mathrm{so}} \left\{ C_{\mathrm{p,so}} \left(T_{\mathrm{mp}} - T_{\mathrm{so}} \right) + L_{\mathrm{so,l}} + C_{\mathrm{p,l}} \left(T_{\mathrm{l}} - T_{\mathrm{mp}} \right) \right\}, \tag{4.4}$$

where m_{so} is the mass of the solid, $C_{\mathrm{p,so}}$ and $C_{\mathrm{p,l}}$ are respectively the specific heat capacities of the solid and liquid phases, T_{mp} is the melting point, T_{so} and T_{l} are respectively the initial and final temperatures, and $L_{\mathrm{so,l}}$ is the latent heat of fusion. The latent heat of fusion is a key component in LHS systems.

Phase change materials (PCMs) used in LHS can be organic, inorganic, or eutectic mixtures. Organic PCMs, like paraffin, offer benefits such as stable phase change

Table 4.1 Major characteristics of sensible heat storage liquid materials

Storage medium	Temperature		Average density (kg/m^3)	Average heat conductivity (W/m.K)	Average specific heat capacity (J/kg.K)	Volume specific heat capacity (kWh/m^3)	Cost per unit mass (US$/kg)	Cost per unit of energy (US$/kWh)
	Cold (K)	Hot (K)						
Solar salt	493	–	1899	–	1495	–	1.3	–
Hitec	415	–	1860	–	1560	–	1.93	–
Hitec XL	393	406	1990	0.6	1495*	–	1.66	–
Mineral oil	473	573	770	0.12	2600	55	0.3	4.2
Synthetic oil	523	623	900	0.11	2300	57	3	42
Silicon oil	573	673	900	0.1	2100	52	5	80
Nitrite salts	523	723	1825	0.57	1500	152	1	12
Nitrate salts	538	838	1870	0.52	1600	250	0.5	3.7
Carbonate salts	723	1123	2100	2	1800	430	2.4	11
Liquid sodium	543	803	850	71	1300	80	2	21

*At 573 K

Source of data Fernandes et al. (2012); González-Roubaud et al. (2017)

without supercooling, but suffer from low thermal conductivity and reduced volumetric storage. Inorganic PCMs, including salt hydrates, molten salts, and metallic alloys, are more common in CSP technology due to their higher energy densities and thermal conductivities. Eutectic PCMs, formed by mixing two or more substances, offer customizable melting points and improved thermophysical properties, making them ideal for specific applications.

Molten salts are particularly versatile, serving both as a storage medium and a heat transfer fluid. A popular example is solar salt, a mixture of 60% sodium nitrate and 40% potassium nitrate, which is stable across a wide temperature range (533–894 K) and is commonly used in CSP plants. However, it has limitations such as a high freezing point (511 K). The freezing of molten salt can cause damage to the pipes and pumps in the plant (Kim et al. 2018). Therefore, auxiliary heating is required to prevent solidification of the molten salt. Alternatives like Hitec salt (a ternary mixture) and Hitec XL offer lower melting points (393–415 K) but have lower maximum operating temperatures compared to solar salt, which may require further optimization for widespread use. Overall, while molten salts are the most expensive component in thermal storage systems, their exceptional thermal properties make them the most widely used storage medium in commercial CSP applications. The

Table 4.2 Major characteristics of sensible heat storage solid materials

Storage medium	Temperature		Average density (kg/m^3)	Average heat conductivity (W/m.K)	Average heat capacity (J/kg.K)	Volume specific heat capacity (kWh/m^3)	Cost per unit mass (US$/kg)	Cost per unit of energy (US$/kWh)
	Cold (K)	Hot (K)						
Sand-rock-mineral oil	473	573	1700	1	1300	60	0.15	4.2
Reinforced Concrete	473	673	2200	1.5	850	100	0.05	1
NaCl (solid)	473	773	2160	7	850	150	0.15	1.5
Cast iron	473	673	7200	37	560	160	1	32
Silica fire bricks	473	973	1820	1.5	1000	150	1	7
Magnesia fire bricks	473	1473	3000	1	1150	600	2	6

Source of data Fernandes et al. (2012)

advantages of different PCMs are summarized in Table 4.3, showcasing the diverse options available for optimizing thermal storage systems.

The Latent Heat Storage (LHS) system offers several advantages: isothermal operation, high energy density and low system pressure.

- *Isothermal operation*: Solid–liquid phase transitions occur at nearly constant temperatures, allowing for efficient energy storage when the phase change temperature aligns with the power block's maximum operating temperature. This is particularly beneficial for achieving high power cycles.
- *High Energy Density*: LHS systems provide higher energy density than sensible heat storage because the latent heat of fusion is significantly greater than specific heat capacity. For instance, aluminium silicon has a specific heat capacity of 1038 J/kg·K and a latent heat of fusion of 560,000 J/kg (Fernandes et al. 2012). This means that, for the same mass, LHS can store more energy, resulting in a more compact storage system.
- *Low System Pressure*: The solid–liquid phase change involves lower volumetric expansion compared to gas-phase transitions, leading to lower system pressures. This is advantageous for high-temperature applications, such as those found in gas turbines, where there is a growing interest in developing PCMs for these environments. However, one challenge is that most PCMs have low thermal conductivity, which can reduce the efficiency of heat transfer during charging and discharging. Research and development are focused on enhancing thermal conductivity while maintaining the desirable thermal properties of PCMs.

Table 4.3 Advantages and disadvantages of different types of solid–liquid PCMs

Type of PCM	Advantages	Disadvantages
Organic	High chemical and thermal stability Non-corrosive Low or non-subcooling Compatibility with conventional construction materials	Low phase change enthalpy Low thermal conductivity Flammability
Inorganic	Higher phase change enthalpy High thermal conductivity Cheap	Supercooling Corrosive Phase segregation Lack of thermal stability
Eutectic	High volumetric energy density (slightly higher than organic PCMs) Sharp melting temperature	Limited tested data available due to thermophysical properties

Adapted from Zeng et al. (2017)

4.3.2.3 Thermochemical Energy Storage

Thermochemical energy storage involves using an energy source to drive a reversible chemical reaction, where the energy is stored in the chemical bonds of the reaction products. A chemical reaction is a process that results in the rearrangement of atoms within molecules, either of the same substance or different substances, to form one or more new substances. These reactions can be classified as either reversible or irreversible.

In reversible reactions, the reaction can proceed in both directions, reaching a state of equilibrium where the rate of the forward reaction equals the rate of the reverse reaction. Conversely, irreversible reactions proceed predominantly in one direction, consuming the reactants until they are fully converted into products. For example, consider the following reactions:

$$A \leftrightarrow B + C \tag{4.5}$$

$$D + E \rightarrow F \tag{4.6}$$

where *A*, *D*, and *E* are reactants, while *B*, *C*, and *F* are products. Equation (4.5) represents a reversible reaction, whereas Eq. (4.6) is irreversible. Depending on the nature of the reaction, it may either absorb heat from the surroundings (endothermic reaction) or release heat into the surroundings (exothermic reaction).

The source of energy for these reactions can be thermal (thermochemical reactions), electrical (electrochemical reactions in batteries), or electromagnetic (photochemical reactions like photosynthesis). In thermochemical energy storage, heat is applied to a storage medium to drive a forward chemical reaction. The stored heat can later be recovered by reversing the reaction, which may sometimes require the addition of a catalyst. The reverse reaction, typically exothermic, releases the stored energy, usually at a lower temperature than the forward reaction.

Thermochemical storage methods are broadly classified into two categories: (a) chemical reaction storage and (b) sorption storage.

(a) Chemical reaction storage

This storage method relies on the principle of reversible chemical reactions, which involves an endothermic decomposition process to store heat and an exothermic synthesis process to release it. The reactants and products are separated and stored, potentially allowing for indefinite storage periods without energy loss. For instance, when heat is added to ammonia (NH_3), a reversible reaction occurs where nitrogen (N_2) and hydrogen (H_2) gases are formed:

$$\mathrm{NH_3} + \text{ heat} \leftrightarrow \mathrm{N_2} + \mathrm{H_2} \tag{4.7}$$

In applications like solar gas turbines, the heat source on the left side of Eq. (4.7) could be a solar receiver. The separated gases (N_2 and H_2) can be stored for extended periods, and the stored heat is recovered by mixing these gases to induce the reverse reaction that produces NH_3.

(b) Sorption storage

Sorption involves capturing or fixing a gas or vapour (called the sorbate) onto a condensed substance (solid or liquid), known as the sorbent. This process occurs at the interface between two phases, and the forces involved include van der Waals and hydrogen bonding. Unlike chemical reactions, sorption does not change the molecular structure of the substances involved and encompasses both absorption (where the sorbent is a liquid) and adsorption (where gas molecules attach to the surface of a solid).

For example, when heat is added to magnesium sulphate heptahydrate ($MgSO_4.7H_2O$), it decomposes into magnesium sulphate ($MgSO_4$) and water (H_2O):

$$MgSO_4 \cdot 7H_{2O} + \text{heat} \leftrightarrow MgSO_4 + 7H_{2O} \tag{4.8}$$

Research in sorption processes has often focused on metal salts with various sorbates such as water, ammonia, methanol, methylamine, and metal alloys with hydrogen (Cot-Gores et al. 2012).

(c) Advantages and challenges of thermochemical energy storage

Thermochemical materials offer several advantages, including:

- High heat storage density, which allows for compact storage systems.
- Significant temperature increases during the exothermic reaction.
- The possibility of storing reactants (both sorbent and sorbate) at ambient temperatures without self-discharge.

However, despite its high energy density and efficiency, thermochemical heat storage is perceived as costly and remains at an early stage of development (Fernandes et al. 2012). Further research and technological advancements are required to make it commercially viable on a large scale.

4.4 Compressor

Some engineering processes require gases to be compressed to elevated pressures, and compressors are the devices designed to achieve this by increasing the pressure of a gas through volume reduction. Compressors are primarily classified into two main types: (a) positive displacement compressors and (b) dynamic compressors, as shown in Fig. 4.6.

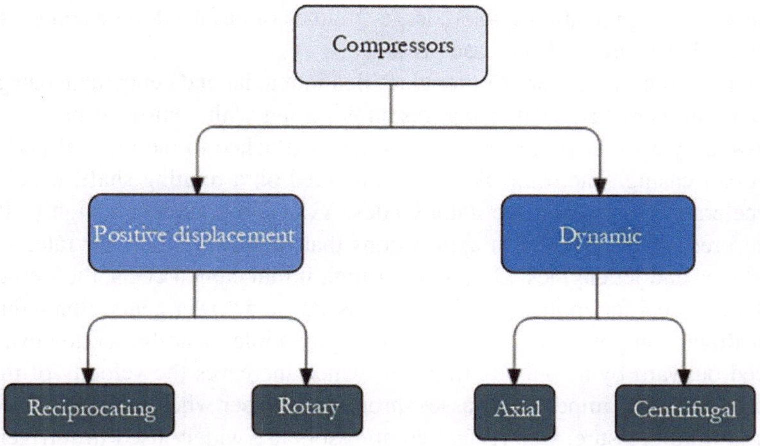

Fig. 4.6 Classification of compressors

4.4.1 Positive Displacement Compressors

Positive displacement compressors function by trapping a fixed volume of gas within a chamber and then mechanically reducing the volume to increase its pressure. They typically employ mechanisms such as lobes, pistons, screws, or vanes to achieve this compression process. When the volume of the gas is decreased, the pressure naturally increases due to the inverse relationship between volume and pressure (as described by Boyle's law). Once the desired pressure is achieved, the chamber opens, and the compressed gas is discharged. An everyday example of a positive displacement compressor is a simple bicycle pump, where the air is compressed by the downward movement of a piston inside a cylinder.

Positive displacement compressors can further be categorized into: Reciprocating and rotary compressors. Reciprocating compressors use a piston in a cylinder to compress the gas. These are suitable for high-pressure applications and are widely used in industrial gas processes, refrigeration, and pneumatic tools. Rotary compressors utilize rotating elements such as screws, vanes, or lobes to compress the gas. These are often used in continuous operation applications, such as in air conditioning systems or automotive superchargers.

4.4.2 Dynamic Compressors

Dynamic compressors increase the gas pressure by converting the kinetic energy (velocity) of the gas into potential energy (pressure) using axial or centrifugal forces. Unlike positive displacement compressors, dynamic compressors are designed for continuous flow applications, where large volumes of gas need to be compressed at a relatively lower pressure increase per stage.

Dynamic compressors are further classified into axial and centrifugal categories. Axial compressors have a rotor that resembles a fan with contoured blades, which is followed by a set of stationary blades (stator) attached to the internal part of the compressor casing. The rotor blades are mounted on a rotating shaft, which spins and accelerates the gas over the stator blades. Axial compressors are highly efficient and are predominantly used in applications that require a high flow rate, such as gas turbines and jet engines. Despite their high initial capital costs, their efficiency makes them ideal for applications in the aerospace and power generation industries. In centrifugal compressors, a gas enters through an inlet near the suction eye and it is forced outward by a spinning impeller, which increases the velocity of the gas. As the gas exits the impeller, it passes through a diffuser where the high velocity is converted into pressure. Centrifugal compressors are widely used in refrigeration, natural gas processing, and various industrial applications where moderate pressure increases and large flow rates are required.

4.4.3 *Applications and Performance of Compressors*

Compressors are integral components in many industries, including gas turbines, industrial gas processing, refrigeration, HVAC systems, and various other mechanical and chemical processes. They are selected based on key parameters such as the desired gas flow rate, the inlet pressure and the discharge pressure. Compressors can operate in single-stage or multi-stage configurations, depending on the pressure requirements. Single-stage compressors compress the gas to the desired pressure in a single compression cycle before discharging it. Multi-stage compressors pressurize the gas multiple times, where the discharge from one stage becomes the intake for the next stage. This staged compression is more efficient for applications requiring very high-pressure outputs.

4.4.3.1 Performance Metrics

The performance of a compressor is often measured by its compression ratio (r_{co}), which is the ratio of the discharge pressure to the inlet pressure:

$$r_{co} = \frac{p_{dis}}{p_{in}} \tag{4.9}$$

where p_{dis} is the discharge pressure and p_{in} is the inlet pressure.

A higher compression ratio indicates better performance, as it means the compressor can achieve a higher pressure differential. However, achieving a high compression ratio usually requires more energy, making efficiency a critical factor in compressor design, particularly in applications like solar gas turbines, where minimizing power consumption is crucial for the overall system performance.

4.4.3.2 Energy Considerations

Compressors are known for their significant power consumption, which can account for a substantial portion of the total energy use in an SGT system. Thus, selecting a compressor with high efficiency and optimal performance characteristics is essential to maximize the conversion of solar energy to mechanical or electrical energy.

4.5 Combustor

A combustor (or combustion chamber) is a critical component in many engineering systems, such as gas turbines, where it facilitates a rapid chemical reaction between a fuel and an oxidant, typically air. Combustion is an exothermic process, releasing a significant amount of heat energy, predominantly in the form of hot gases. This process also produces light, free ions and electrons, and, in some cases, noise due to the turbulent flow of gases (American Society of Heating, Refrigerating and Air-Conditioning Engineers 2001).

In most combustors, air acts as the oxidant and is mixed with a fuel, such as natural gas, diesel, or kerosene, within the combustion chamber. The two primary functions of a combustor are: (a) efficient mixing of air and fuel to ensure a homogeneous mixture, and (b) maintaining stable combustion to maximize heat output while minimizing pollutant emissions. Achieving complete combustion is the goal, wherein all the hydrogen and carbon in the fuel react fully with oxygen to form water (H_2O) and carbon dioxide (CO_2), thus maximizing energy output. This is often accomplished by supplying excess air above the stoichiometric (theoretical) requirement to ensure complete oxidation and reduce the production of harmful emissions such as carbon monoxide (CO) and unburned hydrocarbons.

Excess air can also lead to increased production of nitrogen oxides (NO_x) that are significant pollutants. Therefore, modern combustor designs must balance these competing needs to achieve optimal performance and adhere to environmental regulations. The design and operational parameters of combustors (including airflow patterns, fuel injection rates, and flame stabilization methods) are critical for maintaining efficient and clean combustion under varying environmental conditions.

Historically, two main types of combustion systems were considered: constant volume and constant pressure (Cohen et al. 1996). Theoretically, the efficiency of a constant volume combustion process is higher than that of a constant pressure process. This is because, at constant volume, the entire energy release from the fuel is used to increase the temperature and pressure of the gases without any loss in expansion work, resulting in a higher thermodynamic efficiency. Nevertheless, implementing a constant volume combustion system posed significant mechanical challenges. It required valves to maintain a fixed volume, leading to complex mechanical designs. Furthermore, the combustion process in a constant volume system would be intermittent, causing pulsations and uneven engine operation. These challenges eventually led to the discontinuation of constant volume combustion systems in favor of constant pressure systems that offer continuous combustion without the need for complex valving.

Constant pressure combustion, as used in most modern gas turbine engines, allows for steady and uninterrupted combustion, resulting in smoother engine operation and reduced mechanical complexity. However, research into constant volume combustion may resume in the future as advancements in materials and engineering techniques potentially mitigate the earlier limitations.

4.5.1 Types of Combustors

The design of combustors for gas turbines varies significantly based on the intended application, fuel type, and environmental regulations. Modern considerations, such as emissions reduction and fuel efficiency, add complexity to combustor design. There are three primary types of combustor configurations used in gas turbine technology:

4.5.1.1 Can Combustors

The can combustor consists of multiple separate combustion chambers, or "cans," arranged around the engine shaft (Fig. 4.7). Each can has its own fuel injector (fed from a common fuel manifold), ignition plug, liner, and casing. The air from the compressor is split and delivered to each can, where it mixes with fuel for combustion inside the liner. Can combustors are relatively easy to maintain and are often used in older or smaller gas turbine designs. However, they tend to have higher pressure losses compared to other configurations and are less efficient in terms of space utilization.

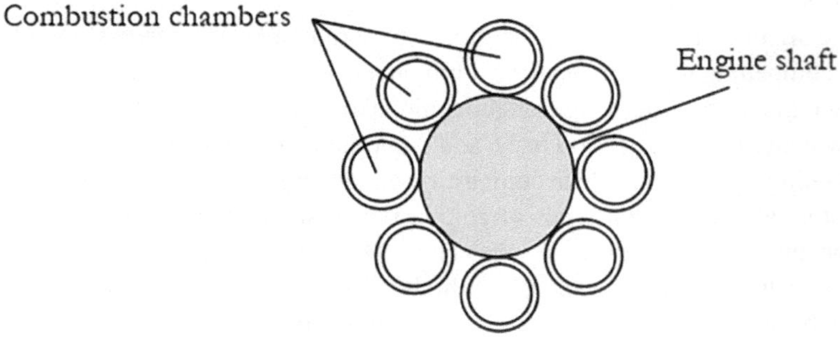

Fig. 4.7 Cross-section of a can combustor

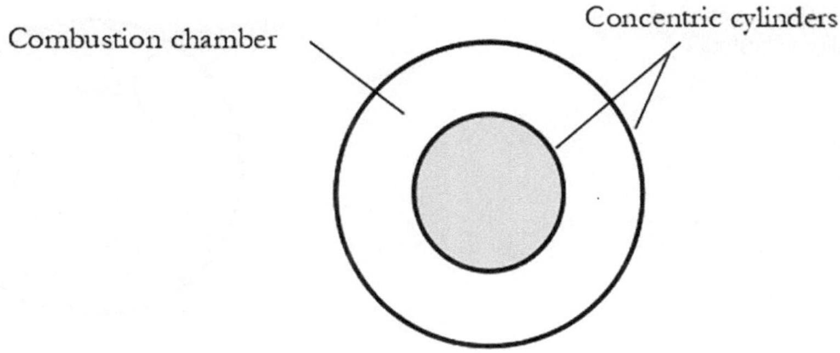

Fig. 4.8 Cross-section of annulus combustors

4.5.1.2 Annular Combustors

Annular combustors feature a continuous, ring-shaped combustion chamber formed between two concentric cylinders (Fig. 4.8). In this design, compressed air flows through the annular space and mixes with fuel injected into the chamber. The annular design is highly efficient, offering the lowest pressure loss among the three types due to its streamlined flow path. It also enables more uniform combustion and temperature distribution, which enhances overall performance and reduces thermal stresses on the turbine components. This design is common in modern high-performance jet engines and large industrial gas turbines.

4.5.1.3 Cannular (Can-Annular) Combustors

The cannular combustor is a hybrid between the can and annular designs (Fig. 4.9). It consists of multiple can-like chambers, similar to the can combustor, but these chambers share a common outer annular liner. This design retains some of the maintenance advantages of the can combustor while providing some of the performance benefits of the annular combustor, such as reduced pressure losses and better temperature distribution. Cannular combustors are often used in medium-sized gas turbines and applications where both performance and ease of maintenance are important.

4.5.2 Requirements for Operation

The efficient operation of a gas turbine combustor requires precise control of several critical factors to ensure sustained combustion, optimal performance, and minimal emissions. In a standard gas turbine, combustion is initiated by an electric spark, which ignites the fuel–air mixture. Once started, the combustion process is self-sustaining under continuous operation, provided there is a consistent supply of fuel

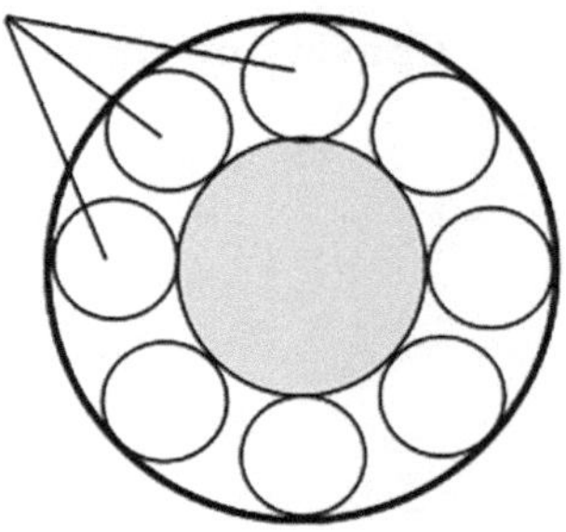

Fig. 4.9 Cross-section of cannulus combustors

and air. To achieve high efficiency, the combustor must operate at high temperatures while minimizing pressure losses. It also plays a crucial role in maintaining the correct temperature distribution at the turbine inlet, which is essential for reducing thermal stresses on turbine components and minimizing pollutant formation.

One of the most important factors affecting the efficiency of combustion in a gas turbine is the fuel–air ratio, the proportion of fuel to air in the mixture. This ratio directly influences the completeness of combustion and the production of emissions like nitrogen oxides (NO_x) and unburned hydrocarbons. The fuel–air ratio is affected by the temperature of the air entering the compressor. Since the compressor draws in air at ambient conditions, variations in weather (such as changes in temperature, humidity, or pressure) can impact combustor performance. For example, as the temperature of the inlet air increases, its density decreases, reducing the mass flow rate of air into the compressor. This lower density, in turn, requires adjustments in fuel flow to maintain the desired fuel–air ratio. Similarly, altitude affects the performance of a combustor. As altitude increases, the density of the air decreases due to lower atmospheric pressure. This reduction in air density necessitates careful adjustment of the fuel flow rate to ensure efficient combustion. To manage such environmental variations, the design of the combustor must include robust control mechanisms that can adapt to changes in air density and maintain the required turbine inlet temperature.

Efficient mixing of fuel and air is also critical to achieve complete combustion. Proper mixing is typically achieved by generating turbulent flow within the combustor, which enhances the contact between fuel and air molecules, thereby increasing the combustion efficiency. One common method to stabilize the flame and promote efficient combustion is the use of an inlet air swirler (Torkzadeh et al. 2016). A swirler, which may consist of tilted blades or grooves, is placed at the entrance of the combustor. As the air flows through the swirler, its direction is altered, creating a rotational or swirling flow upon exiting the swirler. This swirling motion enhances mixing and stabilizes the flame by creating a central recirculation zone that holds the flame in place. Similarly, a swirl fuel injector can be used to induce rotational flow of the fuel, further promoting mixing and enhancing combustion stability (Choi et al. 2017).

In addition to airflow management, modern stationary gas turbine combustors are often designed for dual-fuel capability. This allows the combustor to operate on natural gas as the primary fuel while retaining the flexibility to switch to liquid fuels, such as diesel, in situations where natural gas is unavailable or impractical. This dual-fuel capability ensures operational flexibility and reliability in various environments and fuel supply conditions.

To understand and optimize combustor flow characteristics, engineers often employ computational fluid dynamics (CFD) simulations. CFD tools enable detailed analysis of fluid flow, heat transfer, and chemical reactions within the combustor, allowing for improved designs that maximize efficiency and minimize emissions. However, such simulations are complex and go beyond the scope of this discussion.

In summary, the successful operation of a gas turbine combustor hinges on the following key requirements:

- Accurate control of the fuel–air ratio to maintain efficient combustion and minimize emissions.
- Adaptability to environmental changes, such as variations in temperature, pressure, and altitude, to ensure consistent turbine inlet temperatures.
- Efficient mixing mechanisms, such as swirling flow, to enhance combustion efficiency and flame stability.
- Dual-fuel capability for operational flexibility.

4.5.3 Gas Turbine Emissions

Gas turbine emissions are a significant concern due to their environmental and health impacts. The major pollutants produced by gas turbines originate from both the fuel used and the air involved in the combustion process. Conventional fuels, such as natural gas, diesel, and other hydrocarbons, primarily consist of carbon (C) and hydrogen (H), with smaller amounts of nitrogen (N), sulphur (S), and trace elements. The combustion of these fuels with atmospheric air, which contains about 78% nitrogen, typically results in carbon dioxide (CO_2) and water vapour (H_2O) as the primary products. However, trace amounts of other harmful pollutants, including nitrogen oxides (NO_x), sulphur dioxide (SO_2), carbon monoxide (CO), and unburned hydrocarbons (UHCs), are also produced.

While renewable energy sources generally have a lower carbon footprint compared with fossil fuels, emissions of NO_x can still occur even when biofuels are used. This is primarily due to the high temperatures required for combustion in gas turbines, which facilitate the formation of nitrogen oxides. For stoichiometric combustion of a hydrocarbon fuel, the required amount of oxygen can be determined from the following equation:

$$4C_xH_y + (4x + y)O_2 \rightarrow 4xCO_2 + 2yH_2O \tag{4.10}$$

This equation assumes complete combustion, where all carbon is converted to carbon dioxide (CO_2). However, in reality, incomplete combustion may occur, leading to the production of carbon monoxide (CO) and UHCs in the exhaust gases. CO_2 emissions from fossil fuels are a major contributor to climate change (see

Chap. 11). For example, combined cycle (CCGT) and open cycle (OCGT) gas turbines in Singapore have reported emission values of 384 kg CO_2/MWh and 789 kg CO_2/MWh, respectively (Finenko and Cheah 2014). Although systems for carbon capture and storage (CCS) can be integrated into fossil-fuelled gas turbines to mitigate CO_2 emissions, this approach significantly increases the levelized cost of electricity (Canepa and Wang 2015).

Nitrogen Oxides (NO_x) are primarily formed when combustion temperatures exceed 1800 K (Shalaj et al. 2015). Controlling the flame temperature is one of the most effective methods for reducing NO_x emissions. However, gas turbines typically operate at high temperatures to maximize efficiency, which creates ideal conditions for NO_x formation. These oxides are harmful to human health, particularly affecting the respiratory system, and can lead to problems like asthma, bronchitis, and reduced lung function (Shalaj et al. 2016). Additionally, NO_x compounds react with ozone (O_3) in the atmosphere, contributing to the depletion of the ozone layer. The stratospheric ozone layer is vital for blocking harmful ultraviolet (UV) radiation from reaching the earth's surface; depletion of this layer is linked to an increased incidence of skin cancer and other health issues.

Sulphur Dioxide (SO_2) is another pollutant generated during the combustion of fuels containing sulphur. SO_2 and NO_x respectively contribute to the formation of sulfuric acid (H_2SO_4) and nitric acid (HNO_3) in the atmosphere. These acids result in acid rain, which has numerous detrimental effects, including lowering the pH of lakes and streams, harming aquatic life, damaging forests, and degrading buildings and cultural monuments (Livingston 2016). The acidic deposition can also disrupt soil chemistry, affecting plant health and biodiversity in terrestrial ecosystems. More details about the environmental performance of solar gas turbines are reported in Chap. 11.

4.5.4 Some Techniques for Emissions Reductions in Gas Turbines

Reducing emissions from gas turbines can be achieved through various techniques at both the combustion and post-combustion stages. While post-combustion methods, such as carbon capture and desulfurization, are commonly applied in coal-fired power plants, they are less frequently used for gas turbines. The primary focus in gas turbines is on minimizing pollutants during the combustion process itself. There are three key methods for emission reduction in the gas turbine industry: (a) water/steam injection, (b) selective catalytic reduction, and (c) dry low-NO_x technology.

4.5.4.1 Water/Steam Injection

Water or steam injection is a proven method to reduce flame temperature, thereby limiting the formation of nitrogen oxides (NO_x). By lowering the peak combustion temperatures, the production of NO_x is curtailed. Additionally, injecting steam into the combustion chamber increases the mass flow rate through the turbine, thereby enhancing the turbine's efficiency. However, an unintended consequence of this method is the potential increase in carbon monoxide (CO) and unburned hydrocarbons (UHCs) if the water/fuel ratio is too high. The steam used in this technique is often sourced from waste heat recovery systems, but it can also be produced using renewable energy, such as solar radiation in solarized gas turbines. To mitigate the formation of CO and UHCs, modern dry-low-emission combustors incorporate a premixing chamber with swirl injection of liquid fuels, which enhances the air–fuel mixing and improves combustion efficiency. Recent CFD simulations indicate that the location and direction of liquid fuel injection significantly affect fuel droplet breakup, evaporation, and combustion efficiency (Asgari and Amani 2017).

4.5.4.2 Selective Catalytic Reduction

Selective catalytic reduction (SCR) is a post-combustion technique that uses a catalyst to promote a chemical reaction that reduces NO_x emissions. In this process, controlled amounts of ammonia (NH_3) are injected into the exhaust gases, which then pass over a catalyst. This reaction converts nitrogen oxides (NO_x) into harmless nitrogen (N_2) and water vapour (H_2O). The effectiveness of SCR is temperature-dependent, typically operating optimally within a range of 558–673 K. Although SCR is highly effective at reducing NO_x emissions, its implementation can be expensive due to the costs associated with catalysts and the infrastructure required to handle and inject ammonia. Additionally, SCR systems may produce ammonia slip, a phenomenon where excess ammonia is released into the atmosphere, which itself can pose environmental risks.

4.5.4.3 Dry Low-NO_x Combustion

Dry low-NO_x (DLN) technology aims at reducing NO_x emissions without the use of water or steam injection. This approach involves achieving low emissions by optimizing the air–fuel mixture to either lean or rich conditions. In the "lean burn" method, a fuel–air mixture is burned with an excess of air, which lowers the combustion temperature and reduces NO_x formation. Conversely, the "rich burn/quick quench" technique involves burning the fuel-rich mixture initially, followed by a rapid introduction of dilution air to cool the combustion products, thereby minimizing NO_x emissions. However, lean burn is the more widely adopted approach due to its effectiveness and simplicity.

A significant advancement in lean combustion technology is the development of flameless combustion (FC), also known as flameless distributed combustion or colourless distributed combustion. In FC, the combustion process is distributed throughout the combustion chamber, eliminating visible flames and achieving a more uniform temperature distribution at the turbine inlet (Khidr et al., 2017). This uniformity helps to reduce thermal stresses on turbine blades, thereby prolonging the engine's lifespan. FC is controlled by specific mixture ratios and temperature conditions and can be achieved under carefully managed flow and thermal environments (Wünning and Wünning 1997). One of the primary benefits of FC is its ability to reduce NO_x and CO emissions to ultra-low levels, often less than 10 parts per million (ppm).

4.6 Turbine

A turbine is a mechanical device that extracts energy from a fluid flow and converts it into rotational energy, which is then used to drive a load such as an electric generator or other machinery. The core of a turbine is the rotor assembly, consisting of a shaft (or drum) with attached blades. When a fluid flows over these blades, it imparts kinetic energy, causing the rotor to spin. This rotation can be harnessed to produce electricity or perform other mechanical work.

In a gas turbine, the inlet temperature is directly proportional to the rate of fuel flow into the combustor. However, there is an upper limit to this temperature, dictated by the material properties of the turbine blades, which can only endure a specific level of thermal stress. To ensure the safe and efficient operation of the turbine, parameters like the fuel–air ratio and combustion temperature are carefully controlled.

4.6.1 Types of Turbines

Turbines are generally classified into two main types: radial flow and axial flow. A radial turbine is similar in design to a radial compressor, but with some key differences. In a radial turbine, the diffuser vanes are replaced by a ring of nozzle or guide vanes (called a stator), which direct the fluid inward towards the axis of rotation (Fig. 4.10). The working fluid enters the rotor with a high tangential velocity and exits close to the axis of rotation with a low rotational velocity. Radial turbines are particularly effective at extracting a significant amount of work from low mass flow rates. Their compact size, robustness, and ability to manage a wide range of operating conditions make them suitable for applications such as turbochargers, small gas turbines, and waste heat recovery systems.

In an axial turbine, the working fluid flows parallel to the axis of rotation (Fig. 4.11). The design of an axial turbine closely resembles that of an axial compressor but operates in the reverse direction, converting the kinetic energy of

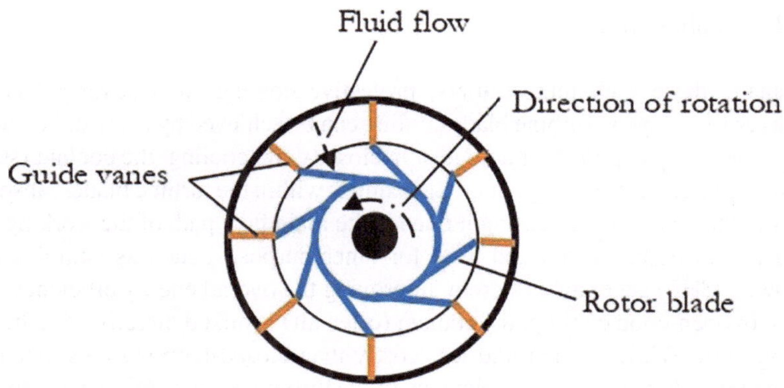

Fig. 4.10 Cross-section of a radial turbine

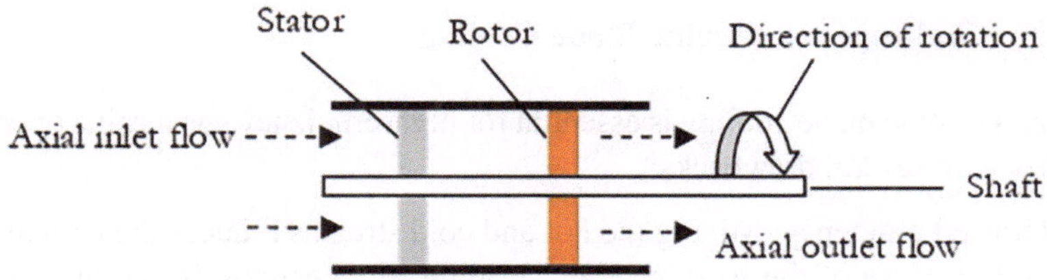

Fig. 4.11 Cross-sectional view of an axial turbine

the fluid into rotational kinetic energy. Axial turbines are commonly used in aircraft engines and large power plants due to their ability to manage high mass flow rates and provide higher efficiency than radial turbines. They can operate with single or multiple stages, depending on the available pressure drop. A single stage is used when the pressure drop is small, while multiple stages are required for larger pressure drops to maximize energy extraction.

4.6.2 Turbine Blade Cooling

The thermodynamic efficiency of a gas turbine increases with the inlet temperature, but higher temperatures also subject the turbine components, especially the blades, to extreme thermal stresses. This is a significant concern in large gas turbine power plants, where temperatures can exceed the thermal limits of the materials used. The primary constraint in raising the gas temperature is the mechanical strength of the turbine materials. Exceeding these limits can lead to material degradation, reduced turbine efficiency, and even structural failure.

4.6.2.1 Cooling Methods

To manage these high temperatures, protective coatings and advanced cooling systems are employed. Turbine blade cooling can be achieved by using either closed-loop or open-loop cooling technologies. In closed-loop cooling, the coolant (such as air or liquid) circulates through internal channels within the turbine blades, absorbing heat from the blades and releasing it outside the main flow path of the working fluid. This heat can be recovered and used for other purposes, such as combined heat and power (CHP) generation, thereby improving the overall energy efficiency of the system. In open-coop cooling, the coolant (often air) is mixed directly with the main working fluid. While simpler and less costly than closed-loop systems, open-loop cooling may reduce the overall efficiency of the turbine because mixing hot and cool streams decreases the enthalpy and total pressure of the working fluid.

4.6.2.2 Challenges of Turbine Blade Cooling

Although turbine blade cooling is essential for high-efficiency gas turbine operation, it comes with several drawbacks:

(a) Reduced efficiency: Mixing the hot and cold streams reduces the enthalpy and total pressure of the working fluid, leading to a decrease in the turbine shaft work and overall efficiency.
(b) Lower exhaust gas temperature: The temperature of the exhaust gases is decreased, reducing the potential for heat recovery in downstream processes, such as in heat recovery steam generators (HRSGs).
(c) Higher production costs: The inclusion of cooling systems significantly increases the production costs of turbines due to the complexity of the design and the need for advanced materials capable of withstanding high temperatures and thermal cycling.

4.6.2.3 Advances in Turbine Cooling Technologies

Recent advancements in turbine cooling include transpiration cooling and the use of ceramic matrix composite (CMC) materials. Transpiration cooling involves porous turbine blades through which a coolant (typically air) is allowed to permeate, creating a protective film on the blade surface. This method can provide more uniform cooling and reduce thermal gradients, which minimizes the risk of thermal fatigue. CMC materials have a high resistance to thermal stress and can withstand higher temperatures than traditional metal alloys. The use of CMCs in turbine components allows for higher operating temperatures, increasing efficiency and reducing the need for complex cooling systems.

4.7 Basic Electric Generator

A turbine develops rotational mechanical energy which needs conversion to produce electricity. An electric generator is a device that converts mechanical energy from a prime mover to electric energy as an output (Fig. 4.12). Prime movers are sources of mechanical energy, and they include gas, steam, hydro and wind turbines. Electromechanical generation of electricity can be achieved via two concepts: electrostatic and magnetic induction.

4.7.1 *Electric-Field Generators*

Electrostatic induction is achieved by using electrets (dielectric materials with embedded charge) (Wada et al., 2016). This trapped charge is quasi-permanent, and it produces an electrostatic field around itself. The electret generator is a device that converts mechanical energy into electrical energy by using an electric field created by the electrets. An example of an electrostatic generator is presented in Fig. 4.13. It comprises rotor and stator electrodes, and electrets with an air gap separating the rotor electrode from the electrets. The stator electrode and electrets are stationary. An external electric field is induced between the rotor and stator electrodes when the rotor spins about its axis.

Wada et al. (2016) found a maximum power output of 3.6 μW for electret generator (using ceramic electret). For vibrational electret generators, output power values in the range $1.0 \times 10^{-4} - 1.2 \times 10^{-1}$ μW have been reported (Tao et al. 2015), which are relatively low (μ-scale). This magnitude of power is suitable for low-power electronic devices such as microphones. Low-power prime movers are exploited to harvest ambient energy. Considering the wide spectrum of needs for electric power, it is also desirable to generate electricity at higher capacity scales, and magnetic generators are popular in this regard.

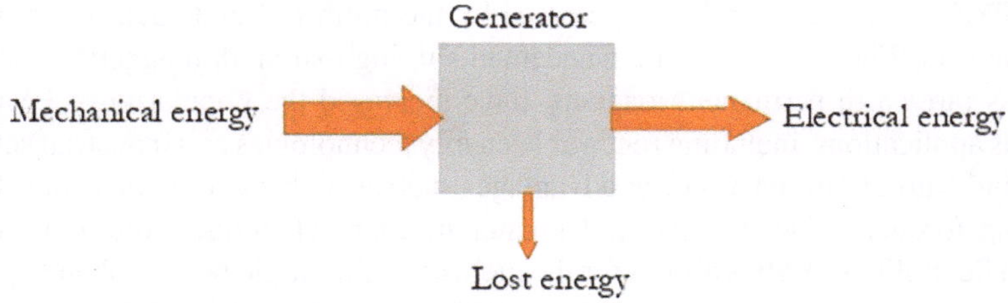

Fig. 4.12 Block diagram of electromechanical energy conversion by a generator

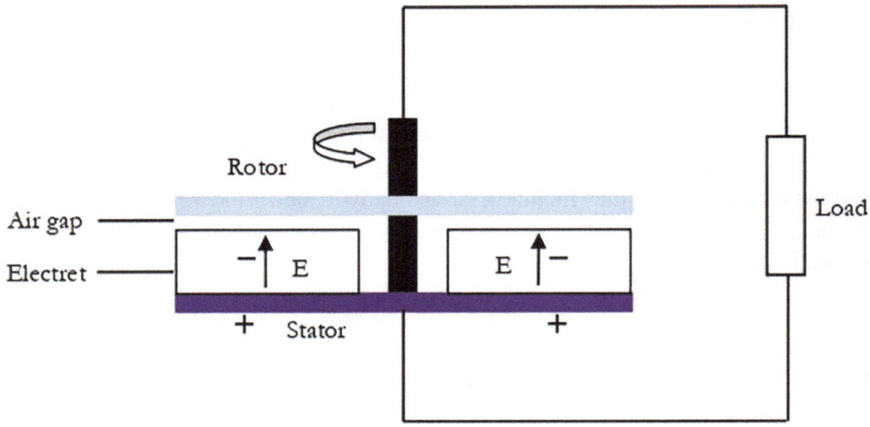

Fig. 4.13 An example of a simple electrostatic generator, showing electric field E inside electrets

4.7.2 Magnetic-Field Induction

Magnetic-field generators are the most prevalent type in the power sector and are based on electromagnetic induction, a principle discovered by Michael Faraday in the nineteenth century. According to Faraday's Law, a voltage is induced in a conductor when it either moves relative to a constant magnetic field or when the magnetic field around a stationary conductor changes. The magnitude of the induced voltage is proportional to the rate of change of the magnetic field. A typical magnetic-field generator consists of two main components: a stator (which provides the magnetic field) and armature windings (which are the conductors where electricity is generated) (Fig. 4.14). The magnetic field can be either axial (where magnetic flux lines run parallel to the axis of rotation) or radial (where the lines are perpendicular to the axis).

The magnetic field in these generators can be produced by permanent magnets (self-excited) or magnetic induction (field-excited). In a permanent magnet generator (PMG), the magnetic field is created by the north (N) and south (S) poles of the magnets (Fig.4.15). Recent advancements in high-strength magnetic materials, such as rare-earth permanent magnets, have increased the popularity of PMGs in various applications, including renewable energy technologies and industrial settings (Gor and Kurt 2016). PMGs offer advantages such as higher energy densities, lower cogging torque, reduced costs, and greater mechanical torque. Additionally, the design flexibility of PMGs allows for the inclusion of multiple poles, enhancing their efficiency and performance. However, they also generate significant heat, requiring effective cooling systems to maintain optimal operation. In contrast, field-excited generators use an external electric current, typically low-voltage direct current (DC), to create a magnetic field. This method provides more control over the magnetic field strength, which can be adjusted as needed.

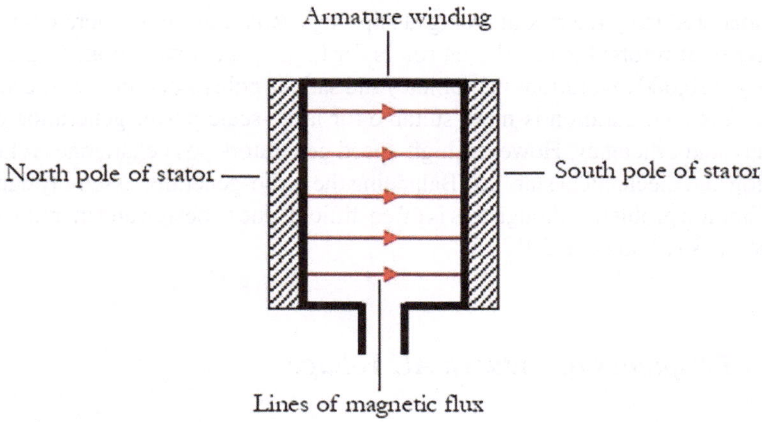

Fig. 4.14 Schematic cross-section of a radial flux permanent magnet generator

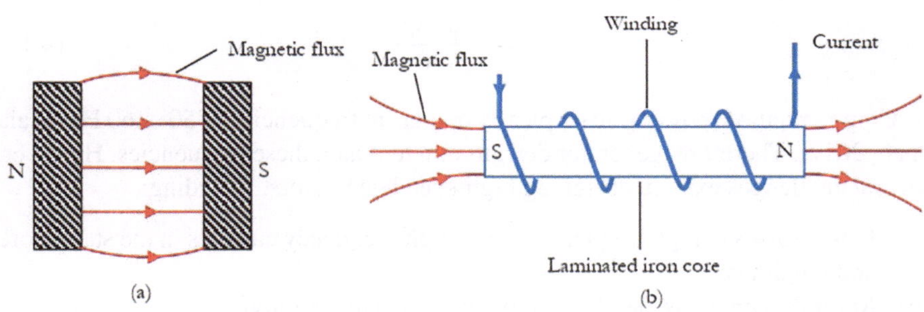

Fig. 4.15 Magnetic field between north (N) and south (S) poles of: **a** permanent magnets and **b** simple inductor

4.7.3 Direct Current and Alternating Current Generators

Generators can produce either direct current (DC) or alternating current (AC). In DC generators, the armature is always the rotor, while in AC generators, the armature can be either the stator or rotor. AC generators are further classified into rotating-armature and rotating-field generators.

In rotating-armature generators, the armature rotates relative to the magnetic field. This configuration requires slip rings and brushes to transfer the induced current from the rotating armature to the stationary load. However, maintaining insulation of the armature, brushes, and slip rings is challenging. Insulation aging, often caused by partial discharges, can lead to faults and potential failures (Danikas and Karlis 2011). High voltages further complicate insulation management, increasing the risk of arcing and short circuits. Thus, rotating-armature configurations are typically used for low-power applications. In rotating-field generators, the magnetic field rotates relative to a stationary armature. The high voltage induced can be directly conducted to the load, reducing the risk of arcing at slip rings. Rotating-field generators use two main types of rotors: turbine-driven rotors for high-speed applications (typically at speeds $\geq$ 120,000 revolutions per minute) and salient-pole rotors for low-speed operations. This configuration is more suitable for large-scale power generation due to its safety and efficiency. However, high-speed generators pose challenges related to vibration and mechanical stability. Balancing the turbo-generator assembly can mitigate vibration problems, though this is often difficult due to design and manufacturing complexities (Maia et al. 2017).

4.7.4 Frequency of Induced AC Voltage

The frequency (f) of the output voltage from an AC generator depends on the number of poles (N_{po}) and the rotational speed in revolutions per minute (N_{rpm}) of the rotor:

$$f = \frac{N_{\text{rpm}} N_{po}}{120} \tag{4.11}$$

Commercial AC power grids typically operate at frequencies of 50 or 60 Hz (Maia et al., 2017). Therefore, generator designs aim to match these frequencies. However, several challenges exist in designing high-speed generators, including:

(a) Power losses at high frequency due to high-frequency currents in the stator core and windings.
(b) Rotor dynamics: Issues related to vibration and stability.
(c) Thermal characteristics: The generation of heat from electrical resistance, core losses, and friction in bearings.
(d) Bearing components: Bearing wear and tear under high-speed operations.

The most common failures in electric rotating machines occur in the stator windings, often due to insulation degradation (Foster et al. 2016). The heat generated in the windings by electrical resistance, core losses in iron laminations, and friction in bearings contribute to insulation breakdown. Additional heat sources include arc-over air gaps and heat transfer from nearby turbines. To prevent excessive temperature rises and ensure reliable operation, generators are equipped with appropriate cooling systems.

Recent innovations focus on improving the reliability, efficiency, and power density of generators. Superconducting materials are being explored for winding applications to reduce resistance losses significantly, while enhanced cooling techniques such as liquid cooling and cryogenic cooling are being developed to manage higher thermal loads. Furthermore, smart monitoring systems that use advanced sensors and artificial intelligence (AI) algorithms are being integrated into modern generators to predict faults, optimize performance, and reduce maintenance costs.

4.8 Control System

Control systems are crucial in various industrial applications, including the power generation sector. An automatic control system maintains a variable, such as temperature, pressure, humidity, airflow, or fuel flow, at a set point or within a specified range of values. A basic control system consists of three main components: sensors, controllers, and actuators. The sensor monitors the value of the controlled variable, compares it with a predefined set point, and sends a signal to the actuator to initiate the necessary action to maintain or adjust the variable within the desired range (American Society of Heating, Refrigerating and Air-Conditioning Engineers 2001).

There are two types of control systems: open-loop and closed-loop. An open-loop control system does not use feedback from the regulated variable but it relies instead on a predefined sequence of actions to manipulate a control agent. In a closed-loop control system, the controller continuously monitors the actual changes in the regulated variable and adjusts the actuator to achieve the desired set point, stopping once the target is reached. Closed-loop control systems can be further categorized by their action type: two-position, floating, modulating, and fuzzy-logic controls. Two-position control operates at minimum or maximum levels (such as "on" or "off") and is simple and cost-effective, making it widely used in industrial applications. Floating control systems adjust the actuator in discrete steps, either opening or closing at a fixed rate. Modulating controllers provide a continuous range of output, allowing for fine adjustments. They can implement proportional, proportional-integral, or proportional-integral-derivative (PID) actions. Among these, PID controllers offer superior performance by adjusting the control variable based on present, past, and anticipated future errors. Fuzzy-logic controllers use a set of rules (such as 'if–then' statements) to mimic human decision-making processes in controlling complex systems.

Gas turbines require particularly sophisticated control systems due to their rapid response to operational and environmental changes and high-speed operation. For example, an aircraft gas turbine can accelerate from idle to take-off speed in less than 60 s (Giampaolo 2003). In electricity generation, gas turbines often drive components like compressors and generators, whose mass inertia limits the rate of acceleration. Environmental factors like ambient temperature and pressure can significantly impact turbine performance; typically, efficiency declines with increasing ambient temperatures. Rapid changes in conditions may lead to issues such as compressor surge, overheating, or excessive speeds, necessitating robust control mechanisms.

Gas turbine control functions generally fall into three main categories, each with some overlap: sequencing, routine operation and protection controllers. A sequencing controller controls the start, load, unload, and stop sequences of the turbine. Once the turbine is started correctly, the sequencing controller passes control to the routine operation controller. It resumes control before loading to set parameters like inlet and discharge valves and synchronizes the generated power with the grid. Routine operation controllers are responsible for maintaining steady-state operation, including management of various parameters such as speed, acceleration, temperature, and pressure. They also adjust fuel flow to match power demands while avoiding overloading or damaging components. The protection controller is designed to prevent faults or damage to the turbine by monitoring critical parameters and shutting down or adjusting operations in response to hazardous conditions. A simple controller of a solar gas turbine for electricity generation is presented in Fig. 4.16.

4.8.1 Control of Power Output and Safety

Controlling a gas turbine's power output is the primary function of its control system, involving both steady-state and dynamic operations. Increasing shaft power requires raising the fuel flow rate, which raises combustion temperatures and enhances shaft power. However, a rapid increase can exceed turbine temperature limits or cause compressor surge, leading to potential damage. Similarly, a rapid reduction in fuel flow can extinguish the flame or cause a surge.

Ambient temperature changes, though generally slower than the turbine's response time, can still have significant effects. For example, low temperatures combined with high humidity can cause ice formation upstream of the compressor, necessitating anti-icing strategies like heating the air intake.

Solar gas turbines also present unique control challenges due to the intermittent nature of solar radiation and the use of sun-tracking concentrators. The input heat from the solar field varies with time, requiring the control system to manage fluctuations in solar input and adjust fuel flow accordingly to maintain optimal turbine inlet temperatures. The control system must also ensure a smooth transition between solar and backup fuel sources, as depicted in Fig. 4.16.

4.8.2 Additional Control Considerations

Modern control systems incorporate advanced algorithms and sensors to enhance performance and reliability. For instance, predictive maintenance tools are now integrated into control systems to monitor equipment health, detect early signs of wear or failure, and schedule maintenance before a failure occurs. Advanced AI-based controllers can optimize turbine operations under varying conditions, improving efficiency and extending equipment life. Moreover, safety mechanisms are more sophisticated, using redundancy and fault-tolerant designs to prevent single-point failures.

Given the high speed and dynamic response requirements of gas turbines, vibration control is also critical. Balancing the turbo-generator assembly helps reduce vibration but can be challenging due to the separate development of turbine and generator components (Maia et al. 2017). Additionally, the control system must consider thermal dynamics, electrical harmonics, and grid stability, particularly in applications involving renewable energy integration.

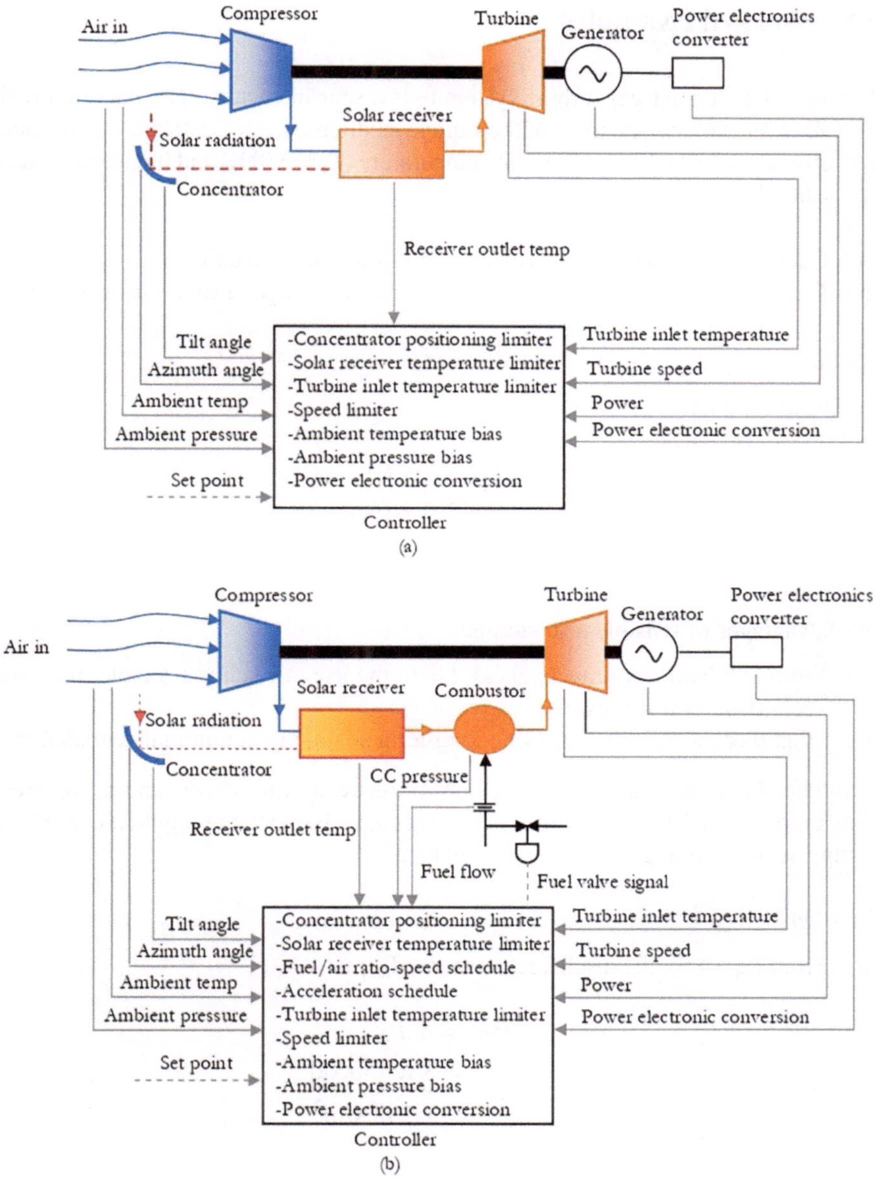

Fig. 4.16 A simplified control system of: **a** solar-only gas turbine and **b** solar gas turbine with backup heating. CC = combustion chamber

4.9 Worked Examples

Example 4.1 A solar gas turbine system uses a sensible heat storage medium with the following parameters: mass of the storage medium ($m_{sm} = 2000$ kg), specific heat capacity ($C_{p,sm} = 2.0$ kJ/kg·K), initial temperature ($T_i = 350$ K) and final temperature ($T_f = 600$ K).

(a) Calculate the amount of thermal energy stored in the medium.
(b) Discuss two advantages of using sensible heat storage in solar power systems.

Solution

(a) Using Eq. (4.1):

$$
\begin{aligned}
Q &= m_{sm} C_{sm} (T_f - T_i). \\
&= 2000 \times 2.0 \times (600\text{–}350). \\
&= 1.0 \text{ GJ}.
\end{aligned}
$$

(b) Advantages of sensible heat storage

(i) Simplicity and cost-effectiveness due to the use of readily available materials such as water and rocks.
(ii) High thermal stability, enabling long-term storage with minimal degradation.

Example 4.2 A gas turbine has an inlet pressure of 150 kN/m^2 and a compression ratio of 8. Calculate the discharge pressure. Explain the significance of the compression ratio in gas turbine performance.

Solution

(a) From Eq. (4.5), the discharge pressure (d_{dis}) is:

$$
\begin{aligned}
P_{dis} &= r_{co} P_{in} \\
&= 8 \times 150000 \\
&= 1.2 \text{ MN/m}^2
\end{aligned}
$$

(b) Significance of compression ratio:

(i) A higher compression ratio improves the thermal efficiency of the turbine cycle.
(ii) It increases the power output by raising the pressure and temperature of the working fluid entering the turbine.

Example 4.3 An AC generator with 6 poles operates at a speed of 1500 rpm. Determine the frequency of the generated AC voltage. Discuss the impact of increasing the number of poles on the frequency and generator size.

Solution

(a) From Eq. (4.6), frequency (f) is:

$$f = \frac{N_{po} \cdot N_{rpm}}{120}$$
$$= \frac{6 \cdot 1500}{120} = 75\,\text{Hz}$$

(b) Impacts of increasing the number of poles are:

(i) Higher frequency output can be achieved, beneficial for specific applications.
(ii) Larger pole numbers increase the generator size, making it more complex and expensive.
(iii) High-pole configurations are better suited for low-speed prime movers like hydro turbines.

Example 4.4 A concentrating solar power plant uses a binary molten salt to store heat. In solid form, the salt has specific latent heat and solid specific heat capacity of 161 kJ/kg and 1250 J/kg.K, respectively. On a certain day, the salt is heated from 350 to 900 K, and 1.0 GJ of heat is absorbed by the storage medium. If the melting point of the salt is 800 K, and the specific heat capacity the salt in liquid form is 1500 J/kg.K. Calculate the mass of the storage medium, assuming there is no heat loss from the storage system.

Solution

From Eq. (4.4), mass of the salt (m_{so}) is:

$$m_{so} = \frac{Q}{C_{p,so}\left(T_{mp} - T_{so}\right) + L_{so,l} + C_{p,l}\left(T_l - T_{mp}\right)}$$
$$= \frac{1000000000}{1250 \times (873 - 350) + 161000 + 1500 \times (900 - 800)}$$
$$= 1404\,\text{kg}$$

Example 4.5 For the system described in Example 4.4, suppose there is insufficient solar radiation, resulting in a temperature drop to 810 K. It is a requirement to maintain the temperature of the storage medium above its melting point. So, an auxiliary heater is used to elevate the temperature of the storage medium to 900 K in 2 h. Calculate the nominal power output from the heater, neglecting heat losses.

Solution

Power output = Q/time,
where Q is the heat absorbed by the liquid molten salt.

$$Q = m_{so} C_{ll} (T_2 - T_1)$$

Therefore,

$$\text{Power output} = 1404 \times 1500 \times (900 - 810)/7200$$
$$\approx 26\ \text{kW}$$

Nomenclature

A	Area (m^2)
C	Specific heat capacity (J/kg K)
f	Frequency of the output voltage from an AC generator
L	Specific latent heat (J/kg)
m	Mass (kg)
N_{po}	Number of poles (dimensionless)
N_{rpm}	Number of revolutions per minute (dimensionless)
p	Pressure (N/m^2)
Q	Stored energy (J)
$\dot{Q}$	Charging rate (W)
r_{co}	Compression ratio (dimensionless)
T	Temperature (K)

Subscripts

a	Ambient/air
dis	Discharge
f	Final
i	Initial
in	Inlet
l	Liquid
mp	Melting point
p	Constant pressure
sm	Storage material
so	Solid

References

Alami AH, Olabi AG, Mdallal A, Rezk A, Radwan A, Rahman SMA, Shah SK, Abdelkareem MA (2023) Concentrating solar power (CSP) technologies: status and analysis. Int J Thermofluids 18:100340

Almanza R, Hernández P, Martín I, Mazari M (2009) Development and mean life of aluminium first-surface mirrors for solar energy applications. Sol Energy Mater Sol Cells 93:1647–1651

American Society of Heating, Refrigerating and Air-Conditioning Engineers (2001) Fundamentals Handbook. American Society of Heating, Refrigerating and Air-Conditioning Engineers, Atlanta

Asgari B, Amani E (2017) A multi-objective CFD optimization of liquid fuel spray injection in dry-low-emission gas-turbine combustors. Appl Energy 203:696–710

Ávila-Marín AL (2011) Volumetric receivers in solar thermal power plants with central receiver system technology: a review. Sol Energy 85:891–910

Besarati, S.M., Goswami, D.Y., Stefanakos, E.K., 2015. Development of a solar receiver based on compact heat exchanger technology for supercritical carbon dioxide power cycles. ASME J Sol Energy Eng 137 (3)

Blanco MJ, Santigosa LR (2017) Advances in concentrating solar thermal research and technology. Elsevier, Amsterdam

Canepa R, Wang M (2015) Techno-economic analysis of a CO_2 capture plant integrated with a commercial scale combined cycle gas turbine (CCGT) power plant. Appl Therm Eng 74:10–19

Choi M, Sung Y, Won M, Park Y, Kimi M, Choi G, Kim D (2017) Effect of fuel distribution on turbulence and combustion characteristics of a micro gas turbine combustor. J Ind Eng Chem 48:24–35

Cohen H, Rogers GF, Saravanamuttoo HI (1996) Gas turbine theory,4th edn. Longman Group Limited, Essex

Cot-Gores J, Castell A, Cabeza LF (2012) Thermochemical energy storage and conversion: a-state-of-the-art review of the experimental research under practical conditions. Renew Sustain Energy Rev 16:5207–5224

Cuddihy EF (1980) Theoretical considerations of soil retention. Sol Energy Mater Sol Cells 3:21–33

Dähler F, Ambrosetti G, Steinfeld A (2017) Optimal solar dish field layouts for maximum collection and shading efficiencies. Sol Energy 144:286–294

Danikas MG, Karlis A (2011) A review on electrical machines insulation aging and its relation to the power electronics arrangements with emphasis on wind turbine generators. Renew Sustain Energy Rev 15:1748–1752

Dimitrakellis P, Gogolides E (2018) Hydrophobic and superhydrophobic surfaces fabricated using atmospheric pressure cold plasma technology: a review. Adv Colloid Interface Sci 254:1–21

Dutta P (2017) High temperature solar receiver and thermal storage systems. Appl Therm Eng 124:624–632

Fernandes D, Pitié F, Cáceres G, Baeyens J (2012) Thermal energy storage: how previous findings determine current research priorities. Energy 39(1):246–257

Fernández-García A, Cantos-Soto ME, Röger M, Wieckert C, Hutter C, Martínez-Arcos L (2014) Durability of solar reflectror materials for secondary concentrators used in CSP systems. Sol Energy Mater Sol Cells 130:51–63

Finenko A, Cheah L (2014) Carbon dioxide reduction potential in Singapore's power generation sector. Energy Proceedia 61:527–532

Foster SN, Cintron-Rivera JG, Strang EG (2016) Detection of incipient stator winding faults in PMSMs with single-layer fractional slot concentrated windings. Electr Pow Syst Res 131:231–243

García IL, Álvarez JL, Blanco D (2011) Performance model for parabolic trough solar thermal power plants with thermal storage: comparison to operating plant data. Sol Energy 85:2443–2460

Giampaolo T (2003) The gas turbine handbook: principles and Practices, 2nd edn. Fairmont, Lilburn

González-Roubaud E, Pérez-Osorio D, Prieto C (2017) Review of commercial thermal energy storage in concentrated solar power plants: steam vs. molten salts. Renew Sustain Energy Rev 80:133–148

Gor H, Kurt E (2016) Preliminary studies of a new permanent magnet generator (PMG) with the axial and radial flux morphology. Int J Hydrog Energy 41:7005–7018

Ho CK, Iverson BD (2014) Review of high-temperature central receiver designs for concentrating solar power. Renew Sustain Energy Rev 29:835–846

Ho C (2017) Advances in central receivers for concentrating solar applications. Sol Energy :38–56

Khidr K, Eldrainy Y, EL-Kassaby M (2017) Towards lower gas turbine emissions: flameless distributed combustion. Renew Sustain Energy Rev 67:1237–1266

Kiejna A (1999) Stabilized jellium—simple model for simple-metal surfaces. Prog Surf Sci 61(5–6):85–125

Kim H, Kim H, Lee S, Kim J (2018) A study on heat transfer characteristics of quinary molten salt mixture. Int J Heat Mass Transf 127:465–472

Korzynietz R, Brioso JA, del Río A, Quero M, Gallas M, Uhlig R, Ebert M, Buck R, Teraji D (2016) Solugas—comprehensive analysis of the solar hybrid Brayton plant. Sol Energy 135:578–589

Kumaresan G, Sudhakar P, Santosh R, Velraj R (2017) Experimental and numerical studies of thermal performance enhancement in the receiver part of solar parabolic trough collectors. Renew Sustain Energy Rev 77:1363–1374

Kurt E, Gör H, Demirtaş M (2014) Theoretical and experimental analyses of a single phase permanent magnet generator (PMG) with multiple cores having axial and radial directed fluxes. Energy Convers Manag 77:163–172

Le Roux WG, Bello-Ochende T, Meyer JP (2014) The efficiency of an open-cavity tubular solar receiver for a small-scale solar thermal Brayton cycle. Energy Convers Manag 84:457–470

Livingston RA (2016) Acid rain attack on outdoor sculpture in perspective. Atmospheric Environment. Atmos Environ 146:332–345

Maia TA, Faria OA, Barros JE, Porto MP, Filho BJ (2017) Test and simulation of an electric generator driven by a micro-turbine. Electr Pow Syst Res 147:224–232

Moya E (2012) Parabolic-trough concentrating solar power (CSP) systems. In Lovegrove K and Stein W (eds) Concentrating solar power technology. Woodhead Publishing Limited, pp 197–239

National Renewable Energy Laboratory (2017) Concentrating solar power projects. https://www.nrel.gov/csp/solarpaces/power_tower.cfm. Accessed 8 Aug 2017.

Okoroigwe E, Madhlopa A (2016) An integrated combined cycle system driven by a solar tower: a review. Renew Sustain Energy Rev 57:337–350

Saghafifar M, Gadalla M (2016) Thermo-economic analysis of conventional combined cycle hybridization: United Arab Emirates case study. Energy Convers Manag 111:358–374

Sarver T, Al-Qaraghuli A, Kazmerski LL (2013) A comprehensive review of the impact of dust on the use of solar energy: History, investigations, results, literature and mitigation approaches. Renew Sustain Energy Rev 22:698–733

Schwarzbözl P, Buck R, Sugarmen C, Ring A, Crespo MJ, Altwegg P, Enrile J (2006) Solar gas turbine systems: design, cost and perspectives. Sol Energy 80:1231–1240

Shalaj VV, Mikhailov AG, Slobodina EN, Terebilov SV (2015) Issues on nitrogen oxides concentration reduction in the combustion products of natural gas. Procedia Eng 113:287–291

Shalaj VV, Mikhajlov AG, Novikova EE, Terebilov SV, Novikova TV (2016) Gas recirculation impact on the nitrogen oxides formation in the boiler furnace. Procedia Eng 152:434–438

Spelling JD (2013) Hybrid solar gas-turbine power plants: a Thermoeconomic analysis. Doctoral Thesis, KTH Royal Institute of Technology, Stockholm. KTH Royal Institute of Technology, Stockholm

Stein WH, Buck R (2017) Advanced power cycles for concentrated solar power. Sol Energy 152:91–105

Tagle-Salazar PD, Nigam KDP, Rivera-Solorio CI (2020) Parabolic trough solar collectors: a general overview of technology, industrial applications, energy market, modelling, and standards. Green Process Synth 9(1):595–649

Tao K, Miao J, Lye SW, Hu X (2015) Sandwich-structured two-dimensional MEMS electret power generator for low-level ambient vibrational energy harvesting. Sens Actuators A 228:95–103

Torkzadeh MM, Bolourchifard F, Amani E (2016) An investigation of air-swirl design criteria for gas turbine combustors through a multi-objective CFD optimization. Fuel 186:734–749

Wada N, Horiuchi N, Mukougawa K, Nozaki K, Nakamura M, Nagai A, Okura T, Yamashita K (2016) Electrostatic induction power generator using hydroxyapatite ceramic electrets. Mater Res Bull 74:50–56

Wünning JA, Wünning JG (1997) Flameless oxidation to reduce thermal NO-formation. Prog Energy Combust Sci 23:81–94

Zeng C, Liu S, Shukla A (2017) Adaptability research on phase change materials based technologies in China. Renew Sustain Energy Rev 73:145–158

Zhang HL, Baeyens J, Degrève J, Cacères G (2013) Concentrated solar power plants: review and design methodology. Renew Sustain Energy Rev 22:466–481

Chapter 5
Thermodynamic Cycles of Solar Gas Turbines

5.1 Introduction

A solar gas turbine (SGT) converts heat from a solar receiver to mechanical work. Depending on system design, the solar gas turbine may be backed up by a conventional heater. Solar gas turbines operate on thermodynamic cycles. A thermodynamic cycle involves a working fluid that changes from one state to another within the cycle. Heat is transferred to or from the working fluid throughout the cycle (Çengel and Boles 2006), and the fluid may be in sub-cooled liquid, liquid–vapour or superheated vapour states (Fig. 5.1). A critical state is attained when the fluid is at a critical pressure and temperature. For instance, the critical temperature and pressure of steam are respectively 647.3 K and 22.1 MN/m^2 (Dunham and Iverson 2014). Above the critical temperature, the vapour of the fluid is not liquefiable regardless of the magnitude of the pressure.

Based on the mode of heat transfer between heat source and sink, thermodynamic cycles fall into two categories: power and refrigeration cycles. In a power cycle, the working fluid transfers heat (input) from a heat source to a heat sink and converts part of the input heat into a mechanical work output. A refrigeration cycle transfers heat from low- to high-temperature reservoirs by using mechanical work input. Devices that produce refrigeration effects are known as refrigerators, air conditioners or heat pumps, and they operate on refrigeration cycles.

Solar gas turbines produce mechanical work output from input heat and so they operate on power cycles. This mechanical work can be used to drive an electric generator for electricity production. Many power cycles have been reported in literature but only a selected number of relevant cycles (Carnot, Rankine and Brayton cycles) are briefly presented in this book. It is perceived that raising the temperature difference between the heat source and heat sink has great potential of improving the thermal efficiency of power cycles (Zheng et al. 2023). The Carnot cycle is included as an ideal cycle for bench-marking. This cycle yields the highest thermal efficiency when operated under the same heat source and sink as any other cycle. No other engine

A. Madhlopa and R. Nkhoma, *Principles of Solar Gas Turbines for Generating Electricity*, Green Energy and Technology,
https://doi.org/10.1007/978-3-031-84096-8_5

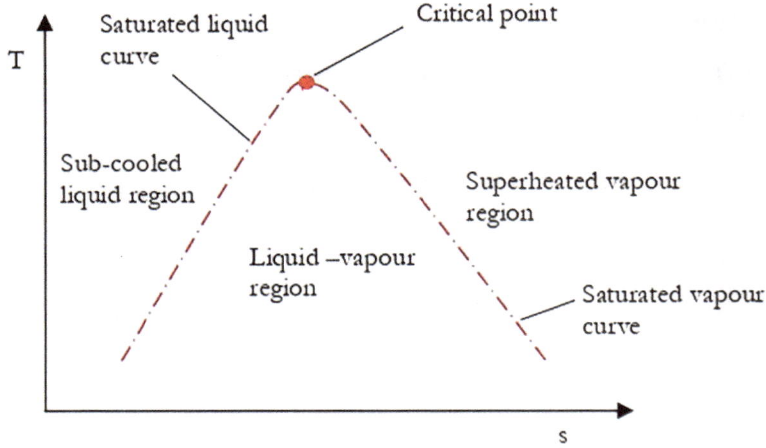

Fig. 5.1 Temperature-entropy (T-s) diagram of a generalized fluid

has higher thermodynamic efficiency than the Carnot engine operating between the same heat source and sink. The Rankine cycle is the most widely applied power cycle in practical concentrating solar power (CSP) plants but it has lower cycle efficiency. Consequently, research attention is also given to the possible exploitation of advanced thermodynamic cycles. In this context, the gas turbine (Brayton) cycle is one of the promising cycles for integration with CSP systems.

5.2 Carnot Cycle

A Carnot cycle is an ideal engine cycle in which heat is drawn from a hot reservoir at a temperature (T_1) and rejected to a cold reservoir at a temperature (T_2). The Carnot cycle may be a vapour or gas cycle. This thermodynamic cycle was proposed by a French scholar named Sadi Carnot in 1824, and provided the foundation for classical thermodynamic theory (Carnot 1824). The heat loss between the working fluid and heat reservoir, and the limitations of time and heat exchanger area were ignored in classical thermodynamics (Wang et al. 2022). In this context, the Carnot cycle is reversible: it comprises two reversible isothermal processes at T_1 and T_2, and two reversible isentropic processes. For a Carnot vapour cycle, the two isothermal processes can be achieved through isobaric heating and cooling, within the region in which the vapour is wet (Fig. 5.2). In an isothermal process, the temperature of the working fluid remains constant. An isentropic process takes place at constant entropy of the working fluid, while the pressure of the working fluid does not change in an isobaric process. The efficiency of a Carnot engine (η_{car}) can be given by (Shang et al. 2014):

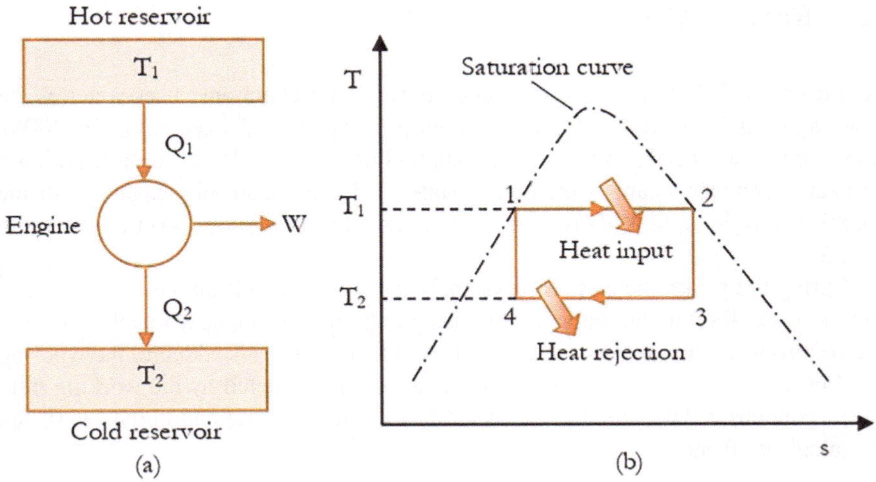

Fig. 5.2 Schematic representation of a Carnot vapour cycle: **a** energy flow and **b** temperature-entropy (T-s) diagram

$$\eta_{\text{car}} = 1 - \frac{T_2}{T_1} \tag{5.1}$$

where $T_1 > T_2$.

This is the maximum efficiency at which heat extracted from a hot reservoir can be converted into work by an engine that is operating in thermodynamic contact with a cold reservoir. Thus, no other engine (operating between the same reservoirs) can have a thermal efficiency greater than that of the Carnot engine.

In case of the CSP technology, the efficiency of the engine is a product of two efficiencies (Blanco and Santigos 2017): (a) the thermal efficiency of the conversion of solar radiation into useful thermal energy, and (b) the efficiency of the subsequent conversion of the useful solar thermal energy into work by a heat engine. The efficiency of solar-to-thermal conversion decreases with the operating temperature. A higher proportion of solar radiation is transformed into heat when the receiver is at a lower temperature. In contrast, the conversion of thermal energy to work increases with the operating temperature but it is limited by the second law of thermodynamics. As the operating temperature (T_1) in the Carnot engine increases to infinity, the efficiency of the engine approaches unity. The operating temperature of practical heat engines does not exceed 6000 K (blackbody temperature at which the sun emits radiation).

5.3 Rankine Cycle

A conventional Rankine cycle is an engine cycle that commonly uses water as the working fluid. This cycle is ideal for vapour power plants (Çengel and Boles 2006), and it comprises: (a) isentropic liquid compression (states 1–2), (b) isobaric addition of heat to produce steam in the boiler (states 2–3), (c) isentropic expansion of the steam (states 3–4), and (d) isobaric steam condensation (states 4–1) as shown in Fig. 5.3b.

During the isentropic expansion of the steam in an ideal Rankine cycle, there is output work ($\dot{W}_{34}$) in the process. The feed pump provides input work ($\dot{W}_{12}$) during the isentropic compression of the working fluid. Heat ($\dot{Q}_{23}$) is added to the working fluid between states 2 and 3 while some heat ($\dot{Q}_{41}$) is rejected by the working fluid in the condenser. The added and rejected heats, and the input and output work can be calculated from:

$$\dot{Q}_{41} = \dot{m}_{\mathrm{v}}(h_1 - h_4) \tag{5.2}$$

$$\dot{Q}_{23} = \dot{m}_{\mathrm{v}}(h_3 - h_2) \tag{5.3}$$

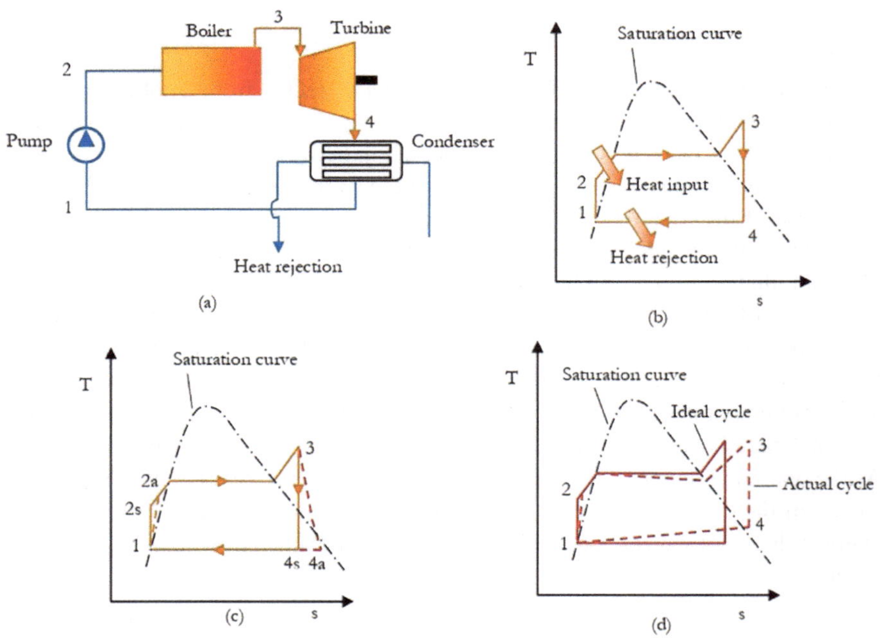

Fig. 5.3 Schematic representation of a standard Ranken cycle **a** Energy flow, **b** temperature-entropy (T-s) diagram of an ideal cycle, **c** temperature-entropy (T-s) diagram with ideal (2 s and 4 s) and actual (2a and 4a) states and **d** deviation of ideal and actual cycles

$$\dot{W}_{12} = \dot{m}_v(h_2 - h_1) \tag{5.4}$$

$$\dot{W}_{34} = \dot{m}_v(h_4 - h_3) \tag{5.5}$$

where $\dot{m}_v$ is rate of vapour mass flow, and h_1, h_2, h_3 and h_4 are the values of the enthalpy of the working fluid at the specified points. Enthalpy is a thermodynamic property that results from a combination of three properties: internal energy, pressure and volume of a system (American Society of Heating Refrigerating and Air-Conditioning Engineers 2017). A change in the enthalpy of a system depends only on the initial and final states of the system. The net output work ($\dot{W}_{rc}$) of the ideal Rankine cycle is given by:

$$\dot{W}_{rc} = \dot{W}_{34} - \dot{W}_{12} = \dot{m}_v(h_4 - h_3) - \dot{m}_v(h_2 - h_1) = \dot{m}_v(h_1 - h_2 + h_3 - h_4) \tag{5.6}$$

Thermal efficiency is an important measure of the performance of a power plant, and it is defined as the ratio of the net output work to the total heat added to the cycle. For the same input power, an energy technology with higher efficiency is more preferable. The thermal efficiency (η_{rc}) of the ideal Rankine cycle can be computed from:

$$\eta_{rc} = \frac{\dot{W}_{rc}}{\dot{Q}_{23}} = \frac{(h_1 - h_2 + h_3 - h_4)}{(h_3 - h_2)} \tag{5.7}$$

Ideal and actual steam cycles differ because of irreversibilities arising mainly from: (a) heat loss from various components to the surroundings and (b) fluid friction. Heat is lost from the steam to the ambient environment as the steam flows through different components (such as pipes and turbine) of an actual steam engine. So, more heat must be transferred to the steam from the boiler to compensate for these heat losses to maintain the same level of net output work. This decreases the efficiency of the power cycle. Notable irreversibilities occur within the pump and the turbine. Due to irreversibilities, a pump demands a higher input work while a turbine produces a lower output work. The flow through these two devices proceeds at constant entropy in the ideal cycle. Nevertheless, actual pumps and turbines do not operate at constant entropy (Fig. 5.3c). In this vein, states 2a and 4a are respectively the exit states of the actual pump and turbine, and 2 s and 4 s are the corresponding states for the isentropic processes. Isentropic efficiencies are used to account for the departure of actual pumps and turbines from the isentropic ones. The pump ($\eta_{p,is}$) and turbine ($\eta_{tu,is}$) isentropic efficiencies are given by:

$$\eta_{p,is} = \frac{W_{p,is}}{W_{p,a}} = \frac{h_{2s} - h_1}{h_2 - h_1} \tag{5.8}$$

$$\eta_{tu,is} = \frac{W_{tu,a}}{W_{tu,is}} = \frac{h_3 - h_{4a}}{h_3 - h_{4s}} \tag{5.9}$$

Saturated steam contains liquid water but it is desirable to produce dry steam because of metallurgical limitation. A moisture content of more than 10% leads to the degradation of turbine blades. In this vein, dry steam is produced by superheating the wet steam in the super heater. Superheating raises the average temperature of the steam to T_4 (Fig. 5.4). Superheating also increases the thermal efficiency of the cycle. Other developments have included re-heating the steam. In this case, two turbines are used: low and high pressure (Fig. 5.5). Reheating the steam elevates the temperature of the working fluid, and therefore, its enthalpy.

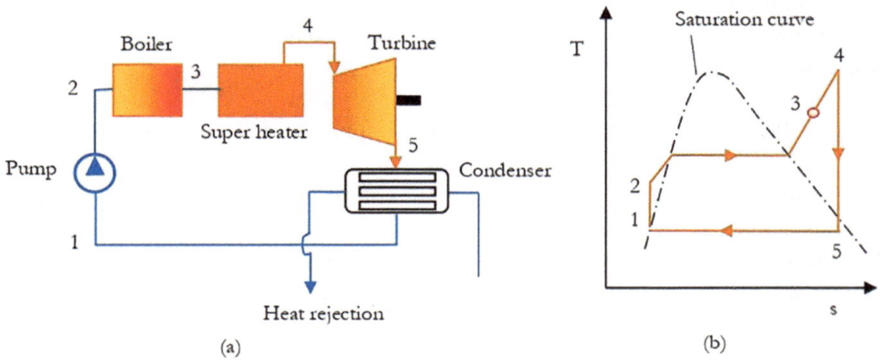

Fig. 5.4 Ideal Ranken cycle with superheating: **a** Energy flow and **b** temperature-entropy (T-s) diagram

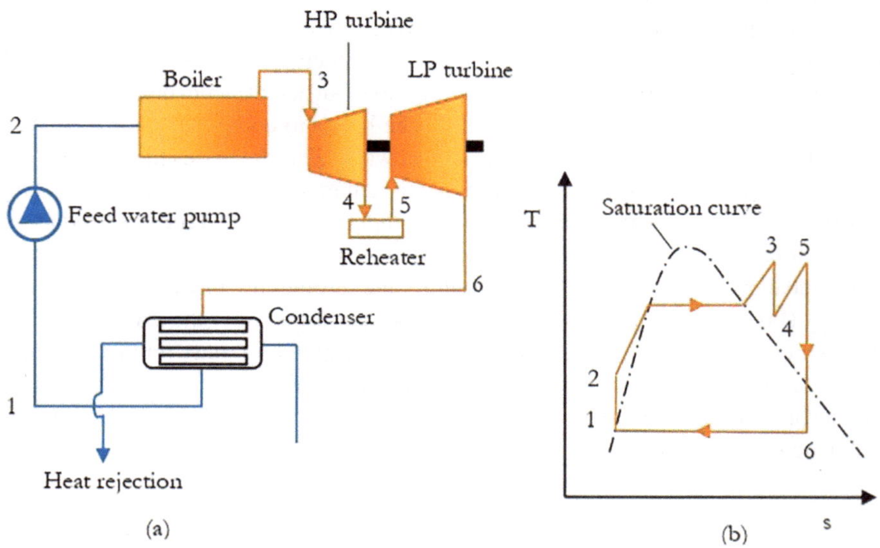

Fig. 5.5 Ideal Rankine cycle with reheating: **a** Energy flow and **b** temperature-entropy (T-s) diagram. HP = high pressure, and LP = low pressure

The net output work $\left(\dot{W}_{rc,sh}\right)$ of the ideal Rankine cycle with superheating is therefore given by:

$$\dot{W}_{rc,sh} = \dot{W}_{45} - \dot{W}_{12} \tag{5.10}$$

$$\dot{Q}_{23} = \dot{m}_v(h_3 - h_2) \tag{5.11}$$

$$\dot{Q}_{34} = \dot{m}_v(h_4 - h_3) \tag{5.12}$$

$$\dot{W}_{12} = \dot{m}_v(h_2 - h_1) \tag{5.13}$$

$$\dot{W}_{45} = \dot{m}_v(h_5 - h_4) \tag{5.14}$$

For the Rankine cycle with superheating, the thermal efficiency $(\eta_{rc,sh})$ can be computed from:

$$\eta_{rc,sh} = \frac{\dot{W}_{rc,sh}}{\dot{Q}_{23} + \dot{Q}_{34}} = \frac{(h_1 - h_2 + h_4 - h_5)}{(h_4 - h_2)} \tag{5.15}$$

The net output work of the Rankine cycle with re-heating can therefore be given by:

$$\begin{aligned}\dot{W}_{rc.rh} &= \dot{W}_{45} + \dot{W}_{67} - \dot{W}_{12} = \dot{m}_v(h_5 - h_4) + \dot{m}_v(h_7 - h_6) - \dot{m}_v(h_2 - h_1)\\ &= \dot{m}_v(h_1 - h_2 - h_4 + h_5 - h_6 + h_7)\end{aligned} \tag{5.16}$$

$$\dot{Q}_{23} = \dot{m}_v(h_3 - h_2) \tag{5.17}$$

$$\dot{Q}_{34} = \dot{m}_v(h_4 - h_3) \tag{5.18}$$

$$\dot{Q}_{56} = \dot{m}_v(h_6 - h_5) \tag{5.19}$$

$$\text{Total input heat} = \dot{Q}_{23} + \dot{Q}_{34} + \dot{Q}_{56} = \dot{m}_v(-h_2 + h_4 - h_5 + h_6) \tag{5.20}$$

For the Rankine cycle with re-heating, the thermal efficiency $(\eta_{rc,rh})$ can be computed from:

$$\eta_{rc,rh} = \frac{\dot{W}_{ra}}{\dot{Q}_{23} + \dot{Q}_{34} + \dot{Q}_{56}} = \frac{h_1 - h_2 - h_4 + h_5 - h_6 + h_7}{-h_2 + h_4 - h_5 + h_6} \tag{5.21}$$

Recovery of waste heat from power plants is a good strategy for promoting sustainable energy production. However, most of the available waste heat falls within the low-temperature range (<373 K). On the other hand, the critical temperature of water is 374 K (Madenoğlu 2017), which implies that the conventional Rankine cycle (with water as the working fluid) can be driven by a narrow spectrum of waste heat sources. This leads to the concept of the organic Rankine cycle (ORC) in which the working fluid is an organic substance. From a thermodynamic stand point, many organic substances are available for ORCs with low-temperature heat sources (Xu and Yu 2014). Some ORC working fluids have low critical temperatures than water as shown in Table 5.1.

Physical properties of an organic fluid influence the thermodynamic performance of an ORC. Nevertheless, other factors (such as environmental impacts, safety, compatibility, availability and cost) have to be considered when selecting an ORC working fluid. Chlorofluorocarbons (CFCs) contribute to the depletion of the stratospheric ozone layer while highly flammable organic substances (isobutane and others) are risky (American Society of Heating Refrigerating and Air-Conditioning Engineers 2017). It is also important to take into account the toxicity level of the ORC working fluid. Some organic working fluids are corrosive and may be incompatible with some structural materials in the components of the ORC engine. Organic Rankine cycles have efficiencies of 10–20%, and are usually used as bottoming cycles (Dunham and Iverson 2014). The cost of the working fluid would affect the competitiveness of the technology on the market even if the thermodynamic performance of an ORC technology is good. Consequently, cheaper substances are preferable options for exploitation in the ORC technology.

Rankine steam cycles are well established and there is extensive experience with this category of heat engines gained from the development of conventional power plants driven by fossil and nuclear sources of energy. In view of this, principles of steam cycle power plants have provided a good foundation for the advancement of the CSP technology. It is therefore not surprising that all the commercial power plants operating in different parts of the world, as of 2024, used (with minor modifications) a sub-critical steam Rankine cycle which is often exploited in the coal-fired power

Table 5.1 Selected organic working fluids

Working fluid	Critical temperature (K)	Critical pressure (10^6 N/m^2)
Isobutane	407.81	3.630
Octafluoropropane	345.02	2.640
Pentane	469.70	3.370
R124	395.43	3.620
R134a	374.21	4.059
R236fa	398.07	3.200
R245fa	427.16	3.65
R245ca	447.57	3.930

plants. One of the commercial steam-cycle plants of concentrated solar power is the Ivanpah Solar Electric Generating System located at Primm, California in the United States of America. This power plant is driven by a solar tower with a heliostat solar-field aperture of 2,600,000 m^2, capacity of 377 MW, and backup heater fuelled by natural gas. The plant started operating in 2014 and it is expected to produce 1079 GWh/yr. It has been theoretically shown that a wet-cooled steam cycle can attain an efficiency of up to 50% at a temperature and a pressure of 1023 K and 30×10^6 Nm^{-2}, respectively (Dunham and Iverson 2014). Nonetheless, material considerations limit the maximum temperature to 900 K based on the current level of state of the art. It is possible that future steam turbines will be operating in the supercritical region at temperatures above this limit.

One drawback of the steam Rankine cycle is the high loss of low grade heat at the condenser, attributed to the fact that this cycle heavily depends on condensation to enable pumping of water. Apparently, the required power to pump water is relatively low (~2% because water has low compressibility) but a high proportion (~ 50%) of the low grade heat is lost through the condensation process (Stein and Buck 2017). Most of the operating CSP plants are based on the steam cycles. The first and second (current) generations of CSP plants are mostly based on the steam cycle and achieve thermodynamic cycle efficiencies of up to 38% and 44% respectively (Khan et al. 2022). The limitation of the steam engine is leading researchers to investigate advanced cycles. In this vein, the thermodynamic efficiency of the third generation of CSP plants is expected to exceed 50%. One of the most promising advanced cycles for solarisation is the Brayton cycle.

5.4 Brayton Cycle

A Brayton cycle is an engine cycle in which a gas (working fluid) is pressurized by a compressor, heated and then directed against a set of turbine blades where part of the energy of the expanding hot gas is transformed into rotational energy (Fig. 5.6a). After exiting the turbine, the working fluid is cooled before re-entering the compressor to start a new cycle. Theoretically, the Brayton cycle comprises two isentropic and two isobaric processes: (a) isentropic gas compression (states 1–2), (b) isobaric heat addition to the gas (states 2–3), (c) isentropic expansion of gas (states 3–4), and (d) isobaric cooling of gas (states 4–1) as shown in Fig. 5.6b. This cycle was first developed by George Brayton in 1873. Ideal and actual Brayton cycles differ from each other as shown in Fig. 5.6c and d. The exhaust gas from the turbine can be rejected to the ambient environment instead of re-circulating it, thereby forming an open cycle gas turbine (Fig. 5.7). The Brayton cycle is an ideal cycle for the gas turbine engine. Solar radiation is the primary source of heat for a basic solar gas turbine.

Actual gas cycles are complicated, and so, some approximations are made to simplify analyses. Such approximations are referred as the air standard assumptions. These assumptions are: (a) all processes are internally reversible (b) air is the working

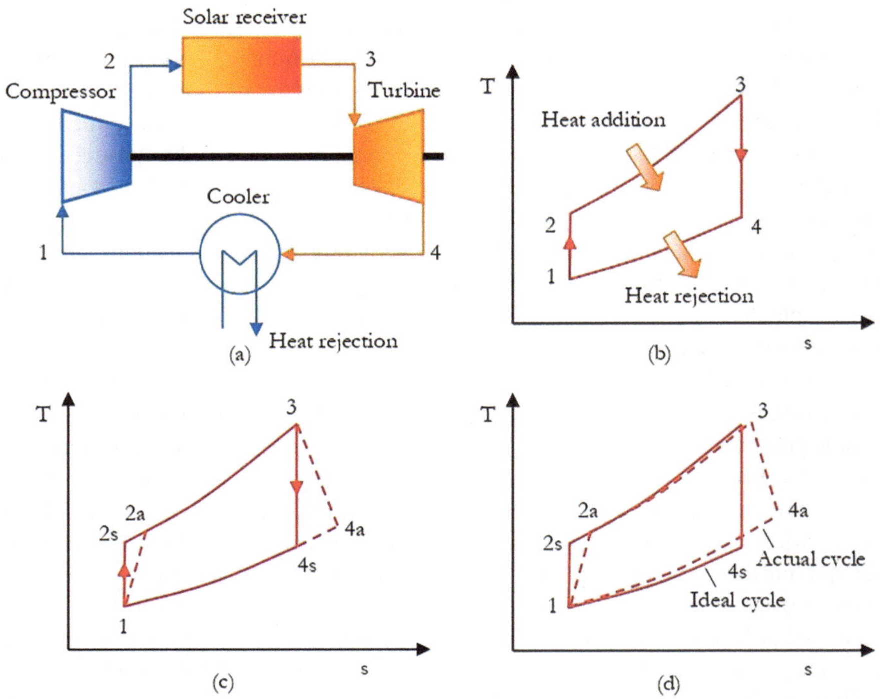

Fig. 5.6 Schematic representation of a closed cycle solar gas turbine: **a** Energy flow, **b** temperature-entropy (T-s) diagram of an ideal cycle, **c** temperature-entropy (T-s) diagram with ideal (2 s and 4 s) and real (2a and 4a) states and **d** deviation of ideal and real cycles

fluid and it behaves like an ideal gas, and (c) combustion and exhaust processes are, respectively, replaced by processes of heat addition and rejection. Air-standard assumptions are known as cold-air-standard assumptions if the specific heats of air are also assumed to be constant at room temperature. Under these assumptions, the thermal efficiency of the ideal Brayton cycle (η_{bc}) is given by:

$$\eta_{bc} = 1 - \frac{\dot{Q}_{out}}{\dot{Q}_{in}} = 1 - \frac{C_p(T_4 - T_1)}{C_p(T_3 - T_2)} = 1 - \frac{1}{r^{(\gamma-1)/\gamma}} \tag{5.22}$$

where r is the compression ratio and γ is the specific heat ratio of air (the ratio of the specific heat at constant pressure to the specific heat at constant volume). Equation (5.22) shows that the thermal efficiency of the Brayton cycle increases with both of these parameters. Actual gas turbines also behave like this.

Parameters that substantially characterize the gas turbine cycle are the pressure ratio and firing temperature. The pressure ratio is the ratio of the compressor discharge pressure to the compressor inlet pressure. The power output of a gas turbine increases with pressure ratio. The firing temperature is defined as the highest temperature attained in the cycle. Again, the output power increases with the firing temperature.

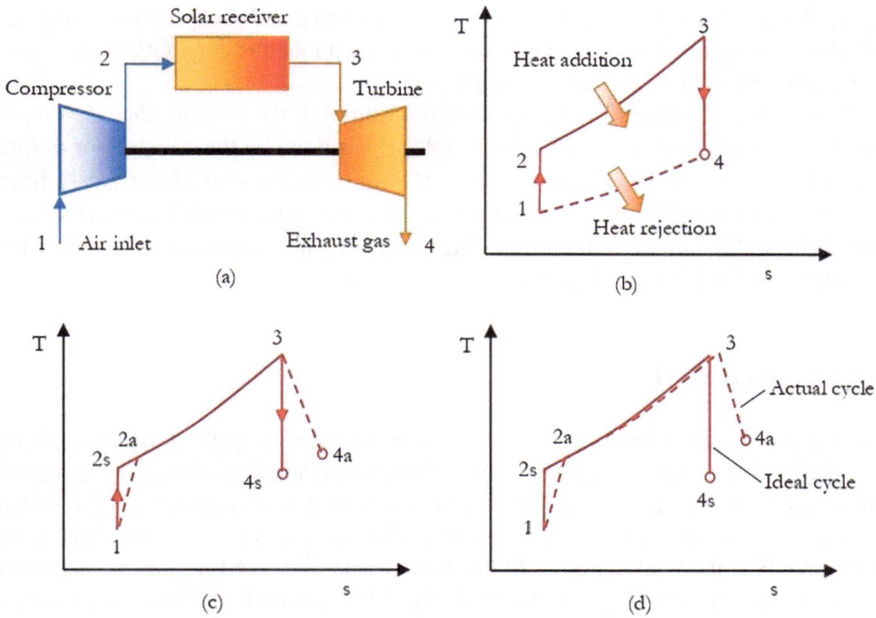

Fig. 5.7 Schematic representation of an open cycle solar gas turbine: **a** Energy flow, **b** temperature-entropy (T-s) diagram of an ideal cycle, **c** temperature-entropy (T-s) diagram with ideal (2 s and 4 s) and real (2a and 4a) states and **d** deviation of ideal and real cycles

Based on flow path, gas turbines can be classified into three basic types: (a) closed cycle gas turbine (CLCGT), (b) open cycle gas turbine (OCGT) and (c) semi-closed cycle gas turbine (SCLCGT). These cycles can be combined to form a combined cycle gas turbine (CCGT).

Solar energy can be the only source of heat for a solar gas turbine. However, solar radiation is intermittent, thereby reducing the reliability of solar-only gas turbines. Therefore, we have two strategies for enhancing the performance of solar gas turbines at present: hybridization and thermal energy storage (see Chap. 4). The process of hybridization can be achieved by using serial and parallel schemes but the former scheme is thermodynamically more advantageous. Therefore, thermodynamic cycles for hybrid solar gas turbines considered in this book are confined to the serial scheme.

5.4.1 *Solar-Only Closed Cycle Gas Turbine*

In a solar-only closed cycle gas turbine, the compressed working fluid is heated by a solar heater and then it expands at constant entropy through the turbine (Fig. 5.6). The exhaust gas from the turbine is cooled in a heat exchanger before inlet to the compressor to repeat the cycle. Air is commonly used as the working fluid in this cycle. Some of the advantages of air are: (a) it is abundant and cheap, and (b) existence of adequate engineering knowledge and experience about this fluid. Other applicable gases are carbon dioxide, helium and nitrogen.

During the expansion of the working fluid through the turbine, there is output work ($W_{tu,a}$) on the turbine shaft. Work $\left(\dot{W}_{com,a}\right)$ is done by the compressor during the compression of the working fluid in the main flow circuit. Heat $\left(\dot{Q}_{23}\right)$ from solar radiation is added to the working fluid between states 2 and 3, and the system rejects heat $\left(\dot{Q}_{41}\right)$ through the cooler. The energy balance can be computed for each component of the system (Fig. 5.6a).

5.4.1.1 Solar Field

In a solar-only gas turbine, concentrated solar radiation is used to heat the working fluid before it expands through the turbine. Only beam radiation (G_b) can be focussed. Incoming beam radiation is incident on the concentrator with aperture (A_{ap}), which focuses the radiation on the solar receiver. The latter component may directly or indirectly heat the working fluid. In the direct mode, the receiver constitutes part of the main flow circuit of the working fluid (Fig. 5.6a). An auxiliary flow circuit, which incorporates the solar receiver, is used for indirect heating of the working fluid. The auxiliary circuit may use a different heat transfer fluid (such as molten salt in the receiver). In this case, useful heat is transferred from the receiver to the working fluid through a heat exchanger.

The optical and thermal characteristics of the solar field vary immensely with the type of CSP technology. Details of governing equations for computing useful solar energy are therefore not identical. So, the following analysis is general for solar input to the working fluid. The total solar power input to the system ($\dot{Q}_s$) can be computed from:

$$\dot{Q}_s = A_{ap} G_b \tag{5.23}$$

where A_{ap} is the aperture area of the solar concentrator and G_b is the beam irradiance intercepted by the concentrator.

Concentrated solar radiation is converted to heat by the solar receiver (collector). If the receiver has a thermal efficiency of η_{rec} and the compressed working fluid is directly heated in the receiver, then it is possible to determine the useful solar thermal energy ($\dot{Q}_{u,s}$) from the receiver:

$$\dot{Q}_{u,s} = \eta_{rec} \eta_{opt} \dot{Q}_s \tag{5.24}$$

where η_{opt} is the optical efficiency of the system and $\dot{Q}_{u,s}$ is the useful energy transferred from the receiver to the working fluid between states 2 and 3. In this case, the useful solar thermal energy can be expressed as:

$$\dot{Q}_{\mathrm{u,s}} = \dot{m}_{\mathrm{wf}} C_{\mathrm{p,wf}} (T_3 - T_2) \tag{5.25}$$

where $C_{\mathrm{p,wf}}$ is the specific heat capacity of the working fluid, $\dot{m}_{\mathrm{wf}}$ is the mass flow of the working fluid in the main flow circuit, and T_2 and T_3 are temperatures of the working fluid at the specified points of the cycle. In terms of enthalpies, the useful solar heat can be expressed as:

$$\dot{Q}_{\mathrm{u,s}} = \dot{m}_{\mathrm{wf}} (h_3 - h_2) \tag{5.26}$$

where h_2 and h_3 are the enthalpies of the working fluid at points 2 and 3 of the cycle, respectively.

5.4.1.2 Compressor

The compressor requires power to compress the working fluid. This power is delivered by the turbine, and minimizing compressor work helps to maximize the efficiency of the solar gas turbine.

$$\dot{W}_{\mathrm{com,a}} = \dot{m}_{\mathrm{wf}} (h_2 - h_1) = \dot{m}_{\mathrm{wf}} C_{\mathrm{p,wf}} (T_2 - T_1) \tag{5.27}$$

where $\dot{m}_{\mathrm{wf}}$ is the mass flow rate of the working fluid, and $C_{\mathrm{p,wf}}$ is the temperature-dependent specific heat capacity of the working fluid. For air, the specific heat capacity can be computed from (Cetin 2006; Kurt et al. 2009):

$$C_{\mathrm{p,a}} = 1000\left(1.0484 - \frac{3.8371 T_{\mathrm{m}}}{10^4} + \frac{9.4537 T_{\mathrm{m}}^2}{10^7} - \frac{5.4903 T_{\mathrm{m}}^3}{10^{10}} + \frac{7.9298 T_{\mathrm{m}}^4}{10^{14}}\right) \tag{5.28}$$

where $T_{\mathrm{m}} = 0.5(T_1 + T_2)$ is the mean of instantaneous inlet and outlet compressor temperatures, T_1 is the air inlet temperature and T_2 is outlet temperature given as (Tsoutsanis et al. 2013).

$$T_2 = \frac{T_1}{\eta_{\mathrm{com,is}}} \left[r^{(\Upsilon - 1)/\Upsilon} - 1 \right] + T_1 \tag{5.29}$$

where $\eta_{\mathrm{com,is}}$ is the isentropic efficiency of the compressor. Different empirical correlations for calculating temperature-dependent properties of heat transfer fluids are available and valid within specific ranges of temperature and pressure. For instance, correlations for specific heat capacity, thermal conductivity, dynamic viscosity, and density of conventional and advanced heat transfer fluids are reported by (Benoit et al. 2016).

The compressor isentropic efficiency ($\eta_{\mathrm{com,is}}$) is defined as the ratio of isentropic work of the compressor to the actual work of the compressor, given by (Effiom et al.

2015):

$$\eta_{com,is} = \frac{\dot{W}_{com,is}}{\dot{W}_{com,a}} = \frac{h_{2s} - h_1}{h_2 - h_1} = \frac{T_{2s} - T_1}{T_2 - T_1} \tag{5.30}$$

where T_1 is the actual temperature of the working fluid at the compressor inlet, T_2 is the actual temperature of the working fluid at the compressor outlet and T_{2s} is the temperature of the working fluid at the compressor outlet if the compression process was adiabatic. The enthalpies of the working fluid at temperatures T_1, T_2 and T_{2s} are respectively h_1, h_2 and h_{2s}. The value of T_{2s} can be computed from:

$$T_{2s} = T_1\left(\frac{P_2}{P_1}\right)^{(\Upsilon-1)/\Upsilon} \tag{5.31}$$

The compression ratio is the ratio of the compressor output pressure P_2 to the ambient pressure P_1. The isentropic compressor work $\left(\dot{W}_{com,is}\right)$ can be computed from:

$$\dot{W}_{com,is} = \frac{\Upsilon RT_1}{\Upsilon - 1}\left[\left(\frac{P_2}{P_1}\right)^{(\Upsilon-1)/\Upsilon} - 1\right] \tag{5.32}$$

5.4.1.3 Turbine

The hot working fluid expands through the turbine, thereby driving the turbine $\left(\dot{W}_{tu,a}\right)$. Part of this actual work is used by the compressor $\left(\dot{W}_{com,a}\right)$. Consequently, the net work output $\left(\dot{W}_{net}\right)$ can be calculated as follows:

$$\dot{W}_{tu,a} = \dot{m}_{wf}(h_3 - h_4) = \dot{m}_{wf}C_{p,wf}(T_3 - T_4) \tag{5.33}$$

$$\dot{W}_{net} = \dot{W}_{tu,a} - \dot{W}_{com,a} \tag{5.34}$$

The turbine isentropic efficiency ($\eta_{tu,is}$) is the ratio of actual work output to the isentropic work output (Olivenza-León et al. 2015). In this case, the turbine isentropic efficiency is computed between thermodynamic states 3 and 4:

$$\eta_{tu,is} = \frac{\dot{W}_{tu,a}}{\dot{W}_{tu,is}} = \frac{h_3 - h_4}{h_3 - h_{4s}} = \frac{T_3 - T_4}{T_3 - T_{4s}} \tag{5.35}$$

where T_3 is the actual temperature of the working fluid at the turbine inlet, T_4 is the actual temperature of the working fluid at the turbine outlet and T_{4s} is the temperature of the working fluid at the turbine outlet if the expansion process was adiabatic.

The enthalpies of the working fluid at temperatures T_3, T_4 and T_{4s} are respectively h_3, h_4 and h_{4s}. Often, an electromagnetic generator is used to convert part of the mechanical power to electricity ($\dot{Q}_e$).

$$\dot{Q}_{\text{e}} = \eta_{\text{gen}}\left(\eta_{\text{tu,is}}\dot{W}_{\text{tu,is}} - \frac{\dot{W}_{\text{com}}}{\text{com,is}}\right) \tag{5.36}$$

where η_{gen} is the generator efficiency.

Considering that the total input energy comes from solar radiation only, the electrical efficiency of the system can be calculated from:

$$\eta_{\text{e}} = \frac{\dot{Q}_{\text{e}}}{\dot{Q}_{\text{s}}} \tag{5.37}$$

5.4.1.4 Cooler

Before the working fluid returns to the compressor (in a closed solar gas turbine), it needs to be cooled to the right level of temperature before inlet to the compressor. This is achieved by circulating a cold heat transfer fluid (htf) through a heat exchanger that removes heat from the working fluid and rejects it to the environment. There is a balance between energy given out by the hot fluid and that gained by the cold fluid, assuming losses from the heat exchanger to the environment are negligible.

$$\dot{Q}_{41} = \dot{m}_{\text{wf}}(h_4 - h_1) = \dot{m}_{\text{wf}}C_{\text{p,wf}}(T_4 - T_1) = \dot{m}_{\text{htf}}C_{\text{p,htf}}(T_6 - T_5) \tag{5.38}$$

5.4.2 Hybrid Closed Cycle Solar Gas Turbine

In a hybrid closed cycle solar gas turbine (CLCSGT), the gas is also pressurized by a compressor and then solar heat is added to the working fluid. The partially-heated working fluid flows into the heat exchanger for further heating (by an external conventional heater) to achieve the required turbine inlet temperature (Fig. 5.8). The hot working fluid expands through a turbine before cooling in the cooler. The use of an external heat source gives flexibility to exploit a wide range of energy resources such as coal, oil, gas, biomass and nuclear in the hybrid closed cycle solar gas turbine.

During the isentropic expansion of the working fluid, there is output work $\left(\dot{W}_{45}\right)$ on the turbine shaft. Part of the output work is used to drive the compressor that does work $\left(\dot{W}_{12}\right)$. during the isentropic compression of the working fluid. Heat from solar $\left(\dot{Q}_{23}\right)$ and conventional $\left(\dot{Q}_{34}\right)$ sources is added to the working fluid between states 2 and 4. The working fluid rejects heat $\left(\dot{Q}_{51}\right)$ in the cooling system. The work done by

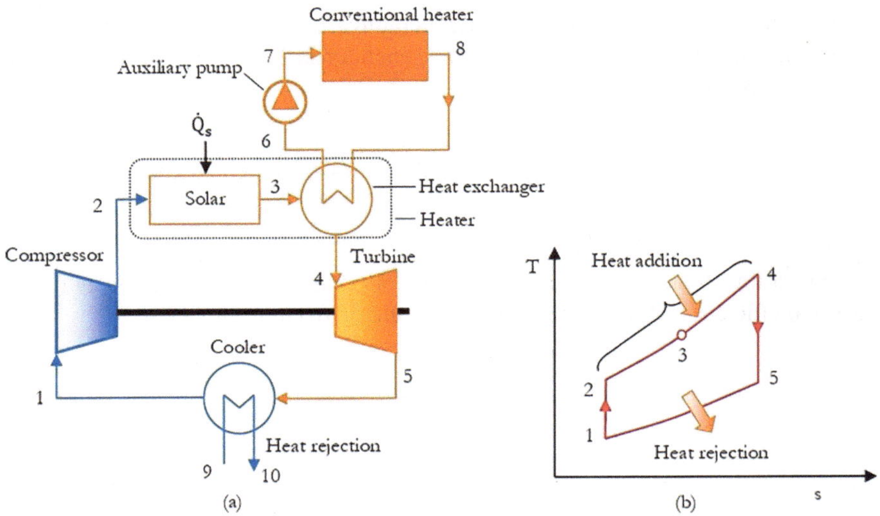

Fig. 5.8 A hybrid closed cycle solar gas turbine

the compressor and the useful power provided by the solar field $\left(\dot{Q}_{\mathrm{u,s}}\right)$, in the hybrid system, can be computed using similar energy balance equations for the solar-only gas turbine.

5.4.2.1 Heat Exchanger for Backup Heating

The working fluid in a hybrid closed cycle solar gas turbine is indirectly heated by the hot fluid (through the heat exchanger for backup heating) from a conventional source of heat such as a combustor or nuclear reactor. The amount of heat gained by the working fluid is equal to the heat given up by the heat transfer fluid in the auxiliary flow circuit, neglecting losses from the heat exchanger to the environment.

$$\dot{Q}_{34} = \dot{m}_{\mathrm{wf}}(h_4 - h_3) = \dot{m}_{\mathrm{htf1}} C_{\mathrm{p,htf1}}(T_7 - T_8) \tag{5.39}$$

Often, the conventional source of heat is a combustor, which uses some fuel such as natural gas, oil, coal or biofuel. The energy content ($\dot{Q}_{\mathrm{f}}$) of the fuel depends on its lower heating value ($H_{\mathrm{f,L}}$) and it can be calculated from:

$$\dot{Q}_f = \dot{m}_f H_{f,L} \tag{5.40}$$

Taking into account the efficiency of the combustor (η_{cc}), the useful energy ($\dot{Q}_{78}$) added to the working fluid from the combustor is:

$$\dot{Q}_{67} = \eta_{\mathrm{cc}} \dot{Q}_{\mathrm{f}} = \dot{m}_{\mathrm{htf1}}(h_8 - h_7) = \dot{m}_{\mathrm{htf1}} C_{\mathrm{p,htf1}}(T_8 - T_7) \tag{5.41}$$

5.4.2.2 Turbine

The hot working fluid from heat exchanger expands through the turbine, thereby driving the turbine. Part of this work is used by the compressor. Consequently the network output can be calculated as follows:

$$\dot{W}_{\mathrm{tu}} = \dot{W}_{45} = \dot{m}_{\mathrm{wf}}(h_4 - h_5) = \dot{m}_{\mathrm{wf}} C_{\mathrm{p,wf}}(T_4 - T_5) \tag{5.42}$$

$$\dot{W}_{\mathrm{net}} = \dot{W}_{\mathrm{tu}} - \dot{W}_{\mathrm{com}} \tag{5.43}$$

For the hybrid closed cycle system, the values of the isentropic efficiency ($\eta_{\mathrm{tu,is}}$) and electricity output power ($\dot{Q}_{\mathrm{e}}$) can be computed from:

$$\eta_{\mathrm{tu,is}} = \frac{h_4 - h_5}{h_4 - h_{5\mathrm{s}}} = \frac{T_4 - T_5}{T_4 - T_{5\mathrm{s}}} \tag{5.44}$$

$$\dot{Q}_{\mathrm{e}} = \eta_{\mathrm{gen}} \left(\eta_{\mathrm{tu,is}} \dot{W}_{\mathrm{tu,is}} - \frac{\dot{W}_{\mathrm{com,is}}}{\eta_{\mathrm{com,is}}} \right) \tag{5.45}$$

Taking into consideration the solar and conventional sources of power input, the overall electrical efficiency of the system can be calculated from:

$$\eta_{\mathrm{e}} = \frac{\dot{Q}_{\mathrm{e}}}{\dot{Q}_{\mathrm{s}} + \dot{Q}_{\mathrm{f}}} \tag{5.46}$$

5.4.2.3 Cooler

Before the working fluid returns to the compressor, it needs to be cooled to the right level of temperature for inlet to the compressor. This is achieved by circulating a cold heat transfer fluid (htf2) through the cooler. The energy given out by the hot fluid is equal to that gained by the cold fluid, when losses to the environment are assumed negligible.

$$Q_{51} = \dot{m}_{\mathrm{wf}}(h_5 - h_1) = \dot{m}_{\mathrm{wf}} C_{\mathrm{p,wf}}(T_5 - T_1) = \dot{m}_{\mathrm{htf2}} C_{\mathrm{p,htf2}}(T_{10} - T_9) \tag{5.47}$$

(d) Performance of hybrid closed cycle solar gas turbine

The compressor work constitutes a substantial fraction of the expansion work done, which tends to lower the ideal efficiency of the gas turbine. It has been estimated that the compressor consumes more than 50% of the work developed by the turbine section (Poullikkas 2005). Nevertheless, application of CO_2 as a working fluid is improving the thermodynamic efficiency of the closed cycle. This gas attains a critical state at a temperature of 304.13 K and pressure of 7.38 MN/m^2. For comparable levels of efficiency, the turbine inlet temperature of a supercritical carbon dioxide (s-CO_2) cycle is about 973 K while that of the conventional air Brayton cycle is above 1273 K (Schöffer et al. 2021). Garg et al. (2013) investigated three different CO_2 Brayton cycles: (a) conventional cycle in which the upper side pressure falls below critical pressure of CO_2, (b) transcritical non-condensing cycle in which the upper side pressure is in the supercritical range and (c) supercritical cycle in which the lower side pressure is also in the supercritical range. The supercritical Brayton cycle exhibits the best thermodynamic performance. Theoretical cycle efficiencies of more than 54% have been reported in recent work (Zheng et al. 2023). This level of performance is superior to that of the Rankine cycle (with a maximum cycle efficiency of about 44%). Some of the advantages of the closed cycle gas turbine are (Olumayegun et al. 2016):

(a) Can attain higher efficiency than the steam cycle at high temperature. The closed cycle gas turbine is less complex than the steam Rankine cycle which has many heat exchangers and pumps, and involves a lot of pipe work.
(b) It is possible to operate a closed cycle gas turbine at higher pressure, which yields more compact components and diminished plant footprint than a steam turbine plant.
(c) Has a higher ratio of power to size, which can lead to reduced capital cost.
(d) Can use solid fuels such as coal and biomass, and nuclear and waste heat unlike the open cycle gas turbine.
(e) Possibility to use different working fluids with suitable thermal and transport properties. The closed cycle gas turbine can use a gas of high density (hence high pressure) as a working fluid, which would result in a smaller turbomachinery system.
(f) There is no fouling and corrosion of system components associated with combustion emissions.
(g) There is no need for air filtration in a contaminated environment because the working fluid is not sucked from the environment.

Some power plants (2–20 MW capacity) have been designed and operated based on the closed cycle gas cycle. Nevertheless, the main disadvantage of this cycle is the need for an external heater, which introduces an ancillary cycle and temperature gradient between the combustion gases and the working fluid (the temperature of combustion gases is higher than that of the working fluid).

5.4.3 *Solar-Only Open Cycle Gas Turbine*

In a simple solar-only open cycle gas turbine, atmospheric air is compressed at constant entropy and then discharged into a pressurized receiver for heat addition (Fig. 5.7). The compressed air may be indirectly heated by the receiver (via a heat exchanger). As in the closed cycle solar gas turbine, the hot air is expanded through the turbine at constant entropy and is rejected into the atmosphere as exhaust gas from the turbine.

During the isentropic expansion of the working fluid, there is output work $(\dot{W}_{34})$ on the turbine shaft. The compressor does work $(\dot{W}_{12})$ during the isentropic compression of the working fluid in the main flow circuit. In this system, heat $(\dot{Q}_{23})$ from solar radiation is added to the working fluid between states 2 and 3. Exhaust air from the turbine is rejected to the ambient environment. The energy balance equations for the various components (compressor, solar receiver and turbine are similar to those of the solar-only closed cycle gas turbine.

5.4.4 Hybrid Open Cycle Solar Gas Turbine

The traditional way of adding heat to thermodynamic cycles is through combustion, where a mixture of fuel and air is burnt to release thermal energy. Consequently, there is a benefit in using air as the working fluid in a gas turbine because air contains oxygen which is required for the combustion process to take place. However, combustion of the fuel–air mixture is an irreversible process which produces by-products such as CO_2, H_2O and other substances that need to be released into the atmosphere. This leads to the concept of an open cycle gas turbine in which the exhaust gases (burnt fuel/flue gas) are released into the atmospheric environment through a stack. Similarly, exhaust gases from a hybrid open cycle solar gas turbine are released into the atmosphere (Fig. 5.9).

Open cycle gas turbine power plants are cheaper and less bulky than equivalent vapour-based plants with large heaters and coolers. The size of a plant is an important factor when considering its transportation from one location to another. The conventional open cycle is more-widely applied in electricity generation than the closed-cycle counterpart. In a hybrid open cycle solar gas turbine, atmospheric air

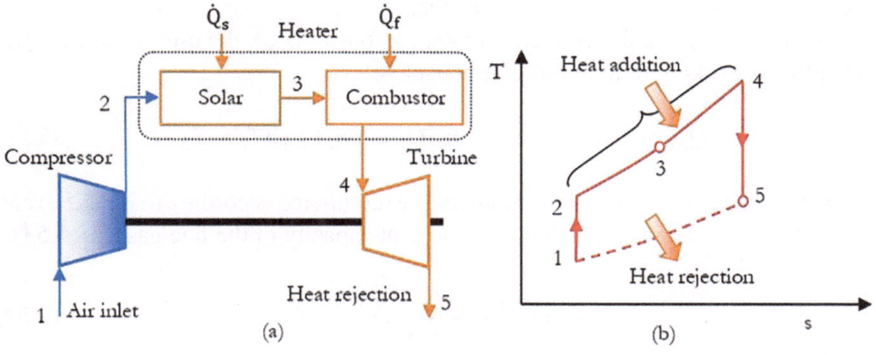

Fig. 5.9 Schematic representation of a hybrid open cycle gas turbine: **a** energy flow and **b** temperature-entropy (T-s) diagram of an ideal cycle

is pressurized at constant entropy by a compressor and then heat is added to the working fluid from a combination of solar and conventional energy resources.

The energy balance equations for the compressor and solar field are similar to those of the solar-only. Nevertheless, the exit hot air from the receiver mixes with fuel in the combustor, thereby increasing the mass flow through the combustor and turbine. Energy balance equations for combustor and turbine can be given as follows:

5.4.4.1 Combustor

The mass flow in the combustor is made up of air ($\dot{m}_a$) and fuel ($\dot{m}_f$) streams, and the amount of air required depends on fuel consumption (Bellos et al. 2017). Therefore, the fuel consumption is multiplied by an air ratio (ξ) and the stoichiometric fuel–air ratio (υ) to determine the air flow rate. Assuming that pressure losses are negligible in the combustor, the air flow rate can be expressed as:

$$\dot{m}_{\mathrm{a}} = \xi \upsilon \dot{m}_{\mathrm{f}} \tag{5.48}$$

In the combustor, the mass flow rate of the gases ($\dot{m}_{ga}$) is computed by considering the total mass flow rate balance:

$$\dot{m}_{\mathrm{ga}} = \dot{m}_{\mathrm{a}} + \dot{m}_{\mathrm{f}} = (\xi \upsilon + 1)\dot{m}_{\mathrm{f}} \tag{5.49}$$

The energy content in the fuel depends on the heating value of the fuel. As shown in Chap. 2, the lower heating value is preferred in the computation of system efficiency. For a given heating value, the energy rate of fuel $\left(\dot{Q}_{\mathrm{f}}\right)$ increases with the fuel mass flow ($\dot{m}_{\mathrm{f}}$):

$$\dot{Q}_{\mathrm{f}} = \dot{m}_{\mathrm{f}} H_{\mathrm{f,L}} \tag{5.50}$$

where $H_{\mathrm{f,L}}$ is the lower heating value of fuel.

Taking into account the efficiency of the combustor (η_{cc}), the useful energy ($\dot{Q}_{34}$) added to the working fluid in this component is:

$$\dot{Q}_{34} = \dot{m}_{\mathrm{ga}}(h_4 - h_3) = \eta_{cc}\dot{Q}_f = \dot{m}_{\mathrm{ga}} C_{\mathrm{p,ga}}(T_4 - T_3) \tag{5.51}$$

where T_4 is the combustion outlet temperature (computed according to Eq. 5.52) and $C_{\mathrm{p,ga}}$ is the temperature dependent specific heat capacity of the flue gas (Eq. 5.53).

$$T_4 = \frac{\dot{m}_{\mathrm{ga}} C_{\mathrm{p,ga}} T_3 + \eta_{\mathrm{Cc}} \dot{m}_{\mathrm{f}} H_{\mathrm{f,L}}}{\dot{m}_{\mathrm{ga}} C_{\mathrm{p,ga}}} \tag{5.52}$$

$$C_{\mathrm{p,ga}} = 1000\left(0.991615 + \frac{6.99703 T_{\mathrm{mg}}}{10^5} + \frac{2.71298 T_{\mathrm{mg}}^2}{10^7} - \frac{1.22442 T_{\mathrm{mg}}^3}{10^{10}}\right) \tag{5.53}$$

where $T_{mg} = 0.5(T_4 - T_3)$ is the mean temperature of the combustion gases.

5.4.4.2 Turbine

The flue gases expand through and drive the turbine, thereby producing shaft work. Part of this work is used by the compressor. Consequently, the network output can be calculated as follows:

$$\dot{W}_{tu,is} = \dot{m}_{ga}(h_4 - h_5) = \dot{m}_{ga} C_{p,ga}(T_4 - T_5) \tag{5.54}$$

$$\dot{W}_{tu,is} = \dot{W}_{45} \tag{5.55}$$

$$\dot{W}_{net,is} = \dot{W}_{tu,is} - \dot{W}_{com,is} \tag{5.56}$$

Only a fraction of the heat from the solar and fuel sources is converted to useful mechanical work in the shaft to drive the compressor and generator. The remaining heat exits the turbine as waste heat in the flue gas. The turbine isentropic efficiency ($\eta_{tu,is}$) and electricity output ($\dot{Q}_e$) can be computed by using the following equations:

$$\eta_{tu,is} = \frac{h_4 - h_5}{h_4 - h_{5s}} = \frac{T_4 - T_5}{T_4 - T_{5s}} \tag{5.57}$$

$$\dot{Q}_e = \eta_{gen}\left(\eta_{tu,is}\dot{W}_{tu,is} - \frac{\dot{W}_{com,is}}{\eta_{com,is}}\right) \tag{5.58}$$

The exhaust gas exit temperature (T_{5s}) can be calculated as follows:

$$\left(\frac{P_4}{P_5}\right) = \left(\frac{T_4}{T_{5s}}\right)^{\frac{\gamma_{ga}}{\gamma_{ga}-1}} \tag{5.59}$$

$$T_{5s} = T_4\left(\frac{P_5}{P_4}\right)^{(\gamma_{ga}-1)/\gamma_{ga}} \tag{5.60}$$

Taking into consideration the solar and fuel energy input to the open cycle gas turbine, the overall electrical efficiency (η_e) can be calculated from:

$$\eta_e = \frac{\dot{Q}_e}{\dot{Q}_s + \dot{Q}_f} \tag{5.61}$$

5.4.5 Semi-closed Cycle Solar Gas Turbine

Carbon emission is one of the main environmental concerns about power plants. The CO_2 concentration is about 4–6% of the exhaust gases from gas turbines (Fiaschi and Baldini 2009). This low concentration of CO_2 creates challenges in efforts to capture and store this gas. There is variation in the cost of carbon capture and storage (CCS) technology but the general consensus is that the CCS technology increases the cost of electricity production. In this vein, a semi-closed cycle gas turbine (SCLCGT) is one of the promising technologies with potential to substantially curtail CO_2 emissions from fossil-fuelled gas turbines.

A conventional semi-closed cycle gas turbine combines the principles of open and closed cycle gas turbines as shown in (Fig. 5.10). The exhaust gas is partly released to the atmosphere through a cooling tower (stack) and partly recirculated to the gas turbine system after extraction of the exhaust heat through a heat recovery steam generator (HRSG) and a separator. Fundamentally, the separator condenses the water vapour from the exhaust gas, thereby concentrating CO_2. This process eases the capturing of CO_2 for disposal. The recirculated exhaust leaving the separator through state 9 is mixed with air at the compressor inlet. In addition, water recovered by the separator can be injected into the combustor, which is also useful for the abatement of NO_x emissions.

Advantages of the semi-closed cycle gas turbine include (Meitner et al. 2000): (a) low emission and specific fuel consumption; (b) permits heavy-duty gas turbines (such as the 501F model) access to recuperative-cycle due to the reduction in the size of the recuperative heat exchanger (RHE) and heat recovery steam generator, which arises from the partial recovery of the exhaust heat in a compact steam turbine bottoming cycle; (c) reduction in water consumption because the semi-closed cycle gas turbine produces water; and (d) amenability to retrofitting an existing OCGT plant with minor modifications because the cycle base pressure is atmospheric and the working fluid remains the same.

Several aspects of the conventional semi-closed cycle have been investigated, and some of them are presented and discussed in this section. For a semi-closed cycle gas turbine with a combined cycle, Fiaschi and Manfrida (1998) found that major sources of exergy losses are from the combustor, heat recovery steam generator, water injection/mixing and water recovery system, with an exergy efficiency of 49–53%. Fiaschi and Manfrida (1999) studied the energy and exergy of a semi-closed cycle gas turbine with recuperator and CO_2 absorption unit. They found cycle efficiencies of 40.0–48.9% for different configurations. The efficiency of CO_2 removal from the exhaust gases was 80%. A mixture of diethanolamine (DEA) and methyldiethanolamine (MDEA) was used to chemically absorb CO_2 from a semi-closed cycle gas turbine (Lombardi 2001). It was observed the CO_2 removal component reduced CO_2 emissions by 85% over the lifespan of the semi-closed cycle gas turbine. The use of an alkali CO_2 absorber (based on sodium carbonate and potassium carbonate) was later suggested (Corti 2004). The thermoeconomic performance of this technology was compared with other systems of CO_2 absorption, namely: (a) conventional amine

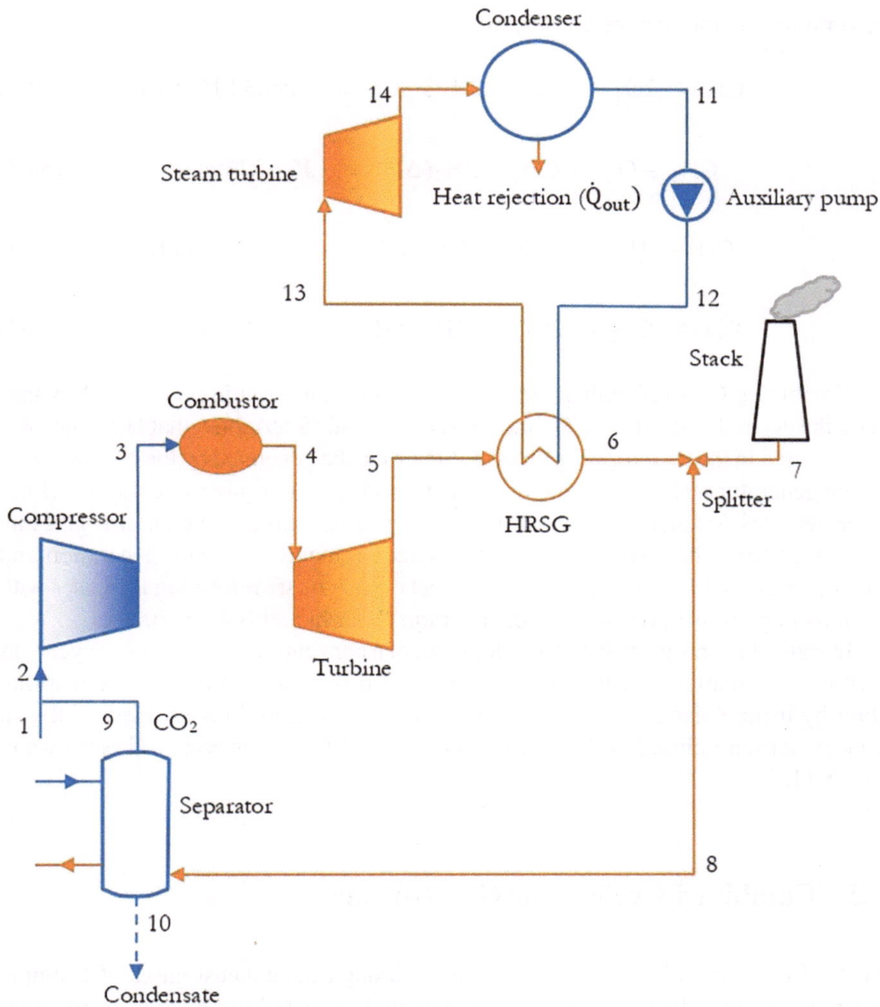

Fig. 5.10 Semi-closed cycle gas turbine with bottoming steam cycle. HRSG = heat recovery steam generator

absorption (DEA and MDEA) and (b) membrane technologies. It was found that the alkali CO_2 absorption technology yielded an intermediate cost between those of the conventional amine absorption process and membrane technologies. As regards the overall costs of CO_2 removal, the amine and alkali processes were distinctly more economical than the membrane process. Fiaschi and Baldini (2009) examined the potential of applying tri-reforming process to the semi-closed cycle. This process links the reactions of CO_2 reforming, steam reforming, partial oxidation and complete oxidation of methane into the same reactor to produce syngas that is rich in hydrogen (with a constant ratio of H_2/CO). In this connection, the following reforming reactions can be considered:

$$CH_4 + 2O_2 \leftrightarrow CO_2 + 2H_2O(\Delta H = -880.3\ \text{kJ/kmol}) \tag{5.62}$$

$$CH_4 + O_2 \leftrightarrow CO + 2H_2(\Delta H = -35.6\ \text{kJ/kmol}) \tag{5.63}$$

$$CH_4 + H_2O \leftrightarrow CO + 3H_2(\Delta H = +206.3\ \text{kJ/kmol}) \tag{5.64}$$

$$CH_4 + CO_2 \leftrightarrow 2CO + 2H_2(\Delta H = +247.3\ \text{kJ/kmol}) \tag{5.65}$$

Combining CO_2-reforming and H_2O-reforming can assist in producing a syngas with the desired level of H_2/CO ratio. Eqs. (5.64) and (5.65) show that H_2O and CO_2 are required in the tri-reforming process. Steam can be generated by the heat recovery steam generator while carbon dioxide is a by-product of flue gas recirculation. Moreover, water is produced by the semi-closed cycle gas turbine, which also gives an advantage from the consideration of water requirements for steam generation and other processes. Findings show that the effect of CO_2 on tri-reforming increases with diminishing levels of the steam/methane ratio (Fiaschi and Baldini 2009).

In spite of the progress in the development of a conventional semi-closed cycle gas turbine, information is scarce on solarisation of this cycle. Pre-heating the working fluid by using solar energy can save fuel and reduce emissions. In view of this, a conceptual semi-closed cycle solar gas turbine (SCLCSGT) is proposed as shown in Fig. 5.11.

5.5 Combined Cycle Solar Gas Turbine

Most of the heat is lost to the environment through the exhaust gases of a simple solar gas turbine. Inclusion of a bottoming cycle assists in recovering waste heat and raising the performance of the system. In view of this, the simple gas turbine cycle can be combined with a bottoming cycle, usually a Rankine cycle or Organic Rankine cycle to form a combined cycle. In general, the combined cycle is defined as a system of one or more gas turbines with heat recovery from the exhaust for driving steam turbines.

5.5.1 Combined Cycle with Closed Topping Brayton Cycle

In the closed topping cycle, the working fluid is compressed between states 1 and 2, and then heated by solar radiation between states 2 and 3 (Fig. 5.12). Depending on the outlet temperature T_3 from the solar receiver, it expands through the gas turbine.

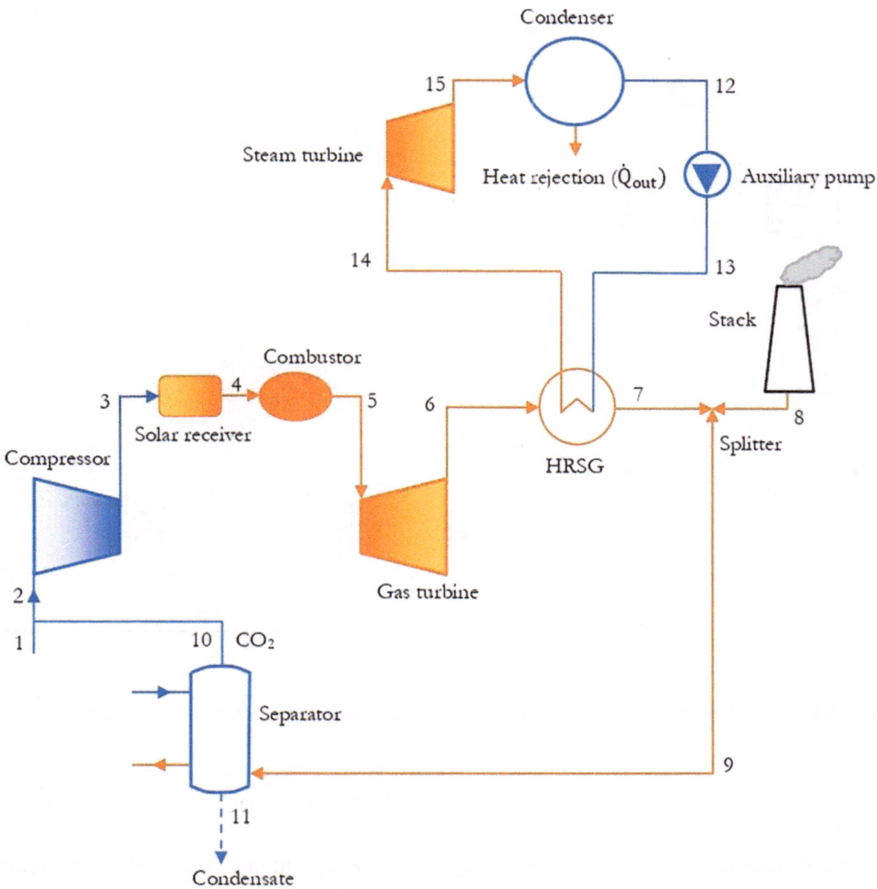

Fig. 5.11 Semi-closed solar gas turbine with bottoming cycle. HRSG = heat recovery steam generator

In the heat recovery steam generator, the hot gas from the gas turbine loses part of its heat to the steam cycle, and it is cooled by the cooler before flowing back into the compressor. Steam from the heat recovery steam generator expands through the steam turbine and then it is condensed in the condenser and the liquid pumped back into the heat recovery steam generator to repeat the cycle.

In the combined cycle with a topping closed gas cycle (Fig. 5.12), the gas turbine has shaft work $\dot{W}_{45}$. Part of the exhaust heat ($\dot{Q}_{56}$) from the gas turbine is transferred to the working fluid in the bottoming cycle which may be a Rankine cycle or an Organic Rankine cycle. Assuming there are no heat losses to the environment, this heat ($\dot{Q}_{65}$) is equal to the amount of thermal energy ($\dot{Q}_{1011}$) gained by the bottoming cycle, thereby producing steam in the heat recovery steam generator. The steam turbine has shaft work of $\dot{W}_{910}$. So, the thermal energy given out by the exhaust gas in the steam generator can be calculated from:

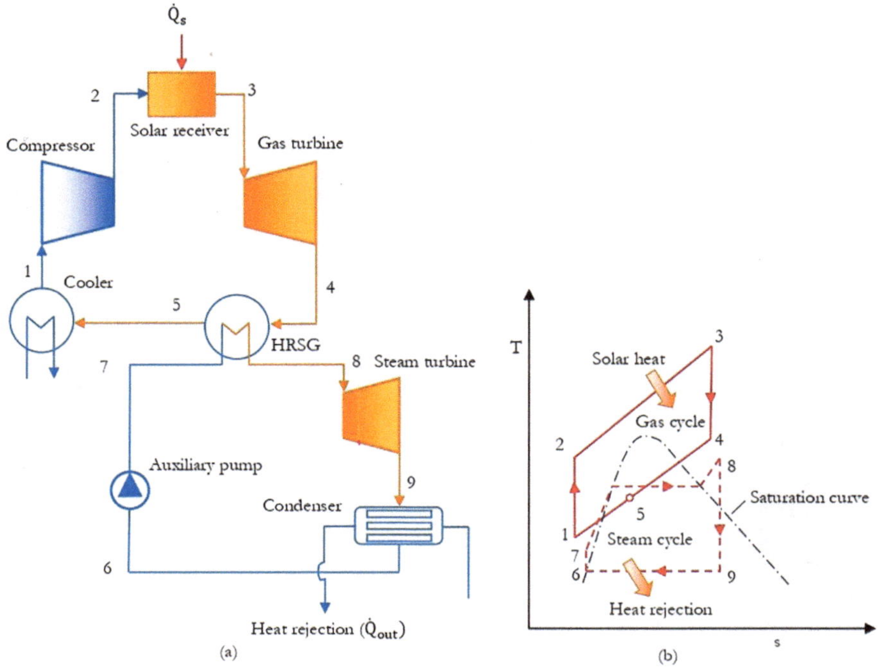

Fig. 5.12 Schematic view of a combined cycle with a closed topping gas turbine cycle

$$\dot{Q}_{65} = \dot{m}_{\mathrm{ga}}(h_5 - h_6) = \dot{m}_{\mathrm{ga}}\, C_{\mathrm{ga}}(T_5 - T_6) \tag{5.66}$$

This heat is gained by the working fluid ($\dot{Q}_{1011}$) in the heat recovery steam generator, raising its specific enthalpy from h_{10} to h_{11}:

$$\dot{Q}_{1011} = \dot{m}_{\mathrm{v}}(h_{11} - h_{10}) = \dot{m}_{\mathrm{v}} C_{\mathrm{v}}(T_{11} - T_{10}) \tag{5.67}$$

Consequently, the energy balance of the heat recovery steam generator can be expressed as:

$$\dot{m}_{\mathrm{v}} C_{\mathrm{v}}(T_{11} - T_{10}) = \dot{m}_{\mathrm{ga}} C_{\mathrm{ga}}(T_5 - T_6) \tag{5.68}$$

There is net work from both the topping $\left(\dot{W}_{\mathrm{net,top}}\right)$ and bottoming $\left(\dot{W}_{\mathrm{net,bot}}\right)$ cycles. The values of these variables can be calculated from the isentropic work of the gas turbine (gt), the compressor (com), the steam turbine (st) and auxiliary pump (ap) as follows:

$$\dot{W}_{\mathrm{net,top}} = \left(\eta_{\mathrm{gt,is}} \dot{W}_{\mathrm{gt,is}} - \frac{\dot{W}_{\mathrm{com,is}}}{\eta_{\mathrm{com,is}}}\right) \tag{5.69}$$

$$\dot{W}_{\text{net,bot}} = \left(\eta_{\text{st,is}} \dot{W}_{\text{st,is}} - \frac{\dot{W}_{\text{ap,is}}}{\eta_{\text{ap,is}}} \right) \tag{5.70}$$

The combined solar-to-electricity efficiency ($\eta_{e,comb}$) of the cycle is made up of the total net work from the two cycles and the input energy to the topping cycle:

$$\eta_{\text{e,comb}} = \eta_{\text{gen}} \left(\frac{\dot{W}_{\text{net,top}} + \dot{W}_{\text{net,bot}}}{\dot{Q}_{\text{s}}} \right) \tag{5.71}$$

The net work for the combined cycle is higher than that for individual cycles. Advanced combined cycle gas turbines for producing electricity have thermal efficiencies in the range 50–58%, which is higher than that (43–45%) of the best oil and coal-fired power plants (Poullikkas 2005). High thermal efficiencies are realised in large power plants with capacities exceeding 300 MWe. The advantages of the combined cycle gas turbine include (Poullikkas 2004):

(a) Higher thermal efficiency compared to other advanced gas turbine cycles.
(b) Diminished emissions because natural gas is frequently used as a fuel in gas turbines. This gas does not produce ash or SO_x, and it emits less amounts of volatile hydrocarbons, carbon monoxide, carbon dioxide and NO_x than coal or oil.
(c) Low capital cost and short construction time (2–3 years).
(d) Smaller space requirement than that occupied by coal or nuclear power plants of equivalent capacity. There are many economic activities that compete for land and other resources. Consequently, compact energy technologies are more preferable.
(e) Flexibility in the size of the plant for a wide range (10–750 MWe) of maximum power output.
(f) Rapid start-up which easies the response to demand fluctuations. In this regard, gas turbines can easily be used during peak demands.

5.5.2 *Combined Cycle Solar Gas Turbine with Open Topping Brayton Cycle*

In the combined cycle solar gas turbine with an open topping Brayton cycle, the exhaust from the heat recovery steam generator is rejected into the environment (Fig. 5.13). It is also possible to pass the exhaust gas through a recuperator before rejection, in order to increase the efficiency of the cycle.

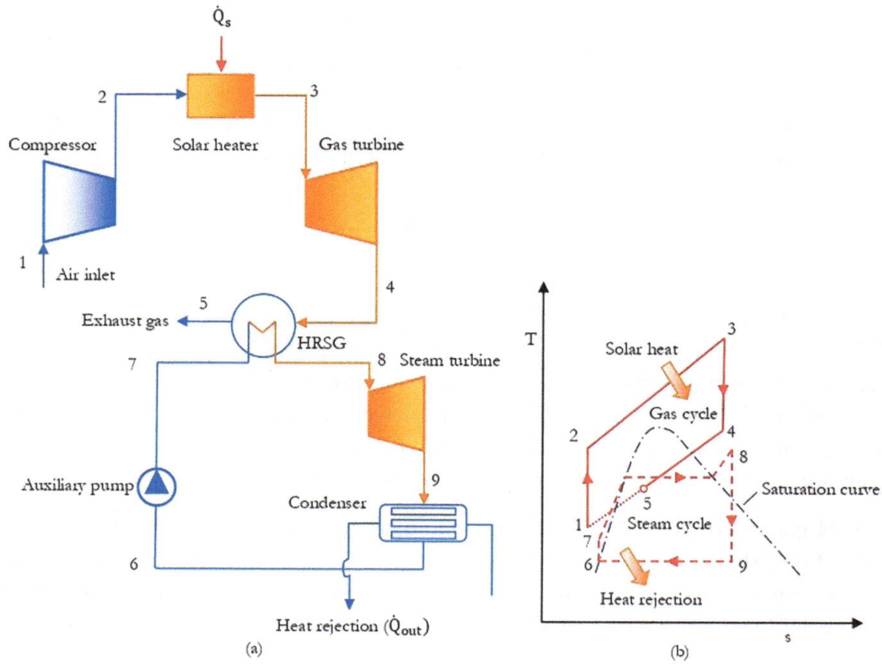

Fig. 5.13 Schematic view of a combined cycle gas turbine with a heat recovery steam generator (HRSG)

5.6 Exergy Analysis of Solar Gas Turbines

The exergy of a system (or subsystem) is equal to the maximum useful work which can be obtained when the system is taken from its given state to the thermodynamic equilibrium with the environment (Querol et al. 2013). A conventional stable reference environment is normally set up at temperature $T_0 = 298$ K (25 °C) and pressure $P_0 = 101{,}325$ N/m^2 (Chen 2015). When the temperature, pressure and chemical potential of a system are equal to those of the surrounding environment, the system is said to be in a dead state. Fundamentally, the natural environment comprises three subsystems (atmosphere, hydrosphere and lithosphere) that are not completely in thermodynamic equilibrium. Properties of each subsystem change with space and time.

Exergy analysis deals with finding the amount of irreversibility that takes place in each component of a thermodynamic cycle and aims at effective parameters (Behbahaninia et al. 2017). Exergy may involve matter and non-matter streams, and direct and indirect methods are available for computation of this variable. Matter-based exergy includes physical and chemical exergy. Exergy of non-matter streams involves work, electricity, heat flow and solar radiation. Both matter and non-matter based exergy are encountered in the thermodynamic analysis of a solar gas turbine.

5.6.1 *Exergy of Matter Streams*

In the context of exergy, matter is a physical substance that occupies space and it possesses mass. The volume of a fixed mass of substance may change with temperature and/or pressure. So, the exergy of matter streams is often expressed as the maximum useful work that can be obtained from a unit mass of the substance. Substances have physical and chemical exergies.

5.6.1.1 Physical Exergy

The physical exergy (e_{ph}) of a substance is the maximum useful work that can be extracted when the state of a unit of mass of the substance changes from its prevailing state (T, P) to the state (T_0, P_0) of the environment through physical processes (Querol et al. 2013). When kinetic and potential energy is neglected, e_{ph} can be calculated by using the specific enthalpy (h) and entropy (s) of the substance at given state points:

$$e_{\text{ph}} = [h(T, P) - h(T_0, P_0)] - T_0[s(T, P) - s(T_0, P_0)] \tag{5.72}$$

Equation (5.72) is used when data is available on the enthalpy and entropy of the system at the specific temperatures and pressures. Indirect methods can be applied in cases where there is scarcity of data on these two variables as functions of temperature and pressure. For incompressible substances, the enthalpy and entropy can be expressed as follows:

$$dh = c_{\text{p}}dT + \left[V - T\left(\frac{\partial V}{\partial T}\right)_{\text{p}}\right]dP \tag{5.73}$$

where V = volume. For condensed matter, the volume V does not change significantly with temperature and so the term $\left(\frac{\partial V}{\partial T}\right)_{\text{p}}$ can be neglected, which reduces Eq. (5.73) to:

$$\mathrm{d}h = c_{\text{p}}\mathrm{d}T + V\mathrm{d}P \tag{5.74}$$

For condensed matter, the entropy can be given by:

$$\mathrm{d}s = C_{\text{p}}\frac{\mathrm{d}T}{T} \tag{5.75}$$

For small changes in temperature, C_{p} can be considered to be constant. In this case, the integration of Eqs. (5.73) and (5.74) yields:

$$h - h_0 = C_{\text{p}}(T - T_0) + V(P - P_0) \tag{5.76}$$

$$s - s_0 = C_p[\ln(T) - \ln(T_0)] = C_p \ln\left(\frac{T}{T_0}\right) \tag{5.77}$$

where $h = h(T, P)$, $h_0 = h(T_0, P_0)$, $s = s(T, P)$ and $s_0 = s(T_0, P_0)$.

Based on Eqs. (5.75) and (5.76), e_{ph} can be estimated from:

$$e_{ph} = h - h_0 - T_0(s - s_0) = C_p(T - T_0) - C_p T_0 \ln\left(\frac{T}{T_0}\right) + V(P - P_0) \tag{5.78}$$

Indirectly, the exergy of an ideal gas can be computed as follows:

$$h - h_0 = \int_{T_0}^{T} C_p \mathrm{d}T \tag{5.79}$$

$$s - s_0 = \int_{T_0}^{T} C_p \frac{\mathrm{d}T}{T} - R \int_{P_0}^{P} \frac{\mathrm{d}P}{P} = \int_{T_0}^{T} C_p \frac{\mathrm{d}T}{T} - R \ln\left(\frac{P}{P_0}\right) \tag{5.80}$$

Assuming C_p does not vary with temperature within the range in question, then:

$$h - h_0 = \int_{T_0}^{T} C_p \mathrm{d}T = C_p(T - T_0) \tag{5.81}$$

$$s - s_0 = C_p \ln\left(\frac{T}{T_0}\right) - R \ln\left(\frac{P}{P_0}\right) \tag{5.82}$$

$$e_{ph} = h - h_0 - T_0(s - s_0) = C_p(T - T_0) - C_p T_0 \ln\left(\frac{T}{T_0}\right) + R T_0 \ln\left(\frac{P}{P_0}\right) \tag{5.83}$$

5.6.1.2 Chemical Exergy

The chemical exergy (e_{ch}) of a substance is the maximum useful work that can be extracted when the state of a unit mass of the substance changes from its prevailing state (T, P) to the state (T_0, P_0) of the environment through chemical processes. It is possible for the substance (such as carbon dioxide) of a stream to be present in the environment. Work can still be done in the expansion of this component where the partial pressure of the substance exceeds the one it has under the conditions of the environment. For instance, consider a stream composed of CO_2 at state point (T_o, P_o) which has no physical exergy. Nevertheless, the pressure of the stream drops down from P_o to the partial pressure of CO_2 (P_{CO_2}) in the environment, which can

Table 5.2 Chemical exergy of selected pure substances at normal temperature and pressure

Substance	Formula	Exergy (kJ/kmol)
Ammonia	NH_3	341,250
Carbon dioxide	CO_2	20,140
Nitrogen	N_2	720
Oxygen	O_2	3970
Water	H_2O	3120

Source of data Querol et al. (2013)

theoretically do some work. This work is also called chemical exergy even though no chemical reaction takes place because CO_2 constitutes a small proportion of the air composition in the atmosphere (0.035%). Values of P_{CO2} vary with time and location. For example, Wrobel (2017) found that the monthly atmospheric P_{CO2} for the Arctic Ocean varied between 37.25 38.91 N/m^2.

The chemical exergy of a substance can directly be obtained from available data. In this respect, the values of exergy for basic substances (such as water, oxygen, carbon dioxide, nitrogen and others) are often tabulated.

Table 5.2 shows the chemical exergy of selected substances.

The chemical exergy of some substances is not tabulated. However, if substance (j) with an unknown exergy ($e_{\mathrm{ch,j}}$) is stoichiometrically related to other substances (1, 2, 3, …i) whose values of chemical exergy ($e_{\mathrm{ch,i}}$) are known at normal temperature and pressure, then the value of $e_{\mathrm{ch,j}}$ can be determined by using an indirect method:

$$e_{\mathrm{ch},j} = g_{\mathrm{f0},j} - \frac{1}{n_j}\sum_{i \neq j} n_i\left(e_{\mathrm{ch},i} - g_{\mathrm{f0},i}\right) \tag{5.84}$$

where g_{f0} is the Gibbs free energy of formation and n_i is the number of moles in the stoichiometric link between substance j and the other substances. This method is particularly handy when computing the exergy of fuels for combustion engines. In the case of a hybrid SGT system, fuel is used to provide a backup source of heat.

The chemical exergy of a fuel stream is of great importance. It can be determined by using the values of lower ($H_{\mathrm{f,L}}$) or higher ($H_{\mathrm{f,H}}$) heating values of the fuel if the other methods described in the preceding paragraphs cannot be applied.

$$e_{\mathrm{f}} = a_{\mathrm{L}} H_{\mathrm{f,L}} = a_{\mathrm{H}} H_{\mathrm{f,H}} \tag{5.85}$$

where a_{L} and a_{H} are empirical coefficients.

5.6.2 *Exergy of Non-matter Streams*

Non-matter includes work, electricity, heat, solar radiation, sound, and other streams. Relevant to the exergy of solar gas turbines are work, electricity, heat, and solar radiation streams.

5.6.2.1 Exergy of Work

The exergy of work (E_w) can be expressed as:

$$E_w = \dot{W} \tag{5.86}$$

where $\dot{W}$ is the rate of doing work.

5.6.2.2 Exergy of Electricity

Electrical energy can be transformed to work, and so the exergy content of electricity (E_e) is equivalent to the energy content of work. Electricity can therefore be considered as a case of exergy of a work stream (Boonnasa et al. 2006):

$$E_e = \dot{W}_{el} \tag{5.87}$$

5.6.2.3 Exergy of Heat Flows

Considering the Carnot engine shown in Fig. 5.2, the maximum output work $\dot{W}_{max}$ that can be obtained from this engine is the available energy (E_{av}). The maximum energy that must be rejected is the unavailable energy (E_{ua}). Based on energy balances, this implies that:

$$\dot{Q}_1 = E_{av} + E_{ua} \tag{5.88}$$

$$\dot{W}_{max} = E_{av} = \dot{Q}_1 - \dot{Q}_2 = \dot{Q}_1 - E_{ua} \tag{5.89}$$

The maximum efficiency (η_{max}) for this engine can be expressed as:

$$\eta_{max} = 1 - \frac{T_2}{T_1} \tag{5.90}$$

Therefore, the maximum work output can be given by:

$$\dot{W}_{\max} = \eta_{\max}\dot{Q}_1 = \left(1 - \frac{T_2}{T_1}\right)\dot{Q}_1 \tag{5.91}$$

As can be seen from Eq. (5.91), the quality of energy increases with the temperature (T_1) of the hot reservoir. The minimum temperature at which a heat engine rejects heat is the ambient temperature (T_a). So, the exergy (E) of heat $\dot{Q}$ at a temperature T is defined as the maximum possible output work at an ambient temperature and E can be computed from (Bellos et al. 2017).

$$E = \left(1 - \frac{T_a}{T}\right)\dot{Q} \tag{5.92}$$

5.6.2.4 Exergy of Solar Radiation

Solar gas turbines are driven by solar radiation from the sun. Unlike thermal reservoirs, the sun is a radiation reservoir. In view of this, different approaches are employed in the computation of exergy for solar radiation (E_s). The model proposed by (Petela 2014) is commonly used in the calculation of exergy for solar energy:

$$E_s = \dot{Q}_s\left\{1 - \frac{4}{3}\left(\frac{T_a}{T_s}\right) + \frac{1}{3}\left(\frac{T_a}{T_s}\right)^4\right\} \tag{5.93}$$

where $\dot{Q}_s$ is the solar heat rate (solar thermal power), T_a is the ambient temperature and T_s is the blackbody temperature of the sun. It is possible to analyse the exergy of each component k of a solar gas turbine. There is exergy destruction ($E_{d,k}$) (in a given component) which can be defined as the difference between its input exergy rates ($E_{in,k}$) and product ($E_{u,k}$) (Petrakopoulou et al. 2017):

$$E_{d,k} = E_{in,k} - E_{u,k} \tag{5.94}$$

Equation (5.94) shows that the destruction of exergy increases with decreasing levels of the product for a fixed amount of input exergy ($E_{in,k}$). The total exergy destruction ($E_{d,tot}$) of the whole plant incorporates all exergy losses:

$$E_{d,tot} = E_{in,tot} - E_{u,tot} - E_{loss,tot} \tag{5.95}$$

For a solar-only gas turbine such as the one shown in Fig. 5.14, the exergy efficiency (Γ) of the plant can be expressed as (Bellos et al. 2017)

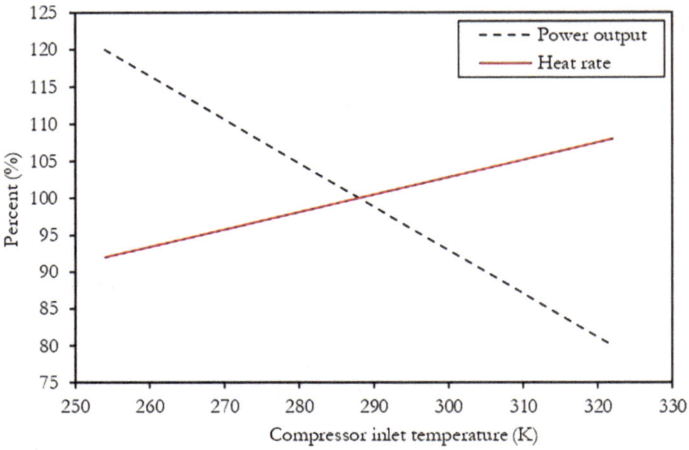

Fig. 5.14 Effect of ambient temperature on power output and heat rate

$$\Gamma = \frac{E_{\mathrm{u,tot}}}{E_{\mathrm{in,tot}}} = \frac{\dot{W}_{34} - \dot{W}_{12}}{E_{\mathrm{s}}} \tag{5.96}$$

where $\dot{W}_{12}$ is compressor work and $\dot{W}_{34}$ is shaft work.

Similarly, the exergy efficiency of a hybrid solar gas turbine shown in Fig. 5.9 can be computed from:

$$\Gamma = \frac{E_{\mathrm{u,tot}}}{E_{\mathrm{in,tot}}} = \frac{\dot{W}_{45} - \dot{W}_{12}}{E_{\mathrm{s}} + E_f} \tag{5.97}$$

where $\dot{W}_{45}$ is shaft work and (E_{f}) is the exergy of fuel. Combustion of fuel requires air and so the fuel–air mixture can fundamentally be regarded as comprising chemical exergy of consumed fuel ($E_{\mathrm{f,ch}}$) and physical exergy of air ($E_{\mathrm{a,ph}}$). In addition to these two exergy components, a steam injected combustion process has a physical exergy component for steam atomisation ($E_{\mathrm{st,ph}}$) (Behbahaninia et al. 2017).

$$E_{\mathrm{f}} = \begin{cases} E_{\mathrm{f,ch}} + E_{\mathrm{a,ph}}, \text{ without steam} \\ E_{\mathrm{f,ch}} + E_{\mathrm{a,ph}} + E_{\mathrm{st,ph}}, \text{ with steam} \end{cases} \tag{5.98}$$

$$E_{\mathrm{f,ch}} = \dot{m}_{\mathrm{f}} e_{\mathrm{f}} \tag{5.99}$$

where e_{f} is the specific exergy of fuel and $\dot{m}_{\mathrm{f}}$ is the mass flow rate of fuel.

The physical exergy of a component can be calculated directly by using Eq. (5.72). If steam and air are assumed to be perfect gases, then the specific exergy (e_{ph}) for these two components can be calculated from:

$$E_{\mathrm{a,ph}} = \dot{m}_{\mathrm{a}} e_{\mathrm{a,ph}} \tag{5.100}$$

$$E_{st,ph} = \dot{m}_{st} e_{st,ph} \tag{5.101}$$

where $\dot{m}_a$ is the mass flow rate of air and $\dot{m}_{st}$ is the mass flow rate of steam.

Some studies have examined the exergy of SGT systems. It has been found that the total exergy destruction in a steam injection solar gas turbine rises with the steam to air ratio and turbine inlet temperature while the exergetic efficiency can increase in the range of 40.0–54.2% (Selwynraj et al. 2015). Most of the exergy destruction occurs in the solar section, especially in the receiver, due to the low conversion of solar radiation to thermal energy that ultimately gets transferred to the working fluid (Giostri 2017). Al-Zahrani and Dincer (2018) also found similar results. In recent work, it has been theoretically found that the use of twisted tapes in parabolic receiver tubes improves the exergetic efficiency of the receiver (Thapa et al. 2023).

5.7 Influential Factors of Solar Gas Turbine Performance

The performance of solar gas turbines is affected by design, environmental and operational factors. For a given set of design parameters, the performance of a solar gas turbine depends on environmental and operational parameters. Environmental factors include solar radiation, ambient air and humidity. Important operational parameters are fuel (for hybrid solar gas turbines), temperature, pressure, air extraction from the compressor and gas turbine degradation.

5.7.1 Solar Radiation

Solar radiation is the most influential environmental factor for solar energy systems (Nafey et al. 2000). It comes from the sun and reaches the surface of the earth in two components: beam and diffuse components. Only beam radiation can be concentrated. Details about the physics of solar radiation are reported in Chap. 3.

A solar gas turbine system utilizes concentrated solar radiation as a source of heat. Theoretically, gas receivers can convert concentrated radiation to heat at high temperatures of up to 1773 K under high levels of beam radiation. However, the levels of solar radiation fluctuate with time which poses a challenge in supplying a base load. One way of enhancing the performance of a solar gas turbine is through integration with conventional sources of energy. This is often achieved through the use of fuels.

5.7.2 Ambient Air Temperature and Pressure

The performance of gas turbines varies with any factor that affects the density of the air intake to the compressor. Climatic conditions vary most frequently from the ISO reference conditions of 288.15 K and 1.013 bars. Nevertheless, the density of air decreases with increasing temperature. Consequently, for the same volume flow rate, the mass flow rate diminishes with increasing air temperature. The effect of ambient air temperature on power output and heat rate is presented in Fig. 5.14. It is observed that the power output diminishes with increasing ambient temperature, which indicates the need for cooling of inlet air to the compressor. Installation of an evaporative cooler or a chiller in the inlet duct downstream of the inlet filters can assist in reducing the compressor inlet temperature. Nonetheless, care is required when implementing the inlet air cooling strategy to avoid condensation or carry-over of moisture which may result in compressor fouling and the deterioration of performance. In view of this, inlet air cooling components are generally followed by moisture separators to curtail moisture crossover into the compressor. Evaporative cooling is most advantageous in hot low-humidity climates (Wang and Chiou 2004).

The density of ambient air decreases with atmospheric pressure. As the altitude of a site increases, the atmospheric pressure of air decreases, thereby decreasing the air density. Consequently, there is a reduction in the mass flow rate for a given volume flow rate as the altitude increases. To make up for this reduction in the mass flow rate of the air, a correction for the altitude is applied to maintain the intended power output.

5.7.3 Humidity

Atmospheric air comprises many gaseous constituents, water vapour and different kinds of contaminants (such as smoke, pollen and other pollutants) (American Society of Heating Refrigerating and Air-Conditioning Engineers 2017). Removal of water vapour and contaminants yields dry air whose composition does not substantially vary with time, geographic location and height above sea level.

Moist air is less dense than dry air, and so the power output decreases with increasing the magnitude of the specific humidity in the air. The effect of humidity was previously neglected but the increasing sizes of gas turbines and the addition of water and steam for purposes of increasing the efficiency of the system and for pollution control have necessitated taking into account the moisture content of inlet air.

5.7.4 Fuel

A fuel is a substance that is burnt to obtain heat, or a nuclear material usable in a reactor to generate thermal energy. Any conventional fuel (coal, oil, gas, biomass, nuclear) can be used in a closed cycle gas turbine. For the open cycle, liquid and gas fuels are appropriate. Frequently, combustible fuels are exploited in gas turbine propulsion. The heat energy depends on the elements of the fuel and products of the combustion process, and natural gas gives higher output than gas oil (Poullikkas 2005). This is attributed to the higher specific heat in the combustion products of natural gas. This gas has a higher ratio of hydrogen to carbon, which accounts for the observed higher specific heats of combustion.

Fuel combustion rate is affected by: (a) the rate at which the combustible fuel constituents react with the oxygen element, (b) the rate at which fuel mixes with air, and (c) the prevailing temperature in the combustor. The rate of combustion increases with the rate of mixing air/fuel and the level of temperature in the combustor. To prevent wastage of fuel, it is better to have excess air (lean mixture). In this connection, the required amount of excess air to be supplied to the combustor is dependent upon the (American Society of Heating Refrigerating and Air-Conditioning Engineers 2017): (a) anticipated changes in fuel properties and in the rates of fuel and air supply, (b) type of combustion equipment, (c) level of required operator supervision, and (d) control requirements. Combustion at a low proportion of excess air is needed for maximum efficiency. Thorough mixing of fuel and air is necessary to eliminate local fuel-rich and fuel-lean zones. Incomplete combustion proceeds in fuel-rich zones thereby reducing the thermal efficiency of the combustor.

In solar gas turbines, variations in the intensity of solar radiation are substantial. Some of these changes are highly stochastic and abrupt. For example, a passing cloud may reduce the beam radiation to zero in a very short time. This implies that the control system has to sense this change in input solar energy and actuate the fuel, air and other relevant components in such a way that maintains the output power from the plant. In this case of declining solar input, the system needs to stimulate a response action that raises the supply of fuel and air to make up for the loss from solar radiation.

5.7.5 Fuel Heating

Combined cycle power plants are designed to operate at maximum efficiency. One way of augmenting efficiency is through the reduction of fuel consumption for the same amount of useful output power. In this regard, combined cycle gas turbines are fitted with fuel heaters to elevate the temperature of the fuel before combustion. Fuel heating reduces the amount of fuel required to increase the temperature of the fuel–air mixture to the level of the firing temperature (lowest temperature at which a substance ignites and goes on burning for a certain period of time after a small flame has been introduced onto its surface under specified conditions).

Fuel heating leads to thermal expansion, which implies that the density of the fuel goes down. As a result of this, the mass flow rate also decreases thereby slightly reducing the output of the gas turbine system. Nevertheless, heat for heating the fuel comes from the feed water for the steam cycle resulting in an overall increase in the efficiency of about 0.6% (Poullikkas 2005). Consequently, exploitation of heat from the steam cycle is advantageous.

5.7.6 Inlet and Exhaust Pressure Losses

A real fluid encounters resistance to flow in the passage (conduit). This resistance may arise from friction produced by conduit-wall shear stresses or friction in conduit-section changes (contractions, expansions and diffusers, entrances, bends, and tees). Conduit-wall shear stresses lead to the conversion of part of the flow energy into internal energy, which is generally lost to the ambient environment.

Air filters, silencers, evaporative coolers or chillers fitted into the compressor inlet cause pressure losses across these components. Similarly, heat recovery devices installed in the exhaust section of the gas turbine contribute to pressure losses in the system. The impact of these losses varies with the design of the system. A pressure drop of about 1000 Nm^{-2} at the inlet to the compressor can reduce the output power by 1.4% (Poullikkas 2005).

5.7.7 Air Extraction from Compressor

Air can be extracted from the compressor discharge for other applications (such as pneumatic instruments) in the gas turbine power plant. Pneumatics is a branch of engineering that applies a pressurized gas to move something. The air extracted from the compressor requires cooling because the temperature of the compressed air is higher compared to ambient air.

Up to 5% of compressor air flow can be extracted from the compressor without changing the casings or on-base piping (Poullikkas 2005). It is possible to extract 6–20% of the air flow from the compressor with some modifications to the casings, piping and controls. Wide-spread changes to the system design are required for air extractions exceeding 20%. It is worth mentioning that air extraction from a compressor entails a decline in the output power due the reduction in mass flow rate.

5.7.8 Degradation of Gas Turbine

The performance of a gas turbine deteriorates with time. This degradation is caused by different mechanisms (Kurtz et al. 2014): (a) fouling, (b) corrosion and oxidation, (c) erosion and abrasion, (d) particle fusing, and (e) mechanical degradation.

Fouling is produced by the adherence of particles to airfoils and annulus surfaces. This adherence is enhanced by oil or water sprays, and it leads to building up of material which augments the roughness of surfaces. The geometry of the airfoils may also be altered by this accumulation of unwanted material. General types of fouling deposits in an axial compressor are (i) salts, (ii) heavy carbons (oils and waxes), (iii) carbon dirt, and (iv) other deposits (sand and other types of dust). Use of an appropriate air filtration system at the inlet can assist in the abatement of fouling in compressors.

Corrosion is the degradation of a substance (often a metal) due to a chemical reaction with its environment. It is caused by inlet air as well as by fuel, water or contaminants produced by the combustion process. The metal often reacts with oxygen in the air, and other chemical reactions may take part in the corrosion mechanism. Some kinds of corrosion (e.g., oxidation, sulphidation and hot corrosion) predominantly damage the hot section of the gas turbine, while crevice corrosion and pitting are prevalent in the compressor section. Hot-section coatings can be applied to control hot corrosion. Proper maintenance and suitable treatment of fuel, air and water are essential factors in the prevention of corrosion because all kinds of corrosion can either be enabled or speeded up by the presence of some contaminants in fuel, air or water supplied to the gas turbine.

Erosion is the wearing down of material from a surface. Material erosion is caused by hard and incompressible particles that impinge on the surfaces of the flow passage. Collisions between airborne solid particles and blades rotating at very high speeds rip off metal fragments from the blades. Abrasion also occurs when a surface rubs against another surface. Blades are designed and manufactured to a high degree of precision. Consequently, minor abrasions can alter their profile and affect the performance of the gas turbine.

Sometimes, particles fuse on hot surfaces. This can cause (a) blockage of cooling passages, (b) modification of surface characteristics, and (c) serious interference with heat transfer. Impedance to heat transfer can cause overheating, leading to thermal fatigue and structural failure. Often, affected surfaces are permanently damaged and so, they must be replaced.

Mechanical degradation is caused by wear in bearings and seals, coupling problems, extreme vibration and noise, or challenges in the lubrication oil system. Lack of proper maintenance can also lead to mechanical degradation. Vibration is a common indicator of mechanical deterioration, and several problems that show vibration symptoms may actually arise from other underlying causes that are related to aerodynamic factors. Usually, bearing problems develop because of diminished pressure of lubrication oil, blockage of oil flow, or extreme loads caused by other factors such as misalignment. It is possible for fuel nozzles to plug up due to factors such as coking, erosion and misassembly. In the hot section of the gas turbine, elevated temperatures may cause distortions in some structures (such as blades), resulting in serious dynamic loads on the blades.

The degradation of gas turbine performance can be categorized as recoverable or non-recoverable loss. Recoverable performance loss is often associated with fouling of the compressor and it can be partly regained by thorough cleaning of the compressor blades and vanes. The magnitude of the effort required to recover the performance loss varies with the nature of the loss. Non-recoverable loss results fundamentally from increases in the clearances of turbine and compressor, and from changes in the surface finish and airfoil contour.

5.8 Worked Examples

Example 5.1 A solar gas turbine comprises a solar tower with a heliostat aperture area of 8500 m^2. The reflector concentrates solar radiation onto a receiver with air as the working fluid. At a certain time of the day, the level of direct normal irradiance intercepted by the reflectors is 800 W/m^2. The inlet temperature to the receiver is 300 K, and the flow rate of the air through the receiver is 8.1 kg/s. The solar field has an optical efficiency of 95%, and receiver thermal efficiency of 85%. Calculate the temperature of the working fluid at the exit from the receiver.

Solution

Solar power intercepted by the reflector, $\dot{Q}_s = A_{ref}\, G_b = 8500 \text{ x } 800 = 6{,}800{,}000$ W.

Useful solar power from the receiver,

$$\dot{Q}_{u,s} = \eta_{rec}\eta_{opt}\dot{Q}_s = 0.85 \times 0.95 \times 6800000 = 5{,}491{,}000\,\text{W}$$

The useful solar power ($\dot{Q}_{s,ue}$) can be estimated from:

$$\dot{Q}_{s,ue} = \dot{m}_a C_{p,a}(T_3 - T_2),$$

where T_2 and T_3 are inlet and outlet temperatures of the solar receiver.

$$T_3 = T_2 + \frac{\dot{Q}_{s,ue}}{\dot{m}_a C_{p,a}}$$

From Eq. (5.26), $C_{p,a}$ is a function of T_2 and T_3. Therefore, T_3 can be found numerically using an iterative method. In this example, the estimated useful solar energy is computed by using an initial guess value of T_3. The error of estimation $\left(\dot{Q}_{s,ue} - \dot{Q}_{s,u}\right)$ is then squared and minimized. The minimization process can be performed in the Microsoft Excel Solver tool, which has a high capability of solving optimization problems (Briones and Escola 2019). This Solver gives values that compare very well with those obtained from more specialized computer programs (Lambert et al. 2012). When the solution converges, $T_3 = 891.82$ K.

Therefore, the temperature of the working fluid at the exit from the receiver is 892 K.

Example 5.2 Pulverised lignite coal is used in a combustor to provide additional heat to a solar gas turbine. Calculate the heat input to the closed cycle, if the rate of fuel feeding into the combustor is 0.004 kg/s, the lower heating value of the fuel is 15 MJ/kg, and the efficiency of the combustor is 75%. Assume combustion is complete and negligible losses from the heat exchanger to the environment.

Solution

Rate of heat input to combustor, $\dot{Q}_\mathrm{f} = \dot{m}_\mathrm{f} H_\mathrm{f,L}$

$$= 0.004 \times 15$$
$$= 60 \text{ kW}$$

Rate of heat input to the cycle $= \eta_\mathrm{cc}\dot{Q}_\mathrm{f} = 0.75 \times 60 = 45$ kW.

Example 5.3 Ethane (C_2H_4) can react with oxygen (O_2) to form carbon dioxide (CO_2) and water (H_2O) as follows:

$$C_2H_6 + 3.5\,O_2 \rightarrow 2CO_2 + 3H_2O$$

Calculate the exergy of C_2H_6.

Solution

The exergy of C_2H_6 can be computed by using Eq. (5.84) and tabulated values of the Gibbs free energy of formation and specific chemical exergy of the reactants and products shown in Table 5.3.

$$e_{C_2H_6} = -32,900 - \frac{1}{(-1)}\{-3.5 \times (3970 - 0) + 2 \times [20,140 - (-394,364)]$$
$$+3 \times [3120 - (-237141)]\}$$
$$= -32,900 - \frac{1}{(-1)}1535896 = 1502996 \text{ kJ/kmol}$$

Table 5.3 Gibbs free energy of formation ($g_\mathrm{fo,i}$) and specific chemical exergy (e_ch) of O_2, CO_2 and H_2O

Substance	n_i	$g_{f0,i}$(kJ/kmol)	e_ch (kJ/kmol)
C_2H_6	− 1	− 32,900	?
O_2	− 3.5	0	3970
CO_2	+ 2	− 394,364	20,140
H_2O	+ 3	− 237,141	3120

Example 5.4 A hybrid open cycle solar gas turbine comprises a solar field and a combustor connected in series. At a certain time of the day, the outlet temperature from the receiver is 400 K, and the flow rate of the air through the receiver is 8.1 kg/s. The preheated air flows into the combustor where it is mixed with natural gas fuel and burnt. During this time, the mass flow rate of fuel into the combustor is 0.135 kg/s. Estimate the specific heat capacity of the combustion gases, assuming complete combustion, negligible pressure losses, a lower heating value of natural gas of 42 MJ/kg, and combustor efficiency of 88%.

Solution

The specific heat capacity of the combustion gases ($C_{\mathrm{p,ga}}$) depends on inlet and outlet temperatures and it can be estimated from Eq. (5.53).

Mass flow rate of combustion gases,

$$\dot{m}_{ga} = \dot{\mathrm{m}}_{\mathrm{a}} + \dot{\mathrm{m}}_{\mathrm{f}} = 8.1 + 0.135 = 8.235\ \mathrm{kg/s}$$

The energy rate of fuel ($\dot{\mathrm{Q}}_f$)($\dot{\mathrm{Q}}_f$)

$$\dot{Q}_{\mathrm{f}} = \dot{m}_{\mathrm{f}} H_{\mathrm{f,L}} = 0.135 \times 42{,}000{,}000 = 5{,}670{,}000\ \mathrm{W}$$

Useful power ($\dot{Q}_{\mathrm{u,f}}$) added to the working fluid in the combustor,

$$\dot{Q}_{\mathrm{u,f}} = \eta_{\mathrm{cc}} \dot{Q}_{\mathrm{f}} = 0.88 \times 5{,}670{,}000 = 4{,}989{,}600\ \mathrm{W}$$

The estimated useful power ($\dot{Q}_{f,ue}$) from the fuel is:

$$\dot{Q}_{\mathrm{f,ue}} = \dot{m}_{\mathrm{ga}} C_{\mathrm{p,ga}} (T_4 - T_3)$$

This implies that:

$$T_4 = T_3 + \frac{\dot{Q}_{\mathrm{f,ue}}}{\dot{m}_{\mathrm{ga}} C_{\mathrm{p,ga}}}$$

$$C_{\mathrm{p,ga}} = \frac{\dot{Q}_{\mathrm{u,fe}}}{\dot{m}_{\mathrm{ga}} (T_4 - T_3)}$$

The values of both $C_{\mathrm{p,ga}}$ and T_4 are unknown. So, T_4 can be found numerically, starting with a guess value. The error of estimation $\left(\dot{Q}_{\mathrm{f},ue} - \dot{Q}_{\mathrm{f,u}}\right)$ is then squared and minimized. Using the Microsoft Excel Solver, $\mathrm{T}_4 = 939.36$ K and $\mathrm{C}_{\mathrm{p,ga}} = 1123.37$ J/kg.K when the solution converges. Therefore, the estimated specific heat capacity of combustion gases is 1123 J/kg K.

Example 5.5 A solar-only closed cycle gas turbine power plant is made up of a compressor, a solar field, a turbine and an electromagnetic generator. At solar noon on a certain day, the input solar energy to the receiver is 4 MW, the gross mechanical power generated by the turbine section is 1.5 MW while the input work to the compressor is 0.4 MW. Calculate the solar-to-electric efficiency of conversion, assuming generator efficiency of 0.98, and isentropic turbine and compressor efficiencies of 0.70 and 0.75 respectively.

Solution

From Eq. (5.36), electrical efficiency (η_e) is:

$$\eta_e = \frac{\dot{Q}_e}{\dot{Q}_s} = \frac{\eta_{gen}\left(\eta_{tu}\dot{W}_{tu} - \frac{\dot{W}_{com}}{\eta_{com}}\right)}{\dot{Q}_s}$$

$$= \frac{0.98 \times \left(0.70 \times 1.5 - \frac{0.4}{0.75}\right)}{4}$$

$$\approx 0.13$$

Nomenclature

A	Area (m^2)
C	Specific heat capacity (J/kg.K)
e	Specific exergy flow (W/kg)
E	Exergy flow (W)
g_{f0}	Gibbs free energy of formation (J/mol)
G	Irradiance (W/m)
h	Specific enthalpy (J/kg)
H	Heating value of fuel (J/kg)
$\dot{m}$	Mass flow rate (kg/s)
n	Number (1, 2, 3…)
P	Pressure (N/m)
$\dot{Q}$	Power (W)
r	Ratio (dimensionless)
R	Gas constant (J/mol)
s	Entropy (J/K)
T	Temperature (K)
V	Volume (m^3)
$\dot{W}$	Work (W)

Greek symbols

γ	Specific heat ratio (dimensionless)

η	Energy efficiency (dimensionless)
υ	Stoichiometric air–fuel ratio (dimensionless)
ξ	Air ratio (dimensionless)
Γ	Exergy efficiency (dimensionless)

Subscripts

0	Reference point
1-n	First to nth thermodynamic points
A	Air/ambient
ap	Aperture
av	Available
ch	Chemical
com	Compressor
comb	Combined
d	Destruction
e	Electrical/electricity
f	Fuel/fuel specific
ga	Gas(es)
gen	Generator
htf	Heat transfer fluid
is	Isentropic
L	Lower
m	Mechanical
max	Maximum
opt	Optical
p	Constant pressure
ph	Physical
rc	Rankine cycle
rec	Receiver
rh	Reheating
s	Solar/sun
sh	Superheating
tu	Turbine
u	Useful
ua	Unavailable
v	Vapour
w	Water
wf	Working fluid

References

Al-Zahrani AA, Dincer I (2018) Energy and exergy analyses of a parabolic trough solar power plant using carbon dioxide power cycle. Energy Convers Manag 158:476–488

American Society of Heating Refrigerating and Air-Conditioning Engineers (2017) ASHRAE handbook—fundamentals. American Society of Heating, Refrigerating and Air-Conditioning Engineers (ASHRAE), Atlanta

Behbahaninia A, Ramezani S, Hejrandoost ML (2017) A loss method for exergy auditing of steam boilers. Energy 140:253–260

Bellos E, Tzivanidis C, Antonopoulos KA (2017) Parametric analysis and optimization of a solar assisted gas turbine. Energy Convers Manage 139:151–165

Benoit H, Spreafico L, Gauthier D, Flamant G (2016) Review of heat transfer fluids in tube-receivers used in concentrating solar thermal systems: properties and heat transfer coefficients. Renew Sustain Energy Rev 55:298–315

Blanco MJ, Santigos LR (eds) (2017) Advances in concentrating solar thermal research and technology. Elsevier, Amsterdam

Boonnasa S, Namprakai P, Muangnapoh T (2006) Performance improvement of the combined cycle power plant by intake air cooling using an absorption chiller. Energy 31(12):2036–2046

Briones L, Escola JM (2019) Application of the Microsoft Excel Solver tool in the solution of optimization problems of heat exchanger network systems. Educ Chem Eng 26:41–47

Carnot S (1824) Reflections on the motive power of fire. Bachelier, Paris

Çengel YA, Boles MA (2006) Thermodynamics: an engineering approach. McGraw Hill, New York

Cetin B (2006) Optimal performance analysis of gas turbines. J Dogus University 7:59–71

Chen Z (2015) On the accuracy of laminar flame speeds measured from outwardly propagating spherical flames: Methane/air at normal temperature and pressure. Combust Flame 162(6):2442–2453

Corti A (2004) Thermoeconomic evaluation of CO_2 alkali absorption system applied to semi-closed gas turbine combined cycle. Energy 29(3):415–426

Dunham MT, Iverson BD (2014) High-efficiency thermodynamic power cycles for concentrated solar power systems. Renew Sustain Energy Rev 30:758–770

Effiom SO, Abam FI, Ohunakin OS (2015) Performance modeling of industrial gas turbines with inlet air filtration system. Case Stud Therm Eng 5:160–167

Fiaschi D, Baldini A (2009) Joining semi-closed gas turbine cycle and tri-reforming: SCGT-TRIREFas a proposal for low CO_2 emissions power plants. Energy Convers Manag 50:2083–2097

Fiaschi D, Manfrida G (1998) Exergy analysis of the semi-closed gas turbine combined cycle (SCGT/CC). Energy Convers Manag 39:1643–1652

Fiaschi D, Manfrida G (1999) A new semi-closed gas turbine cycle with CO_2 separation. Energy Convers Manag 40:1669–1678

Garg P, Kumar P, Srinivasan K (2013) Supercritical carbon dioxide Brayton cycle for concentrated solar power. J Supercrit Fluids 76:54–60

Giostri A (2017) Preliminary analysis of solarized micro gas turbine application to CSP parabolic dish plants. Energy Procedia 142:768–773

Khan MI, Asfand F, Al-Ghamdi SG (2022) Progress in technology advancements for next generation concentrated solar power using solid particle receivers. Sustain Energy Technol Assess 54:102813

Kurt H, Recebli Z, Gedik E (2009) Performance analysis of open cycle gas turbines. Energy Res 33:285–294

Kurz R, Meher-Homji C, Brun K (2014) Gas turbine degradation. 43rd Turbomachinery and 30th Pump Users Symposia (Pump & Turbo 2014), 23–25 September, 2014, Houston TX. Turbomachinery Laboratory, Houston

Lambert RJW, Mytilinaios L, Maitland L, Brown AM (2012) Monte Carlo simulation of parameter confidence intervals for non-linear regression analysis of biological data using Microsoft Excel. Comput Methods Programs Biomed 107:155–163

Lombardi L (2001) Life cycle assessment (LCA) and exergetic life cycle assessment (ELCA) of a semi-closed gas turbine cycle with CO_2 chemical absorption. Energy Convers Manag 42:101–114

Madenoğlu TG (2017) Effect of operating parameters on synthesis of lithium iron phosphate $LiFePO_4$) particles in near- and super-critical water. J Supercrit Fluids 127:103–110

Meitner P, Laganelli A, Senick P, Lear W (2000) Demonstration of a semi-closed cycle. American Society of Mechanical Engineers, New York, Turboshaft Gas Turbine Engine

Nafey AS, Abdelkader M, Abdelmotalip A, Mabrouk AA (2000) Parameters affecting solar still productivity. Energy Convers Manag 41(16):1797–1809

Olivenza-Leόn, Medina A, Hernández AC (2015) Thermodynamic modeling of a hybrid solar gas-turbine power plant. Energy Convers Manag 93:435–447

Olumayegun O, Wang M, Kelsall G (2016) Closed-cycle gas turbine for power generation: a state-of-the-art review. Fuel 180:694–717

Petela R (2014) Exergy analysis of solar radiation. In: Enteria N, Akbarzadeh A, editors. Solar thermal sciences and engineering applications. In: Enteria N, Akbarzadeh A (eds) Solar thermal sciences and engineering applications. CRC Press, London, pp 7–96

Petrakopoulou F, Sánchez-Delgado S, Marugán-Cruz C, Santana D (2017) Improving the efficiency of gas turbine systems with volumetric solar receivers. Energy Convers Manage 149:579–592

Poullikkas A (2004) Parametric study for the penetration of combined cycle technologies into Cyprus power system. Appl Therm Eng 24:1675–1685

Poullikkas A (2005) An overview of current and future sustainable gas turbine technologies. Renew Sustain Energy Rev 9:409–443

Querol E, Gonzalez-Regueral B, Perez-Benedito JL (2013) A practical to exergy and Thermoeconomic analyses of industrial processes. Springer, London

Schöffer SI, Klein SA, Aravind PV, Pecni R (2021) A solid oxide fuel cell- supercritical carbon dioxide Brayton cycle hybrid system. Appl Energy 283:115748

Selwynraj AI, Iniyan S, Polonsky G, Suganthi L, Kribus A (2015) Exergy analysis and annual exergetic performance evaluation of solar hybrid STIG (steam injected gas turbine) cycle for Indian conditions. Energy 80:414–427

Shang R, Zhang Y, Shi W, Wang X, Zhang Y (2014) Fresh look and understanding on carnot cycle. Energy Procedia 61:2898–2901

Stein WH, Buck R (2017) Advanced power cycles for concentrated solar power. Sol Energy 152:91–105

Thapa S, Kumar R, Kumar K, Samir S, Lee D, Singh T (2023) Analytical investigation of solar parabolic trough receiver with multiple twisted tapes using exergy and energy analysis. Case Stud Therm Eng 50:103493

Tsoutsanis E, Meskin N, Benammar M, Khorasani K (2013) Dynamic performance simulation of an aeroderivative gas turbine using the Matlab Simulink environment. American Society of Mechanical Engineers (ASME)

Wang FJ, Chiou JS (2004) Integration of steam injection and inlet air cooling for a gas turbine generation system. Energy Convers Manag 45:15–26

Wang T, Chen L, Liu P, Ge Y, Feng H (2022) Performance analysis and optimization of an irreversible Carnot heat engine cycle for space power plant. Energy Rep 8:6593–6601

Wrobel I (2017) Monthly dynamics of carbon dioxide exchange across the sea surface of the Arctic Ocean in response to changes in gas transfer velocity and partial pressure of CO2 in 2010. Oceanologia 59:445–459

Xu J, Yu C (2014) Critical temperature criterion for selection of working fluids for subcritical pressure Organic Rankine cycles. Energy 74:719–733

Zheng N, Li Z, Luo Y, Fang J, Wei J (2023) Assessment of ternary CO_2 mixtures as working fluids in supercritical Brayton cycles with floating critical points. Case Stud Therm Eng 52:103756

Chapter 6
Configurations of Solar Gas Turbines

6.1 Introduction

Compression, combustion and expansion of the combustion products take place in the same component (usually a cylinder) in a reciprocating engine. In contrast, these processes occur in separate components of a gas turbine, namely: compressor, combustor, and turbine. Components of a gas turbine are designed, fabricated and tested as separate units, and they can be assembled together to form a gas turbine system in different ways (Cohenet al. 1996), thereby yielding a wide variety of configurations. There is no limit to the number of each component, and so multiple compressors, combustors (and/or solar receivers), turbines and shafts (spools) can be used in the system. The arrangement of the components influences the optimal thermal efficiency, and the change of efficiency with output power and of output torque with rotational speed. A given configuration may be suitable for a specific application such as driving: (a) an electric generator under transient load at constant speed and b) a ship's propeller where the power varies as the cube of the speed. It is therefore necessary to identify a configuration that closely matches the application at hand.

The configuration with one shaft is appropriate when a gas turbine is for operation at constant speed and fixed load (such as generation of base-load power) conditions. This configuration has very limited flexibility. Nevertheless, mounting of the compressor and turbine on the same shaft raises the inertia of the engine because the compressor provides significant drag, which ultimately also assists in reducing the risk of over-speeding. In some applications, it is desirable to have operational flexibility, such as driving a variable speed load (Meas et al. 2022). The use of a free turbine and second spool is required in such cases, forming a twin-spool system. Twin-spool systems are applicable in power plants, with the generator connected to the power turbine. Gas turbine engines with up to three spools have been developed to give a higher degree of flexibility (Wang et al. 2017). The engine complexity increases with the number of spools, resulting in higher maintenance costs. However,

A. Madhlopa and R. Nkhoma, *Principles of Solar Gas Turbines for Generating Electricity*, Green Energy and Technology,
https://doi.org/10.1007/978-3-031-84096-8_6

some practical engines with three spools exist, such as the Rolls-Royce RB211 for flight applications. The discussion of triple-spool gas turbine engines falls outside the scope of this book.

6.2 Solar-Only Gas Turbine Systems

6.2.1 Solar-Only Gas Turbine with Recompression and Inter-Cooling

The efficiency of a solar gas turbine increases with the discharge pressure. Consequently, the working fluid can be compressed in multiple stages to raise the ratio of compression (Fig. 6.1). The working fluid is compressed at constant entropy from pressure P_1 at state 1 to a pressure P_x which lies between P_1 and P_4, thereby raising the temperature of the fluid. In view of this, the fluid requires cooling to its original temperature in order to maintain the density of the fluid and, therefore, the rate of mass flow. An intercooler is used to remove waste heat from the working fluid in between the compression stages. It is possible to use wet, dry or hybrid systems of cooling. Wet cooling requires water, and it is the most efficient cooling method because water has a high specific heat capacity. However, the level of direct normal irradiance is generally good in arid regions where water is usually scarce. Consequently, dry cooling can be an alternative option in such conditions. Dry cooling utilizes air and it is less efficient than wet cooling. Two cooling techniques can be combined to obtain a hybrid cooling system, with a water requirement that lies between those of wet and dry cooling techniques. Once the working fluid is cooled to its original temperature (T_1), it is recompressed from pressure P_x to pressure P_4 at constant entropy, with P_4 being greater than P_x.

Although the use of multiple compressors augments the total compressive work, the net output work increases. For instance, Al-Sulaiman and Atif (2015) studied

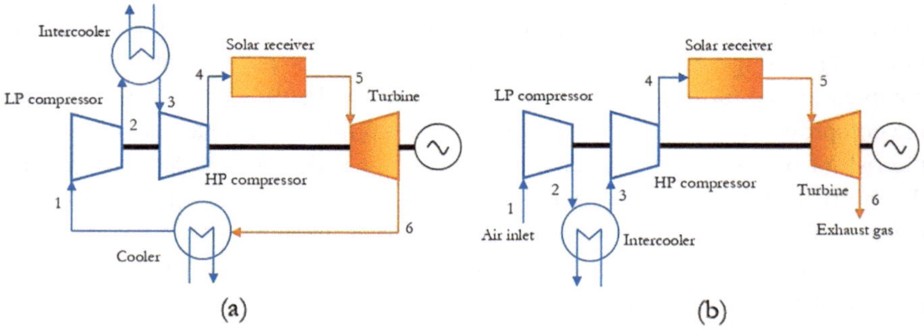

Fig. 6.1 A single-spool solar-only gas turbine with recompression and inter-cooling: **a** closed cycle and **b** open cycle. HP = high pressure and LP = low pressure

the performance of five supercritical carbon dioxide Brayton cycles (simple, regenerative, recompression, pre-compression, and split Brayton cycles) driven by a solar tower in Saudi Arabia. Results show that the recompression process yielded the highest efficiency (52% observed in summer). This level of thermodynamic performance clearly demonstrates the potential of the solar tower technology in driving gas turbine power plants (Okoroigwe and Madhlopa 2016; Madhlopa and Okoroigwe 2017).

The pressure discharge from the second compressor (high pressure compressor) (P_4) is higher than that from the first compressor (low pressure compressor). Assuming that there is no pressure drop across the intercooler, then the pressure at state 2 is equal to that at state 3 ($P_2 = P_3 = P_x$). Equation (6.1) defines the isentropic compressor work ($\dot{W}_{s4}$).

$$\dot{W}_{s4} = \frac{\gamma R T_1}{\gamma - 1}\left[\left(\frac{P_x}{P_1}\right)^{(\gamma-1)/\gamma} + \left(\frac{P_4}{P_x}\right)^{(\gamma-1)/\gamma} - 2\right] \tag{6.1}$$

where R is the gas constant, T_1 is the temperature at state 1, and γ is the specific heat ratio of air (the ratio of the specific heat at constant pressure to the specific heat at constant volume).

Research attention has also been given to solar gas turbines with two spools (Najjar and Zaamout 1992), shown in Fig. 6.2. It has been found that the thermal performance of a twin-spool solar gas turbine is higher than that of a single-spool engine. Moreover, the twin-spool engine is easier to operate and maintain, and adaptable to fluctuations in power. In addition, the twin-shaft engine does not require inter-cooling between compressors, which offers an advantage for its application in arid regions that are generally endowed with good solar resources. The two spools can be arranged in a concentric geometry, with a longer spool connecting the low-pressure compressor to the low-pressure turbine (Langston 2004). This spool is also described as the low spool, and it rotates inside a shorter, larger-diameter spool (also known as the high spool) that connects the high pressure compressor and the high pressure turbine. The high spool rotates faster than the low spool.

6.2.2 *Solar-Only Gas Turbine with Fix and Free Turbines*

In other layouts of the gas turbine engine, the power turbine is delinked from the compressor to yield an engine topology with two spools in which the two turbines are coupled via the flow circuit (Fig. 6.3). The turbine which is not connected to a compressor is said to be free, and it operates at a lower pressure than the fixed turbine.

The use of multiple spools enhances flexibility in the system operation to meet different demands. Nevertheless, the performance of a gas turbine with multiple spools is low when operated in the part-load condition because of its non-optimal

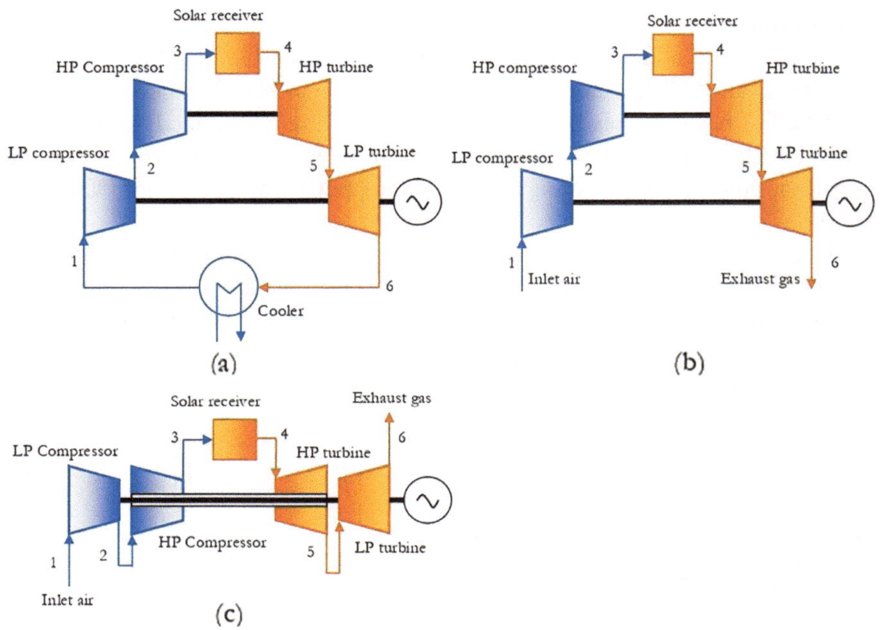

Fig. 6.2 Twin-spool solar-only gas turbine with recompression: **a** closed cycle, **b** open cycle with non-concentric spools and **c** open cycle with concentric spools. HP = high pressure, and LP = low pressure

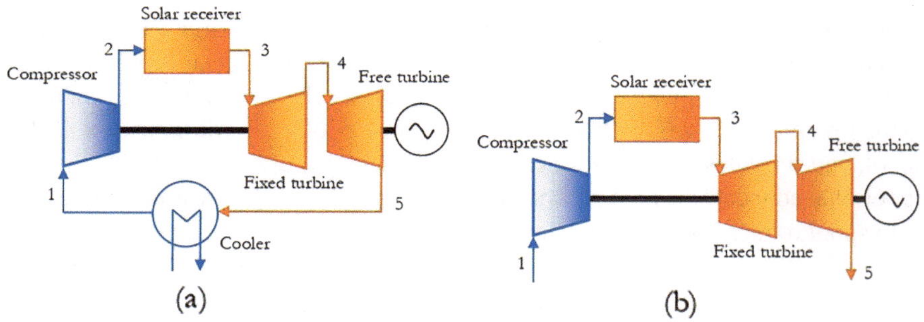

Fig. 6.3 Twin-spool solar-only gas turbines with fixed and free turbines: **a** closed cycle and **b** open cycle

part-load control strategy (Wang et al. 2017). This drawback can be overcome by optimizing the thermodynamic performance of the system in a wide range of the load.

6.2.3 Solar-Only Gas Turbine with Free Turbine and Reheating

The expanding working fluid loses part of its energy in the fixed turbine, resulting in a temperature drop before it flows into the free turbine. So, it is necessary to re-elevate the temperature of the fluid by reheating using a second receiver located between the two turbines. A possible flow configuration of a solar-only gas turbine with a free turbine and reheating is shown in Fig. 6.4. The compressed working fluid is heated by a low-temperature solar receiver before flowing into the fixed turbine section. After exiting this turbine, the fluid is reheated by using a high-temperature solar receiver before it enters the free turbine section.

Recompression may also be applied in twin spool solar gas turbines that have low and high pressure compressors mounted on separate spools with corresponding low and high pressure turbines. In the high pressure turbine, the expanding working fluid loses part of its energy, resulting in a temperature drop before it flows into the low pressure turbine. To re-elevate the temperature of the fluid, reheating can be done by using a second solar receiver located between the two turbines. A possible flow configuration of a twin-spool solar-only gas turbine with high and low pressure turbines with recompression and reheating is shown in Fig. 6.5.

6.3 Hybrid Solar Gas Turbine Systems

The major components of a hybrid solar gas turbine are solar field, compressor, combustor and turbine. These components can also be arranged in a variety of ways to achieve the desired engine performance of the hybrid solar gas turbine.

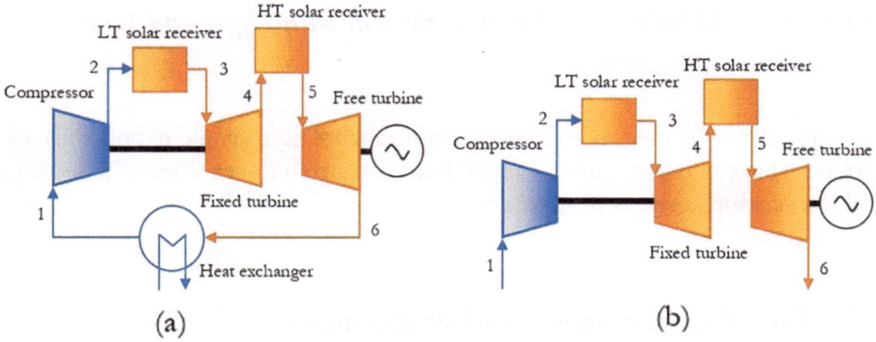

Fig. 6.4 Twin-spool solar-only gas turbines with a free turbine and reheating: **a** closed cycle, and **b** open cycle. HT = high temperature and LT = low temperature

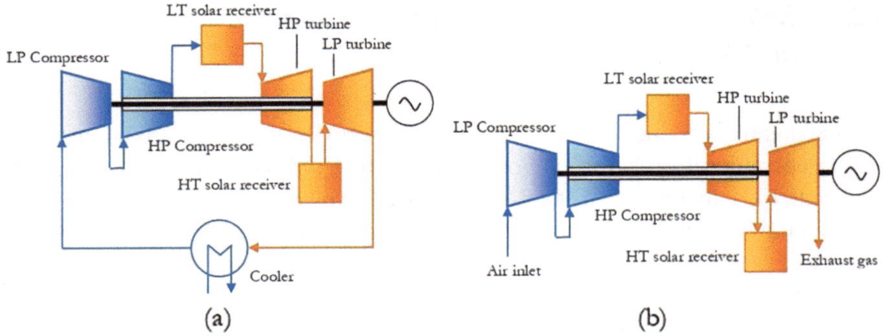

Fig. 6.5 Twin-spool solar-only gas turbines with low and high pressure turbines, concentric spools, and low- and high-temperature solar receivers: **a** closed cycle and **b** open cycle. HP = high pressure, HT = high temperature, LP = low pressure and LT = low temperature

6.3.1 Hybrid Solar Gas Turbine with Recompression and Inter-Cooling

Hybridization is one of the main strategies for enhancing the performance of solar gas turbines (see Sect. 4.3). As in the solar-only gas turbine, recompression can be applied in the hybrid system to increase the efficiency. One possible configuration of this engine is shown in Fig. 6.6.

In recent work, Behar (2018) investigated a hybrid solar gas turbine in which the working fluid was preheated by using a parabolic trough concentrator. Results showed that the compression ratio was more influential than the solar field, which is consistent with findings of Al-Sulaiman and Atif (2015).

6.3.2 Hybrid Solar Gas Turbine System with High and Low Pressure Turbines

Two spools can be used to enhance the performance of a simple hybrid solar gas turbine. This yields configurations with fixed and free turbines (Fig. 6.7), and high and low pressure turbines (Fig. 6.8).

6.4 Solar Gas Turbines with Recuperation

Generally, the temperature of the exhaust gas from the turbine is high (may exceed 673 K), which necessitates the inclusion of a recuperator. In the recuperative design, the compressed working fluid enters the recuperator where it is pre-heated before

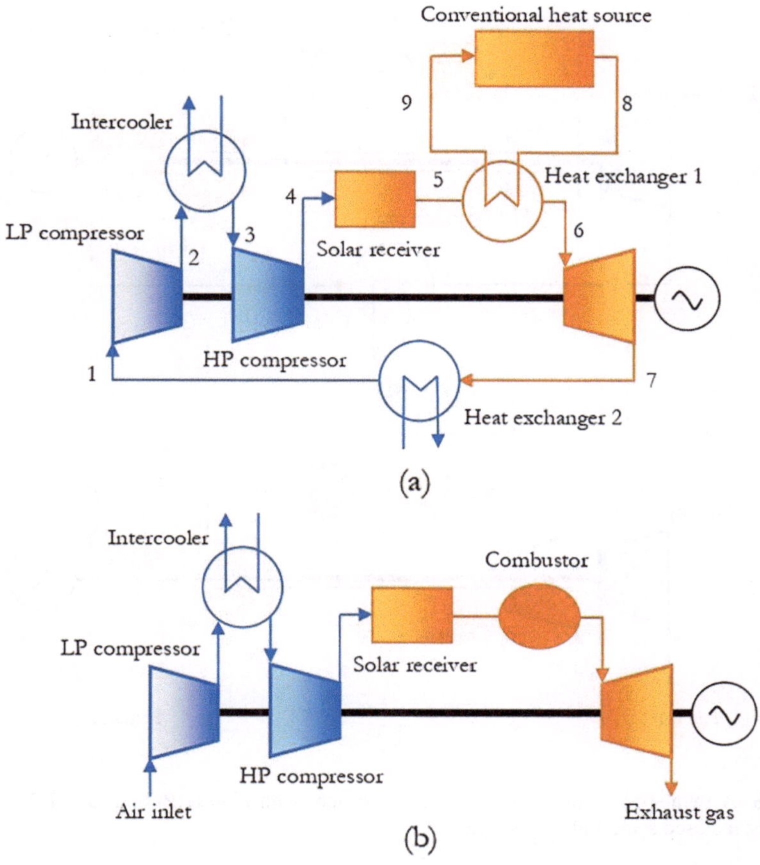

Fig. 6.6 A hybrid solar gas turbine system with two compression stages: **a** closed cycle and **b** open cycle

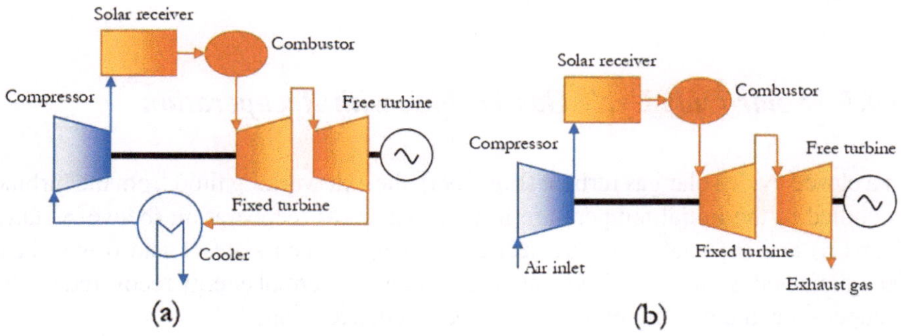

Fig. 6.7 A twin-spool hybrid solar gas turbine system with fixed and free turbines: **a** closed cycle and **b** open cycle

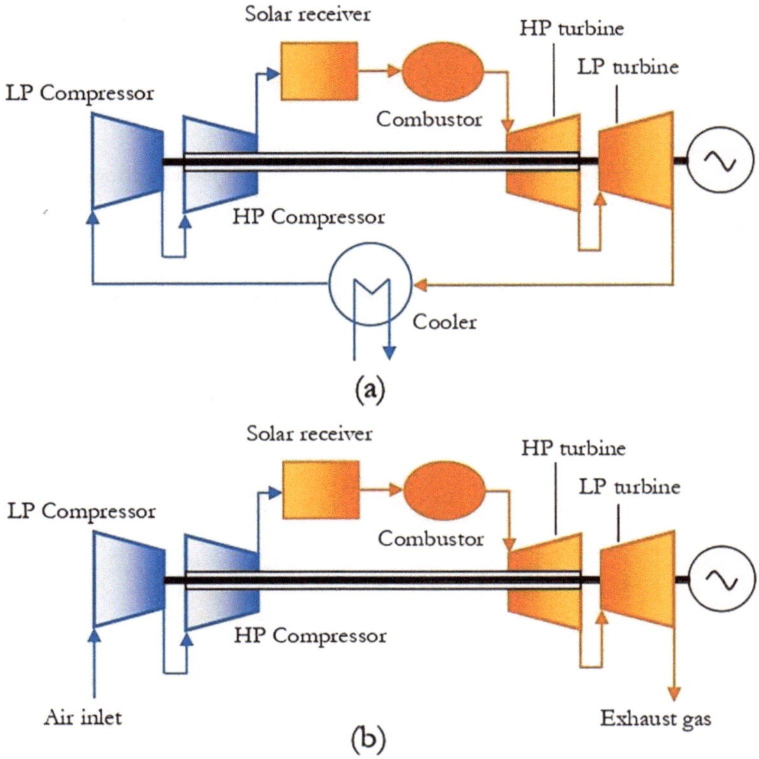

Fig. 6.8 A twin-spool hybrid solar gas turbine system with high (HP) and low (LP) pressure turbines: **a** closed cycle and **b** open cycle

the addition of solar thermal energy. In so doing, the temperature of the compressed fluid rises before flowing into the solar receiver for further heating.

6.4.1 Solar-Only Cycle Gas Turbine with Recuperation

In a closed cycle solar gas turbine (Fig. 6.9a), the exit working fluid from the turbine is cooled to the initial temperature in two stages: (a) recuperation (between states 5 and 6) and (b) cooling in the heat exchanger between states 1 and 6 with heat being rejected from this component. The amount of thermal energy recovered by the recuperation and cooling processes can be calculated from:

$$\dot{Q}_{56} = \dot{m}_{\text{wf}}(h_5 - h_6) = \dot{m}_{\text{wf}}C_{\text{p,wf}}(T_5 - T_6) = \dot{m}_{\text{wf}}C_{\text{p,wf}}(T_3 - T_2) \quad (6.2)$$

$$\dot{Q}_{61} = \dot{m}_{\text{wf}}(h_6 - h_1) = \dot{m}_{\text{wf}}C_{\text{p,wf}}(T_6 - T_1) = \dot{m}_{\text{htf}}C_{\text{p,htf}}(T_8 - T_7) \quad (6.3)$$

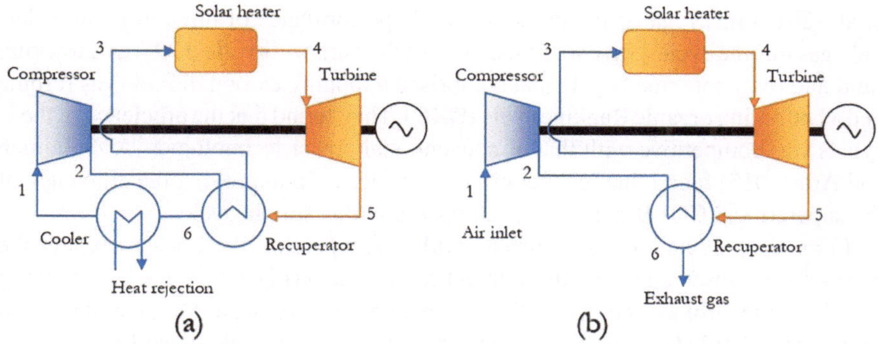

Fig. 6.9 A solar-only gas turbine with recuperation: **a** closed cycle and **b** open cycle

If air is used as the working fluid then the specific heat capacity can be computed from (Cetin 2006; Kurt et al. 2009):

$$C_{p,a} = 1000\left(1.0484 - \frac{3.8371T_m}{10^4} + \frac{9.4537T_m^2}{10^7} - \frac{5.4903T_m^3}{10^{10}} + \frac{7.9298T_m^4}{10^{14}}\right) \quad (6.4)$$

where $T_m = 0.5(T_{in} + T_{out})$ is the mean of instantaneous temperatures at the inlet and outlet points, T_{in} is the inlet temperature and T_{out} is outlet temperature.

It is observed that, for the same solar energy input, the working fluid in the recuperated closed cycle gas turbine gains $\dot{Q}_{61}$ more heat than the un-recuperated cycle. The thermal effectiveness (ε_r) of the recuperator is defined in terms of the inlet (T_{wf_in}) and outlet (T_{wf_out}) temperatures of the cold stream of the working fluid (into and out of the recuperator, respectively) and the inlet temperature (T_{eg_in}) of the exhaust gas entering the recuperator (Wang et al. 2017):

$$\varepsilon_r = \frac{T_{wf\_out} - T_{wf\_in}}{T_{eg\_in} - T_{wf\_in}} \quad (6.5)$$

Recuperators with high thermal effectiveness are preferable to improve the efficiency of the gas turbine. Similarly, the thermal effectiveness (ε_{hex}) of the heat exchanger is defined in terms of the inlet (T_{htf_in}) and outlet (T_{htf_out}) temperatures of the cold stream of the heat transfer fluid (into and out of the heat exchanger respectively) and the inlet temperature (T_{eg_in}) of the exhaust gas entering the heat exchanger:

$$\varepsilon_{hex} = \frac{T_{htf\_out} - T_{htf\_in}}{T_{eg\_in} - T_{htf\_in}} \quad (6.6)$$

The thermodynamic performance of a gas turbine influences its economic competitiveness. In view of this, it has attracted a lot of interest from researchers. Chacartegui et al. (2011) investigated the thermodynamic performance of three different solar-only gas turbine cycles: (a) two closed cycles with carbon dioxide (CO_2) as a working fluid and (b) a combined cycle that comprised a topping carbon dioxide gas turbine and a bottoming organic Rankine cycle (ORC). They found that the efficiency of these cycles was competitive with that of conventional power technologies. Al-Sulaiman and Atif (2015) found that recuperation was effective in improving the efficiency of the supercritical CO_2 Brayton cycle compared with a simple cycle.

In an open cycle solar gas turbine with a recuperator, the hot gas exiting the recuperator is discharged into the ambient environment (Fig. 6.9b). The exit working fluid from the turbine is cooled by the recuperator. In this case, the amount of thermal energy recovered $\left(\dot{Q}_{23}\right)$ by the recuperation process can be calculated from:

$$\dot{Q}_{23} = \dot{m}_{\mathrm{wf}}(h_3 - h_2) = \dot{m}_{\mathrm{wf}} C_{\mathrm{p,wf}}(T_3 - T_2) \tag{6.7}$$

6.4.2 Hybrid Solar Gas Turbine

In the recuperative hybrid closed cycle solar gas turbine, the pressurized working fluid from the compressor enters the recuperator where it is pre-heated before the addition of solar thermal energy (Fig. 6.10a). In so doing, the temperature of the compressed fluid rises before flowing into the solar receiver for further pre-heating. Thereafter, the working fluid is finally heated in the heat exchanger. Thus, the recuperated hybrid closed cycle solar gas turbine has two pre-heating stages.

The exit working fluid from the turbine is cooled to the initial conditions in two stages: (a) recuperation (between states 6 and 7) and (b) cooling in heat exchanger 2

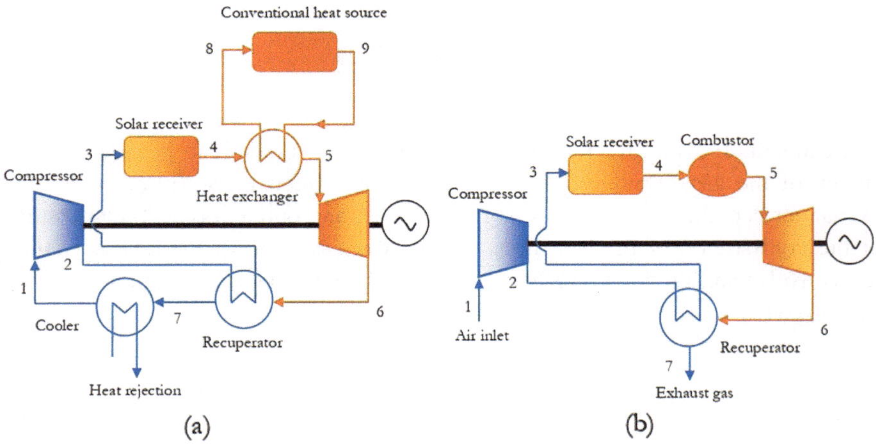

Fig. 6.10 A hybrid solar gas turbine system with recuperation: **a** closed cycle and **b** open cycle

(between states 7 and 1). In the second cooling stage, heat is rejected to the ambient environment. The hot fluid loses heats $(\dot{Q}_{67})$ and $(\dot{Q}_{71})$ in the recuperator and cooler, respectively. The amount of thermal energy recovered by the recuperation process is $(\dot{Q}_{23})$. These quantities of heat can be calculated as follows:

$$\dot{Q}_{23} = \dot{m}_{wf} C_{p,wf}(T_3 - T_2) \tag{6.8}$$

$$\dot{Q}_{67} = \dot{m}_{wf}(h_6 - h_7) = \dot{m}_{wf} C_{p,wf}(T_6 - T_7) \tag{6.9}$$

$$\dot{Q}_{71} = \dot{m}_{wf}(h_7 - h_1) = \dot{m}_{wf} C_{p,wf}(T_7 - T_1) = \dot{m}_{htf2} C_{p,htf2}(T_{11} - T_{10}) \tag{6.10}$$

where $C_{p,htf2}$ and $\dot{m}_{htf2}$ are specific heat and mass flow rate of the heat transfer fluid in heat exchanger 2.

We see that, for the same input solar and fuel energy, the working fluid in the recuperated cycle gains heat Q_{23} over the un-recuperated cycle. No wonder, it has been found that the recuperative process increases the overall thermodynamic efficiency of the cycle (Santos et al. 2016).

The temperature of the exhaust gas from an open cycle gas turbine is also high. A recuperator can also be fitted on the exit side of the turbine to enhance the thermal efficiency of the cycle (Fig. 6.10b).

The hot gas from the turbine is cooled through recuperation (between states 6 and 7) before being ejected into the atmosphere. The amount of thermal energy recovered by the recuperation process can be calculated by:

$$\dot{Q}_{67} = \dot{m}_{wf}(h_6 - h_7) = \dot{m}_{wf} C_{p,wf}(T_6 - T_7) \tag{6.11}$$

$$C_{p,\,ga} = 1000\left(0.991615 + \frac{6.99703 T_{mg}}{10^5} + \frac{2.712987 T_{mg}^2}{10^7} - \frac{1.22442 T_{mg}^3}{10^{10}}\right) \tag{6.12}$$

where $T_{mg} = 0.5(T_6 - T_7)$ is the mean temperature of the combustion gases.

An example of recuperated solar gas turbines is the Solar Mercury 50 model manufactured by Solar Turbines Inc. It is a hybrid single-shaft gas turbine with an ISO thermal efficiency of 40.3% (Schwarzbözl et al. 2006).

In literature, recuperative and regenerative heat exchangers are found in layouts of gas turbine cycles (Olivenza-León et al. 2015; Rajaei et al. 2016). However, these two types of heat exchangers have different heat and mass transfer characteristics. A recuperative heat exchanger permits continuous exchange of heat between the hot and cold streams of the fluids while heat is transferred alternately between hot and cold streams in a regenerative heat exchanger (Rogers and Mahew 1992). Moreover, heat and mass flows are continuous in a gas turbine when the engine is in operation. Thus, there is need for constant heat transfer from the hot exhaust gases (hot stream) to the pressurized working fluid (cold stream) in the heat exchanger of the engine cycle, and the characteristics of a recuperative heat exchanger are more suitable in the layout of a gas turbine with a heat recovery on the exit side of the turbine. Otherwise, the use of a regenerative heat exchanger on this side of the turbine would lead to intermittent pre-heating of the pressurized working fluid from the compressor, which would adversely affect the thermodynamic performance of the engine cycle. The regenerative variety of heat exchangers can be applied in the thermal storage system where heat exchange is recurrent.

6.4.3 Solar Gas Turbines with Thermal Storage

A solar gas turbine can be integrated with a thermal storage unit to store solar energy for use at night or when the level of direct normal irradiance is low. It is difficult to store hot gas such as air, and so a separate storage medium is used, which requires a heat exchanger to indirectly heat the gaseous working fluid. Unfortunately, exploitation of high-temperature heat exchangers is very costly (Blanco and Santigos 2017). In view of this, the use of direct contact heat storage units of the regenerative variety is preferable. These units may include: (a) fixed beds with stacked materials and (b) packed beds with regular or irregular materials, and have often one cycle per day.

Thermal storage units are integrated parallel to the receiver (Fig. 6.11). For purposes of charging, the solar field has to be oversized so that there is excess heat from the receiver. A blower can be included to provide extra air mass flow through the receiver and storage units. This extra air mass flow does not affect the mass flow through the compressor and turbine sections because the loop for the storage unit with blower is essentially closed.

To charge a thermal storage unit (charging mode), excess power is extracted from the receiver by an augmented mass flow that is taken from the compressor and an extra mass flow introduced by the blower downstream the storage unit. Consequently, a proportion of the gas from the receiver outlet passes through the storage unit. Part of the heat in the hot gas is transferred to the storage unit. The receiver outlet temperature is kept at its design value. When the level of irradiation is low (or at night), the receiver is unable to deliver adequate power to drive the gas turbine. In this case, the discharging process starts, and the mass flow from the compressor is split into two streams. One of the streams passes through the receiver while the remainder flows through the storage unit (in a reversed direction). The receiver outlet temperature fluctuates in both charging and discharging modes. These fluctuations are difficult to mitigate in a solar-only gas turbine but they can be compensated by the combustor in the hybrid system.

Storage of solar heat has also been applied in hybrid solar gas turbines. For example, Klein et al. (2015) analysed the performance of a hybrid solar gas turbine with recuperation and thermal storage. Their results show that the system could achieve a storage efficiency of 88% and utilization factor of 85% when solar energy was hybridized with fossil fuels. The combination of hybridization and thermal storage enhances the performance of solar gas turbines. Moreover, biofuels can

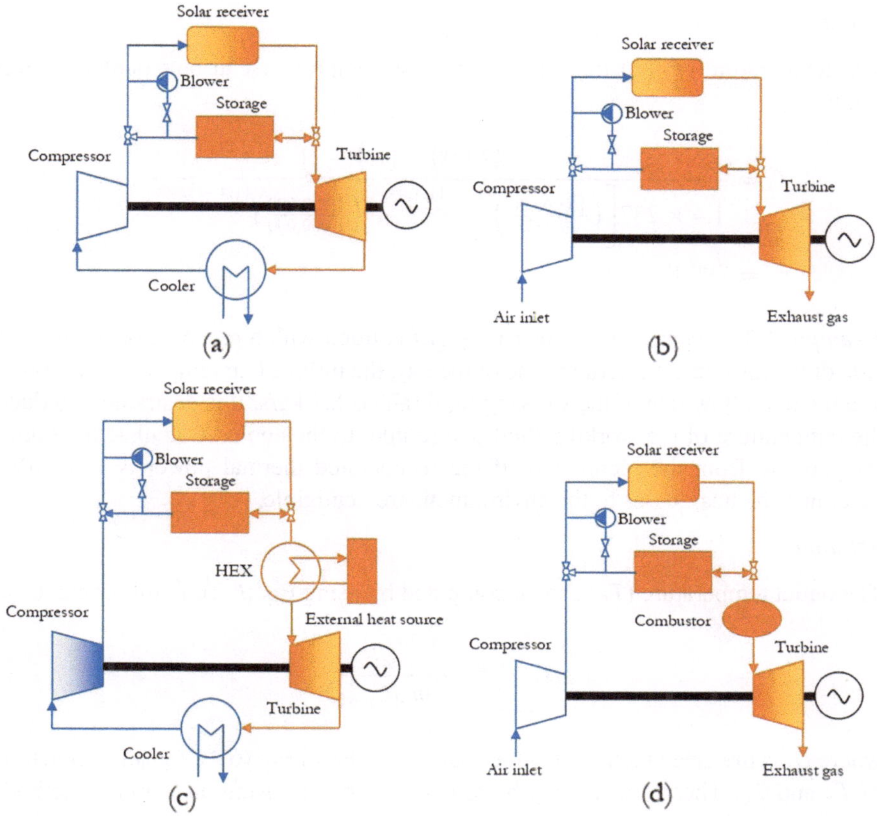

Fig. 6.11 Solar gas turbine with thermal storage component: **a** solar-only closed cycle, **b** solar-only open cycle, **c** hybrid closed cycle and **d** hybrid open cycle. HEX = heat exchanger

be exploited to achieve a hybrid solar gas turbine system that operates on 100% renewable energy with a high level of reliability.

6.5 Worked Examples

Example 6.1 A solar-only open cycle gas turbine uses air as its working fluid which is compressed in two stages (with intercooling). In the first stage, the working fluid is compressed at constant entropy from atmospheric pressure (101,325 N/m^2) to 1.0 MN/m^2. In the second stage, the pressure of the working fluid is increased by 0.8 MN/m^2. Calculate the temperature of the air at the inlet of the first compressor, assuming the values of the total isentropic work done by the compressors, specific heat ratio of air and gas constant are respectively 322 kW, 1.4 and 287 J/kg K.

Solution

The temperature (T_1) at the inlet to the first compressor can be calculated from Eq. (6.1).

$$T_1 = \frac{322{,}000 \times [1.4 - 1]}{1.4 \times 287\left[\left(\frac{1{,}000{,}000}{101{,}325}\right)^{(1.4-1)/1.4} + \left(\frac{1{,}800{,}000}{1{,}000{,}000}\right)^{(1.4-1)/1.4} - 2\right]}$$
$$= 290\ \text{K}$$

Example 6.2 A closed cycle solar-only gas is fitted with a recuperator on the exit side of the turbine. At a certain time of the day, the turbine exit temperature is 500 K and the mass flow rate of the working fluid (air) is 8.1 kg/s. It is desirable to reduce the temperature of the working fluid before inlet to the cooler. Calculate the outlet temperature from the recuperator if the recuperated thermal power is 1.70 MW, assuming the heat losses to the environment are negligible.

Solution

The outlet temperature (T_6) can be computed by using Eq. (6.2). In this context,

$$T_6 = T_5 - \frac{\dot{Q}_r}{\dot{m}_{wf} C_{p,wf}},$$

where $\dot{Q}_r$ is the given recuperated thermal power. From Eq. (6.4), $C_{p,wf}$ is a function of T_5 and T_6. Therefore, T_6 can be found numerically using an iterative method. In this example, the estimated recuperated thermal energy ($\dot{Q}_{r,e}$) is computed by using an initial guess value of T_6. The error of estimation $\left(\dot{Q}_{r,e} - \dot{Q}_r\right)$ is then squared and minimized. The minimization process can be performed in the Microsoft (MS) Excel or other Solver tools. The MS Excel Solver has a high capability of solving optimization problems (Briones and Escola 2019). This Solver gives values that compare very well with those obtained from more specialized computer programs (Lambert et al. 2012). At convergency, $T_6 = 299$ K.

Example 6.3 If the effectiveness of the recuperator in Example 6.2 is 0.80, calculate the inlet temperature of the cold stream of the working fluid into the recuperator, assuming the temperature of this stream is 460 K when exiting the recuperator.

Solution

The inlet temperature of the cold working fluid (T_{htf_in}) can be computed from Eq. (6.6):

$$T_{wf\_in} = \frac{T_{wf\_out} - \epsilon_r\, T_{eg\_in}}{1 - \epsilon_r}$$
$$= \frac{460 - 0.80 \times 500}{1 - 0.80}$$
$$= 300\ \text{K}$$

Example 6.4 In a hybrid open cycle solar gas with a recuperator on the exit side of the turbine, the outlet temperature of the exhaust gases from the recuperator is 350 K with a mass flow rate of the exhaust gases 9 kg/s. Calculate the outlet temperature from the turbine if recuperated thermal power is 2 MW, assuming the heat losses to the environment are negligible.

Solution

The turbine outlet temperature (T_6) can be computed by using Eq. (6.9). In this context,

$$T_6 = T_7 + \frac{\dot{Q}}{\dot{m}_{\text{ga}} C_{\text{p,ga}}},$$

where $\dot{Q}_{\text{r}}$ is the given recuperated thermal power. $C_{\text{p,ga}}$ is a function of T_6 and T_7. Therefore, T_6 can be found numerically using an iterative method. In this example, the estimated recuperated thermal energy ($\dot{Q}_{\text{r,e}}$) is computed by using an initial guess value of T_6. The error of estimation $\left(\dot{Q}_{\text{r, e}} - \dot{Q}_{\text{r}}\right)$ is then squared and minimized. At convergence, $T_6 = 558$ K.

Example 6.5 A solar-only gas turbine with recompression (in two stages) and intercooling uses air as the working fluid. The working fluid is at a temperature of 293 K at the inlet to the low pressure compressor that pressurizes it from 0.1 MN/m^2 to 0.3 MN/m^2. Then, the working fluid is cooled to its original temperature to maintain its density. After cooling, the fluid is recompressed from 0.3 MN/m^2 to 0.5 MN/m^2 at constant entropy. Find the total compressor isentropic work, assuming the gas constant and the specific heat ratio for air are 287 J/kg.K and 1.4 respectively.

Solution

From Eq. (6.1), total compressor isentropic work ($\dot{W}_{14}$) is:

$$\dot{W}_{14} = \frac{\gamma \text{RT}_1}{(\gamma - 1)} \left[\left(\frac{P_{\text{x}}}{P_1} \right)^{(\gamma-1)/\gamma} + \left(\frac{P_4}{P_{\text{x}}} \right)^{(\gamma-1)/\gamma} - 2 \right]$$

$$\frac{1.4 \times 287 \times 293}{(1.4 - 1)} \left[\left(\frac{0.3}{0.1} \right)^{(1.4-1)/1.4} + \left(\frac{0.5}{0.3} \right)^{(1.4-1)/1.4} - 2 \right]$$

$$= 154775 \text{ W}$$

Nomenclature

C	Specific heat capacity (kJ/kg.K)
h	Specific enthalpy (J/kg)

$\dot{m}$	Mass flow rate (kg/s)
P	Pressure (N/m^2)
$\dot{Q}$	Heat rate/power (W)
R	Gas constant (J/kg.K)
T	Temperature (K)
W	Work (W)

Greek Symbols

ε	Thermal effectiveness of a heat exchanger/recuperator (dimensionless)
γ	Specific heat ratio of a gas (dimensionless)

Subscripts

a	Air/ambient
eg	Exhaust gas
hex	Heat exchanger
htf	Heat transfer fluid
in	Inlet
m	Mean
mg	Mean of combustion gases
out	Outlet
n	Nth thermodynamic point (n = 1, 2, 3, …..)
p	Constant pressure
r	Recuperator
wf	Working fluid

References

Al-Sulaiman F, Atif M (2015) Performance comparison of different supercritical carbon dioxide Brayton cycles integrated with a solar power tower. Energy 82:61–71

Behar O (2018) A novel hybrid solar preheating gas turbine. Energy Convers Manag 158:120–132

Blanco MJ, Santigosa LR (2017) Advances in concentrating solar thermal research and technology. Elsevier, Amsterdam

Briones L, Escola JM (2019) Application of the Microsoft Excel Solver tool in the solution of optimization problems of heat exchanger network systems. Educ Chem Eng 26:41–47

Cetin B (2006) Optimal performance analysis of gas turbines. J Dogus University 7:59–71

Chacartegui R, De Escalona JM, Sánchez D, Monje B, Sánchez T (2011) Alternative cycles based on carbon dioxide for central receiver solar power plants. Appl Therm Eng 31:872–879

Cohen H, Rogers G, Saravanamuttoo H (1996) Gas turbine theory. Longman Group Limited, Essex

Klein P, Roos TH, Sheer TJ (2015) Parametric analysis of a high temperature packed bed thermal storage design for a solar gas turbine. Sol Energy 118:59–73

Kurt H, Recebli Z, Gedik E (2009) Performance analysis of open cycle gas turbines. Int J Energy Res 33(3):285–294

Lambert RJW, Mytilinaios I, Maitlan L, Brown AM (2012) Monte Carlo simulation of parameter confidence intervals for non-linear regression analysis of biological data using Microsoft Excel. Comput Methods Programs Biomed 107(2):155–163

Langston L (2004) Turbines, gas. In: Cleveland C (ed.) Encyclopedia of Energy, pp 221–230

Madhlopa A, Okoroigwe E (2017) Solar gas turbine systems. In: Abraham M (ed) Encyclopedia of sustainable technologies. Elsevier, Amsterdam, pp 377–388

Meas MR, von Backström TW, van der Spuy S J (2022) Performance analysis of variable speed solar gas turbine configurations. In: AIP Conference Proceedings 2445(1):090007, SOLARPACES 2020: 26th International Conference on Concentrating Solar Power and Chemical Energy Systems

Najjar YS, Zaamout MS (1992) Comparison of single and twin-shaft solar gas turbine systems. Int J Ambient Energy 13:103–111

Okoroigwe E, Madhlopa A (2016) An integrated combined cycle system driven by a solar tower: a review. Renew Sustain Energy Rev 57:337–350

Olivenza-León D, Medina A, Hernández A (2015) Thermodynamic modeling of a hybrid solar gas-turbine power plant. Energy Convers Manag 93:435–447

Rajaei A, Barzegar A, Eslami E (2016) Exergy and environmental impact assessment between solar powered gas turbine and conventional gas turbine power plant. Int J Chem Eng 2016.

Rogers G, Mahew Y (1992) Engineering thermodynamics: work and heat transfer, 4th edn. Longman, Essex

Santos MJ, Merchán RP, Medina A, Hernández AC (2016) Seasonal thermodynamic prediction of the performance of a hybrid solar gas-turbine power plant. Energy Convers Manag 115:89–102

Schwarzbözl P, Buck R, Sugarmen C, Ring A, Crespo MJ, Altwegg P, Enrile J (2006) Solar gas turbine systems: design, cost and perspectives. Sol Energy 80:1231–1240

Wang H, Li XS, Ren X, Gu C, Ji XX (2017) A thermodynamic-cycle performance analysis method and application on a three-shaft gas turbine. Appl Therm Eng 127:465–472

Chapter 7
Solar Gas Turbine Design

7.1 Introduction

The development of an engineering process or product involves designing and testing stages. Often, a new product or process is designed, tested, modified and re-tested until a satisfactory final product is achieved (Grove and Davis 1992). Design is a process in which a plan is formulated to meet a need or to find a solution to a specific problem. It is necessary that if the plan is aimed at yielding a product, then the product should be functional, safe, reliable, competitive, usable, manufacturable and marketable (Budynas and Nisbett 2011). The process of designing a gas turbine is done in phases: (a) specification, (b) preliminary studies (choice of cycle, type of turbomachinery and layout), (c) studies on thermodynamic design point, (d) aerodynamics of compressor, turbine, intake, exhaust and other, (e) mechanical design, (f) detailed design and manufacture, (g) testing and development, and (h) production presentation. These phases are iterative and inter-related as shown in Fig. 7.1.

7.2 Product Specification

It is necessary to start the design process by identifying the requirements of customers and a potential market. The potential market can be established through a market research, which is a process of acquiring and analysing data on the role companies have in the marketplace, determining potential customers/consumers, and evaluating the outlook on new products (Frazier and Yang 2023). This process is useful in making decisions about the introduction of new product lines.

A. Madhlopa and R. Nkhoma, *Principles of Solar Gas Turbines for Generating Electricity*, Green Energy and Technology,
https://doi.org/10.1007/978-3-031-84096-8_7

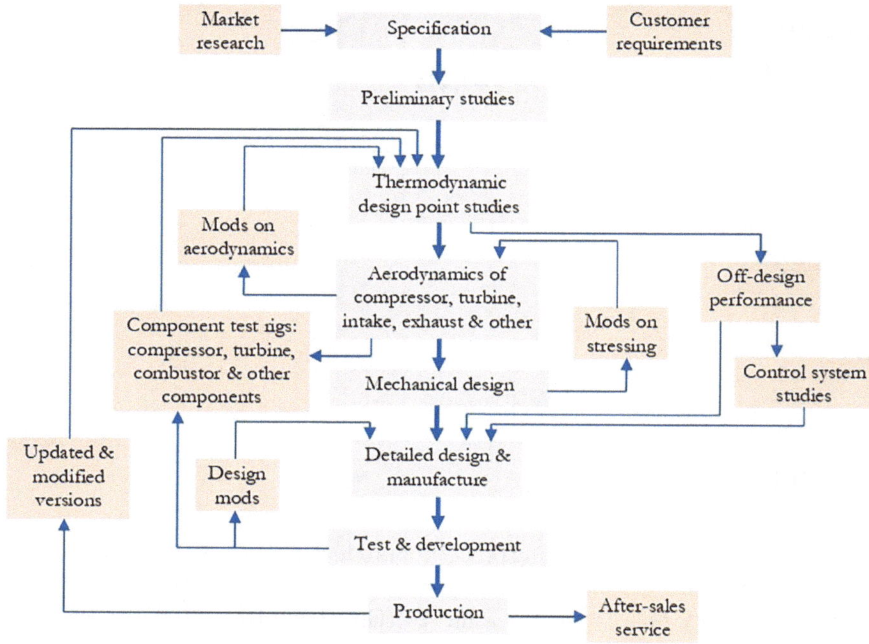

Fig. 7.1 Flowchart showing design phases and iterative links. *Mods* modifications. Adapted from Budynas and Nisbett (2011)

7.2.1 Customer Requirements

Requirements of customers may vary according to region and other factors. For example, the prevailing climate in a given locality may favour a certain engine design. Solar radiation and ambient temperature and pressure are influential environmental factors on the performance of solar gas turbines. So, knowledge about this is important for designing a competitive product on the market.

Specification of a gas turbine includes required power, efficiency, weight, cost, volume, life and noise (Cohen et al. 1996). Some of these factors vary in opposition to one another. A high level of efficiency, for instance, is associated with large investment. Thus, a suitable design can be achieved through the process of system optimization (see Chap. 8).

7.2.1.1 Market Potential of Solar Gas Turbines

A solar gas turbine system utilizes concentrated solar radiation that fluctuates with time, thereby reducing the capability of the system. To increase capability, solar gas turbines can be integrated with thermal energy storage (TES) or hybridized with a conventional heater (Schwarzböзl et al. 2006; Wagner et al. 2018). The TES unit gives concentrating solar power an advantage over other energy systems that are limited to the deployment of batteries to store energy (Kuravi et al. 2013). Integration of concentrating solar power with thermal energy storage provides the potential of full dispatchability of bulk power (Pfenninger et al. 2014). In view of this, recent CSP projects in different parts of the world are equipped with thermal storage units (National Renewable Energy Laboaratory 2021). Integration of thermal energy storage augments the dispatchability of power with negligible or no increase over the levelized cost of electricity (LCOE) of a plant without thermal energy storage (Mehos et al. 2016; Lilliestam et al. 2018).

Hybrid solar gas turbines have a mixture of solar and conventional heaters. It is possible to achieve solar shares of 40–90% in the design and annual solar shares of up to 30% can be achieved in base load (Schwarzbözl et al. 2006). Moreover, the use of modern gas turbine systems in recuperation or combined cycle mode augments conversion efficiencies of the solar heat from about 40% to higher than 50%, which can contribute to cost reduction of solar thermal power. Compared with solar-only power plants, the advantages of hybrid systems are: (a) higher system efficiency due to diminished part load operation and a reduction in losses arising from start-ups and shutdowns, (b) low additional investment because of an adaptable solar share and (c) low technical and economical risks because of supplying fully dispatchable power. In view of the various advantages, solar gas turbine systems have an enormous potential for introduction into the market. However, manufacturers of gas turbines are hesitant to embark on these developments because of market uncertainty. The existing solar gas turbines, predominantly for research and demonstration, are currently developed by modifying gas turbine engines designed for operation with conventional fuels.

The conversion of a conventional gas turbine into a solar gas turbine appears theoretically simple but it is practically challenging. Design assumptions about gas turbines are based on the existing models of conventional gas turbines. Modifications always exist, which have to be considered during the design, analysis and construction stages, regardless of whether a new solar gas turbine plant is to be built or an existing conventional gas turbine plant is to accommodate solar integration. The required changes to the system are often related to interfacing with external heating of the air, modification of the combustor, component cooling and system control.

7.2.2 Preliminary Studies

A preliminary study is an initial exploration of issues involved in a process. It aims at identifying key features that require addressing. The design process involves multiple issues that need initial investigation, once specifications have been established. It is necessary to choose suitable engine cycle, system components and engine layout.

7.2.2.1 Engine Cycle and Components

As reported in Chap. 5, there are many possible power cycles (single or combined cycles) at the disposal of the designer of a solar gas turbine. These cycles have different thermodynamic characteristics. So, the designer has to choose the most suitable cycle for the application at hand.

It is also important to choose suitable components of the engine (such as type of compressor, solar concentrator, receiver, combustor and turbine). Compressors are classified into axial and centrifugal types. Selection of a compressor type is based on gas flow, and inlet and discharge pressures. Consideration needs also to be given to the use of single or multiple stages of compression. As regards components for the solar field, solar concentrators are classified into parabolic trough, linear Fresnel, parabolic dish and solar tower (see Chap. 1). Solar gas turbines of the hybrid variety have a backup heater that is usually in the form of a combustor. To this end, there are different types of combustors. In addition, axial-flow and radial-flow turbines are available for possible consideration during engine design. Again, a suitable type of turbine needs to be established in the preliminary phase of the design process. For electric generation (usually, alternating current), an electric generator is integrated to the gas turbine. Alternating current (AC) generators are classified into rotating-armature and rotating-field types. Rotating-armature AC generators are suitable for low power generation. The type of rotor used in the generator affects the speed at which the generator can be operated. Rotating-field AC generators with turbine-driven rotors are suitable for high speeds as encountered in solar gas turbines. Consequently, the type of generator needs also to be considered when designing a solar gas turbine. Details about the major components of solar gas turbines are reported in Chap. 4. The various components are connected together to form an engine layout.

7.2.2.2 Engine Layout

Components of a gas turbine can be linked in many different ways to obtain an engine layout. It is also possible to consider using a multiple number of components (say, two compressors on one shaft or two different shafts) in a gas turbine system. Various configurations of solar gas turbines, though not exhaustive, have been reported in Chap. 6.

Some of the important factors considered when selecting the engine layout for a given application are: efficiency, structural weight, economic sustainability and environmental sustainability. An unsustainable engine layout cannot be implemented even if it is efficient and light-weight. An optimal combination of these and other critical factors is preferred.

7.2.3 *Thermodynamic Design Studies*

Analysis of the thermodynamic performance of an engine layout is the first major design step. In this step, comprehensive calculations are performed by taking into account of all essential factors such as fluid properties, expected efficiencies of components, air-bleeds and pressure losses. For different values of the cycle parameters listed above, a value for the specific output and specific input of solar energy is computed (inputs for hybrid systems are solar and fuel energies). These calculations are often done on a digital computer using software that enables fast and accurate computation (see Chap. 8). Simulation of the thermodynamic performance of the gas turbine can be performed using software such as Thermoflex and other software (Barigozzi et al. 2012). Based on cycle parameters, the specific output is used to find the airflow required to give the specified magnitude of power.

The choice of cycle parameters is predominantly affected by the engine size, and, in particular, by the air flow required. For instance, a small engine (say, capacity of 500 kW) has very small blades that cannot be cooled due to manufacturing difficulty and cost. The pressure ratio of such an engine would be limited to permit fabrication of blades with a realistic size, and it might require the exploitation of a centrifugal compressor of rather low efficiency. Absence of blade cooling system would restrict the turbine inlet temperature to about 1200 K. In contrast, complex air-cooled blades can be deployed in a large engine (say, capacity of 50 MW) and the engine can be operated at a high turbine inlet temperature (> 1500 K). The 50 MW engine can also use axial compressors with a high pressure ratio.

7.2.3.1 Fluid Properties

A solar gas turbine utilizes heat transfer fluids to distribute heat within the system through convection. Internal flow of heat transfer fluids is achieved by using compressors/pumps that yield forced convention. Heat may be transferred from a hot solid surface to a cold fluid or from a hot fluid to a cold surface. The rate of heat convection $(\dot{Q}_c)$ can generally be expressed as:

$$\dot{Q}_c = h_c(T_2 - T_1) \tag{7.1}$$

where h_c is the coefficient of heat transfer, T_1 is the temperature of the cold fluid or surface and T_2 is the temperature of the hot fluid or surface. For a given specific length and mass flow, the coefficient of forced heat convection depends the thermal conductivity, dynamic viscosity and specific heat capacity (see Chap. 1). The coefficient of heat transfer increases with the thermal conductivity and specific heat capacity of the heat transfer fluid is critical for the augmentation of the heat transfer coefficient. Thus, good properties promote moderate temperature difference between the receiving surface and the heat transfer fluid, thereby permitting the fluid to attain elevated temperatures without creating excessive thermal stresses on the surface.

For solar gas turbine systems with thermal storage, heat storage is achieved by using sensible or latent heat as shown in Chap. 4. The rates of sensible and latent heats increase with the specific heat capacity and the specific latent heat of a storage medium. Even though indirect methods of thermal storage with a secondary medium are available, a high level of the specific heat capacity is desirable to permit direct storage of energy. The rate of heat evacuation by a fluid increases with the mass flow rate. Consequently, density is an import property of the heat transfer fluid. Fluids with high density and thermal conductivity are preferred. Pumping power increases with the viscosity of a fluid. Consequently, a low viscosity is favourable to decrease pressure drop and pumping power.

7.2.3.2 Component Efficiency

A basic solar gas turbine for electricity generation is made up of a solar field (concentrator and receiver) and a power block (compressor, turbine and electric generator). It captures and converts solar radiation to heat, and the generated heat is converted to mechanical energy that drives the generator. Consequently, the overall system efficiency depends on the individual efficiencies of the concentrator, receiver, compressor, turbine and generator (see Chap. 5). In case of the hybrid system, the working fluid is further heated by a conventional heater that has also limited efficiency.

For the basic solar gas turbine, the overall efficiency (η_e) of conversion of solar energy to electricity is given by (Bellos et al. 2017):

$$\eta_e = \frac{\dot{Q}_e}{\dot{Q}_s} = \frac{\eta_{gen}\left(\eta_{is,tu}\dot{W}_{is,tu} - \frac{\dot{W}_{is,com}}{\eta_{is,com}}\right)}{\dot{Q}_s} \quad (7.2)$$

where $\dot{Q}_e$ is electrical power production and $\dot{Q}_s = \left(A_{ap}I_b\right)$ is the solar energy input to the concentrator, A_{ap} is the aperture area of the concentrator, I_b is the beam irradiance intercepted by the concentrator, $\dot{W}_{is,com}$ is the isentropic input work to the compressor, $\dot{W}_{is,tu}$ is the isentropic output work from the turbine, η_{gen} is the efficiency of the electric generator, $\eta_{is,com}$ is the isentropic efficiency of the compressor and $\eta_{is,tu}$ is the isentropic efficiency of the turbine. Computation of the compressor and turbine isentropic works is reported in Chap. 5.

7.2.4 Aerodynamic Design

Analysis of thermodynamic parameters gives information on the flow of the working fluid, pressure ratio and turbine inlet temperature which can be used in the aerodynamic design of the system (Cohen et al. 1996). Aerodynamics deals with fluid flows through ducts or over solid bodies (Fig. 7.2).

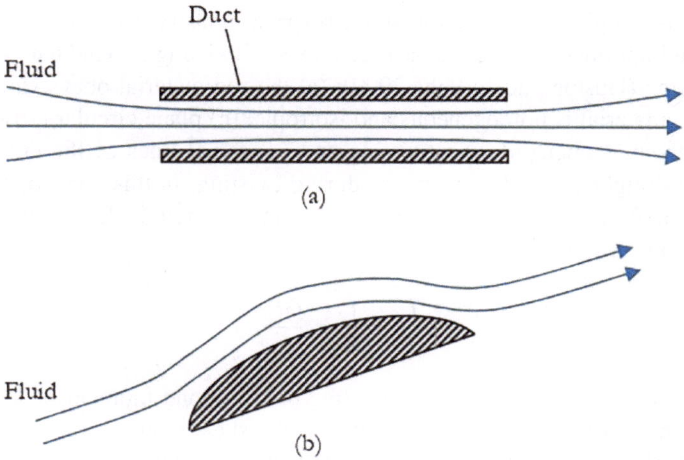

Fig. 7.2 Schematic representation of flow pattern: **a** through a duct and **b** over a surface

7.2.4.1 Fluid Flow

A fluid may flow in an orderly or chaotic manner. An orderly flow is known as laminar flow while a disorderly flow is called turbulent flow. Flow may be over a solid body or through a conduit bounded by solid surfaces. Fluid flow over a body is known as external flow, and in this case, the fluid wholly or partially surrounds the solid body. Internal flows are completely bounded by surfaces of a solid body, and they include flows through pipes, contractions and expansions, valves, and fittings. Flows of working and/or heat transfer fluids through solar receivers are generally internal. Characteristics of fluid flow are used when modelling the dynamics of fluids. This topic is covered in more details in Chap. 8.

7.2.4.2 Shaft Work

Computation of the annulus dimensions, rotational speeds and number of stages can be performed by using the data from the thermodynamic analysis. For example, the geometries and sizes of components are calculated based on thermodynamic data. The isentropic shaft work can be related to the torque produced on the shaft. Thus, the net shaft work (W_{net}) is equal to the shaft torque multiplied by its rotational speed:

$$W_{\text{net}} = T\omega \tag{7.3}$$

where T is torque and ω is rotational speed of the shaft.

Shafts can be solid or hollow, and they may undergo twisting when subjected to twisting moments. Benefits of using hollow shafts include reduction of weight, low rotational inertia, high frequency of resonance, high heat resistance, and low cost. In some applications, such as the aerospace industry, weight reduction is of paramount importance. To derive basic relations of twisting a round bar, we assume the following (Huston and Josephs 2009): (a) the rod material obeys Hooke's law, (b) the rod material is homogeneous and isotropic, (c) plane circular cross sections maintain their geometry during twisting and (d) radial lines of the cross section keep their straight and radial geometry during twisting. In this context, for a solid or hollow shaft of uniform circular cross-section throughout its length, the theory of pure torsion states that:

$$\frac{T}{J} = \frac{\tau}{R} = \frac{G\theta}{L} \tag{7.4}$$

where G is the modulus of rigidity, J is the polar second moment of shaft cross-section area, L is the length of the shaft, τ is the shear stress at radius R and θ is the twist angle. It should be noted that for a given applied torque, the shear stress is zero at $R = 0$ and maximum at $R = R_{\max}$. The polar second moment of area of shaft cross-section is given by:

$$J = \begin{cases} \pi D^4/32, & \text{for a solid shaft} \\ \pi\left(D^4 - d^4\right)/32, & \text{for a hollow shaft} \end{cases} \tag{7.5}$$

where d is the internal diameter and D is the external diameter. However, shear stress and shear strain are related to the angle of twist:

$$\tau = \frac{\mathrm{GR}\theta}{\mathrm{L}} = \mathrm{G}\gamma, \tag{7.6}$$

where γ is the shear strain.

As seen from Eq. (7.4), the shear stress is a function of twist angle. So, we can use this equation to derive an expression for the twisting torque T in terms of the twist angle θ:

$$\mathrm{T} = \frac{\mathrm{G}\theta}{\mathrm{JL}} \tag{7.7}$$

Under transient conditions such as change in compression ratio, fuel mass flow rate, start up and shut down conditions, there is a change in the net power output of the system. This causes fluctuations in the shaft behaviour such as the torque produced.

$$\frac{\mathrm{d}\dot{W}_{\mathrm{net}}}{\mathrm{d}t} = \frac{\mathrm{d}(T\omega)}{\mathrm{d}t} = \omega\frac{\mathrm{d}T}{\mathrm{d}t} + T\frac{\mathrm{d}\omega}{\mathrm{d}t} \tag{7.8}$$

However, torque is a product of the moment of inertia and angular acceleration. Consequently, Eq. (7.8) can be expressed as:

$$\frac{\mathrm{d}(T\omega)}{\mathrm{d}t} = \omega J \frac{\mathrm{d}\Omega}{\mathrm{d}t} + J\Omega^2 \quad (7.9)$$

where J is the inertia and Ω is the angular acceleration of the shaft.

7.2.4.3 Model of Gas Turbine Flow Characteristics

Flow characteristics are often modelled by taking into consideration the governing equations of a given flow, namely: continuity, momentum and energy equations (Anderson 1991). These equations can be expressed in integral or differential forms but the latter form is widely applied because it allows the investigation of the flow field in more detail.

The components of a solar gas turbine have different geometries and functions. So, it is common to model the flow characteristics of each component by using computational fluid dynamics. Details about computational fluid dynamics are covered in Chap. 8.

7.2.4.4 Aerodynamic Losses

Aerodynamic losses in gas turbines are classified into three groups (Ligrani et al. 2017): profile, tip-clearance and end-wall losses. A blade or vane profile generates profile losses that form on the aerofoil surface because of the build-up of boundary layers. Tip leakage losses are caused by the pressure gradient created over the blade tip between the pressure and suction sides of the blade, and often arise in rotors. End-wall losses are caused by viscous effects from the end-wall and the interaction of the end-wall boundary layers with the aerofoils. These viscous effects give rise to secondary flows.

Some challenges may arise when data from the thermodynamic design point does not yield realistic values of aerodynamic parameters. In this case, it may be necessary for experts of aerodynamics and thermodynamics to consult. It is important for the aerodynamic design to take into consideration the viability of manufacturing the system. For example, it may not be practically possible to machine small turbine blades with conduits for cooling (Boyce 2002).

7.2.5 Mechanical Design

Designing of the mechanical system can commence when the progress in thermodynamic and aerodynamic designs are in advanced stage. This allows for better iteration when problems arise in the subsequent design stage. Some important factors to consider in the mechanical design are strength and reliability. It is possible that thermodynamic and aerodynamic analyses may yield a component that cannot withstand the anticipated stress. In addition, vibration of components can lead to fatigue and failure of components or the whole engine structure.

Forces and moments act on different parts of the gas turbine system. For instance, a rotating turbine exerts torque on the engine shaft, and the weight of the turbine produces a bending moment on the shaft. Figure 7.3 is a simplified illustration of a turbine and shaft with bearings resting on two supports. It is noted that there is loading on the bearings and the turbine weight creates deflection of the shaft (Fig. 7.4).

Forces acting on components of the gas turbine result in stresses (tensile, compressive or shear). It should also be noted that heat causes thermal expansion of structures, leading to thermal stress and structural failure. Within elastic limits, the ratio of stress to strain is known as the Young's modulus (Hearn 1997), and the material strength increases with the magnitude of this parameter. Every structural component is fit for a given purpose for only a limited time (Fekete 2023). As the time of service increases, load-bearing components lose their strength to support the same magnitudes of loads as when they were new. Consequently, it is necessary to analyse all

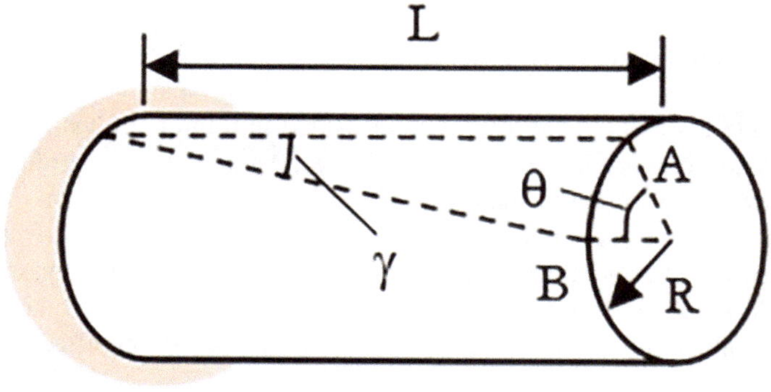

Fig. 7.3 Simplified shaft twisted through an angle θ

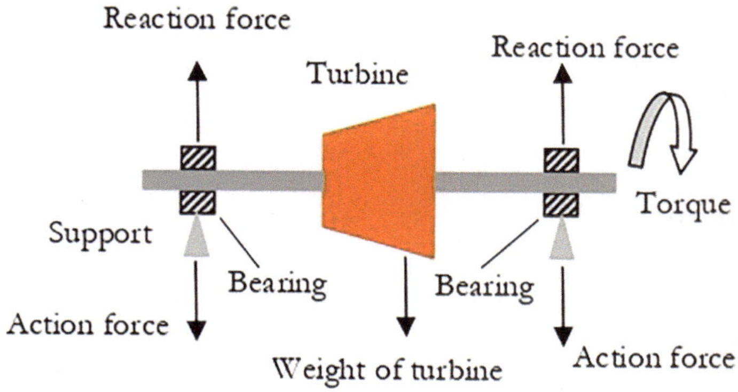

Fig. 7.4 Simplified illustration of possible forces and moments acting on a turbine and bearings

the major loads and stresses that act on the structural components and the whole gas turbine to establish the safety and useful lifetime of the system.

7.2.5.1 Dynamics of Gas Turbines

Dynamics is a branch of mechanics that deals with the motion of a body when forces act on it. When the hot gases expand through the turbine they cause rotational motion of the machine. Hence, a solar gas turbine falls into the category of rotating machines. At present, the trend in gas turbine design is towards increasing the rotational speeds, which apparently is also augmenting the operational challenges pertaining to vibration (Boyce 2002). Vibrating systems can be classified into free and forced categories. A free system vibrates under forces that are intrinsic to the system, and it vibrates at one or more natural frequencies. Forced vibration is caused by an external force, and it takes place at the frequency of the exciting force. The amplitude of vibration is influenced by the frequency of the exciting force, with the largest amplitude observed when the frequency of the applied force is equal (or close) to the natural frequency of the body. At this point, the body undergoes resonant vibration. For example, turbine blades undergo vibration when the gas turbine engine is running. It is reported that up to 42% of gas turbine failures are caused by vibration-induced fatigue (Madhavan et al. 2014). Unfortunately, blade vibration cannot be eliminated from a gas turbine engine but good blade design can assist in mitigating the forces that incite vibration of concern.

The engine layout also affects the natural frequency of the system and therefore the angular speed at which resonance occurs. For instance, Maia et al. (2017) studied various topologies of an electric generator driven by a micro turbine. In this study, prototypes based on topologies shown in Fig. 7.5a and c were fabricated and tested. One shaft was used on the first prototype which was based on the topology shown in Fig. 7.5a while two shafts were loosely coupled in the second prototype (based on the topology in Fig. 7.5c). They found that the angular speed of the first prototype could run up to 100,000 revolutions per minute (rpm). On the other hand, the angular speed of the second prototype could not exceed 50,000 rpm because of excessive vibration. Attempts to align and safely operate the system were not successful. The recommendation was that further work needs to focus on finding a solution for misalignment.

The electric generator converts rotational mechanical energy from the turbine shaft to electrical energy. This component may be mounted on the same shaft with the turbine or connected to the turbine through a mechanical gearbox (Fig. 7.6). The use of a gearbox is economically attractive but it entails the following major disadvantages: (a) increase of fire risk resulting from the use of gearbox lubrication oil, (b) requirement of more protection, (c) heavy weight and (d) excessive levels of noise (Benaboud and Rufer 2016). In view of these drawbacks, power electronics can give the ability to operate with high levels of power, and it augments the efficiency of the gas turbine by employing variable speed (Stemmler 1994; Benaboud and Rufer 2016). Power electronics deals with electric power, electronics and control systems.

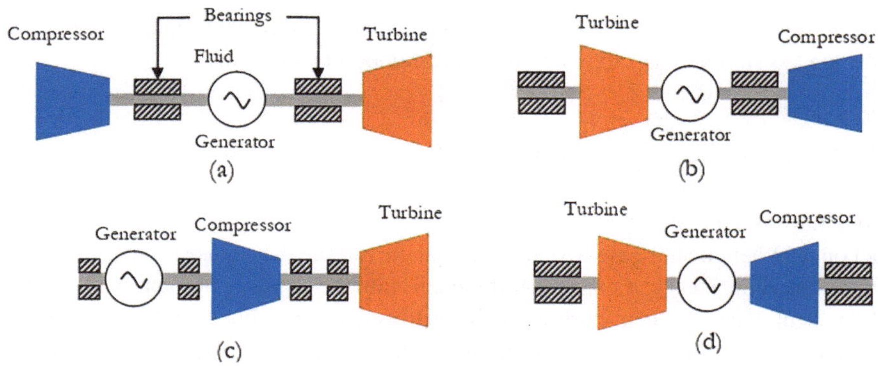

Fig. 7.5 Various configurations of single-shaft gas turbine with generator

Electric power is concerned with static and rotating power devices for generation, transmission and distribution of electric power (Zhang et al. 2018). Electronics jointly deals with solid-state semiconductor devices and circuits with control systems for conversion of electric power. Most of the electric power is converted by using power electronics before it reaches the consumer. In this context, there are four types of power electronics converters: AC–AC, AC–DC, DC–AC and DC–DC that find application in many fields including utilities. The design of a power electronics converter aims at maximizing the efficiency of conversion.

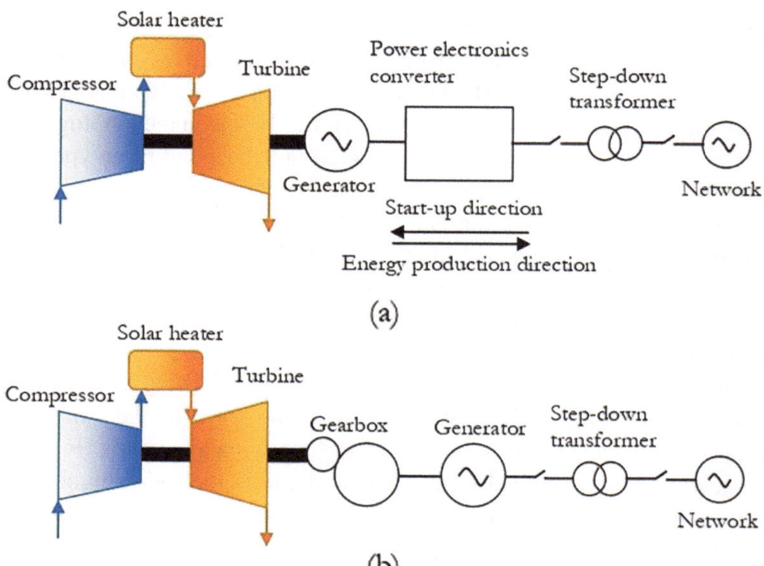

Fig. 7.6 Schematic representation of a solar gas turbine coupling with generator: **a** shafted generator and **b** geared generator

7.3 Design of Control Systems

A gas turbine is a very nonlinear engine that involves multiple inputs and outputs (de Silva et al. 2017). In addition, operating conditions cover extreme temperatures, pressures and loads. As seen in the foregoing discussion, multivariate interactions (thermodynamic, aerodynamic and mechanical variables) occur in the solar gas turbine plant. Some of these interactions are very complicated. Consequently, the plant requires complex multivariate controllers that can attain maximum performance levels while maintaining the stability and safety of operation at minimum overall cost. The three main functions of a gas turbine control system are (Boyce 2002): (a) sequencing of start-ups and shutdowns, (b) controlling the operation of the gas turbine when it is in steady-state conditions and (c) protecting the gas turbine from damage. It is therefore not surprising that all gas turbines are supplied with a control system.

The design of complex multivariate control systems is not easy. In this vein, an optimization approach is commonly applied in designing gas turbine control systems. Optimization aims at maximizing or minimizing a given function relative to a set of available choices under a specified situation. The function being optimized is referred to as the objective function (Cui et al. 2017). Choices may include a number of variables with their feasible ranges. Thus, it is important to answer a number of questions in order to formulate an optimization problem, and some of these questions may pertain to components and processes (Bejan et al. 1996). The problem may comprise one or more objective functions. Details about optimization of solar gas turbines are reported in Chap. 8.

7.3.1 Worked Examples

Example 7.1 A basic solar gas turbine for generating electricity is made up of a reflecting concentrator, solar receiver, compressor, turbine and electric generator. At a certain time of the day, the beam irradiance on the concentrator (with an area of 15,000 m^2 and reflectance of 0.90) is 800 W/m^2. The concentrator reflects part of the irradiance onto the receiver, and 0.85 of the reflected irradiance is absorbed and transferred to the working fluid in the receiver. If the isentropic efficiency of the turbine is 0.60 and the isentropic compressor work is 220 kW, calculate the overall efficiency of the system, assuming the efficiencies of the compressor and generator are respectively 0.75 and 0.80, and all the reflected solar radiation is absorbed and converted to heat by the receiver.

Solution

From Eq. (7.2), the overall efficiency (η_e) is:

$$\eta_e = \frac{\eta_{gen}\left(\eta_{is,tu}\dot{W}_{is,tu} - \frac{\dot{W}_{is,com}}{\eta_{is,com}}\right)}{\dot{Q}_s}$$

Useful solar energy converted to heat is theoretical heat added to the working fluid by the receiver. This hot fluid expands through the turbine, thereby producing turbine isentropic work ($\dot{Q}_{is,tu}$). So,

$$\begin{aligned}\dot{Q}_{is,tu} &= (\alpha\rho)\dot{Q}_s = (\alpha\rho)A_{ap}G_b \\ &= 0.85 \times 0.90 \times 15{,}000 \times 800 \\ &= 9{,}180{,}000\ \text{W} \\ \dot{Q}_s &= A_{ap}G_b = 15{,}000 \times 800 \\ &= 12{,}000{,}000\ \text{W}\end{aligned}$$

This implies that:

$$\begin{aligned}\eta_e &= \frac{0.80 \times \left(0.60 \times 9{,}180{,}000 - \frac{220{,}000}{0.75}\right)}{12{,}000{,}000} \\ &= 0.35\end{aligned}$$

Example 7.2 A turbine is mounted on a solid shaft of diameter 100 mm. The shaft transmits 80 kW at 200 revolutions per minute. Find the maximum shear stress set up in the shaft.

Solution

From Eq. (7.3), torque (T) is:

$$\begin{aligned}T &= \frac{W}{\omega} = \frac{80{,}000}{200 \times 2\pi/60} \\ &= 3819.72\ \text{Nm}\end{aligned}$$

The maximum shear stress can be calculated from Eq. (7.4).

$$\tau_{max} = \frac{TR_{max}}{J} \text{ and } J = \frac{\pi}{32}(0.1)^4 = 9.81748 \times 10^{-6}\ \text{m}^4.$$

Therefore,

$$\begin{aligned}\tau &= \frac{3819.72 \times 0.05}{9.81748 \times 10^{-6}} \\ &= 19.5\ \text{MN/m}^2\end{aligned}$$

Example 7.3 A hollow transmission shaft is 500 mm long, 100 mm external diameter and 60 mm internal diameter. Find the maximum power that may be transmitted at a speed of 200 revolutions per minute if the maximum shear stress is 80 MN/m^2.

Solution

Torque, $T = \frac{I\tau_{\max}}{R_{\max}}$ and $J = \frac{\pi(D^4-d^4)}{32} = \frac{\pi(0.1^4-0.06^4)}{32} = 8.54513 \times 10^{-6}$ m^4.

Therefore,

$$\begin{aligned} T &= \frac{8.54513 \times 10^{-6} \times 80{,}000{,}000}{0.05} \\ &= 13{,}672.21 \text{ Nm} \end{aligned}$$

Maximum power ($\dot{W}_{\max}$)

$$\begin{aligned} \dot{W}_{\max_{\max}} &= T\omega \\ &= 13{,}672.21 \times (200 \times 2 \times \pi/60) \\ &= 286.4 \text{ kW} \end{aligned}$$

Example 7.4 The turbine in Example 7.1 is mounted on a circular solid shaft of length 500 mm and diameter 90 mm. If the shaft rotates at 250 revolutions per minute, calculate the twist angle assuming a rigidity modulus of 240 GN/m^2.

Solution

From Eq. (7.2), net shaft work $\left(\dot{W}_{\text{net}}\right)$,

$$\begin{aligned} \dot{W}_{\text{net}} &= \eta_{\text{is,tu}} \dot{W}_{\text{is,tu}} - \frac{\dot{W}_{\text{is,com}}}{\eta_{\text{is,com}}} \\ &= 0.60 \times 9{,}180{,}000 - \frac{220{,}000}{0.75} \\ &= 5{,}214{,}666.667 \text{ W} \\ T &= \frac{\dot{W}_{\text{net}}}{\omega} = \frac{5{,}214{,}666.667}{250 \times 2 \times \pi/60} \\ &= 199{,}185.6 \text{ Nm} \end{aligned}$$

The twist angle can be computed using Eq. (7.4).

$$\theta = \frac{\text{TL}}{\text{JG}} \text{ and } \text{J} = \frac{\pi \text{D}^4}{32} = \frac{\pi \times 0.09^4}{32} = 6.44125 \times 10^{-6} \text{m}^4$$

Therefore,

$$\theta = \frac{199{,}185.6 \times 0.5}{6.44125 \times 10^{-6} \times 240 \times 10^{9}}$$
$$= 0.06 \text{ radian}$$

Example 7.5 A solar gas turbine power plant comprises a reflecting concentrator, a solar receiver, a compressor, a turbine and an electric generator. At a certain time of the day, the turbine develops 2 MW of actual work. If the net shaft work is 1.0 MW and the compressor isentropic efficiency is 0.70. Calculate the compressor isentropic work.

Solution

From Eq. (7.2), the compressor isentropic work ($\dot{W}_{\text{is,com}}$) can be expressed as:

$$\dot{W}_{\text{is,com}} = \eta_{\text{com,is}}\eta_{\text{tu,is}}\dot{W}_{\text{tu,is}} - \eta_{\text{com,is}}\dot{W}_{\text{net}}$$

But, $\eta_{\text{tu,is}}\dot{W}_{\text{tu,is}} = \dot{W}_{\text{tu,a}}$.

This implies that,

$$\dot{W}_{\text{is,com}} = 0.70 \times 2 - 0.70 \times 1$$
$$= 0.7\,\text{MW}$$

Nomenclature

A	Area (m^2)
d	Internal diameter (m)
D	External diameter (m)
G	Modulus of rigidity (N/m^2)
h	Coefficient of heat transfer ($W/m^2{\cdot}K$)
I	Irradiance (W/m^2)
J	Polar second moment (m^4)
L	Length (m)
R	Radius (m)
t	Time (s)
T_1	Temperature of cold fluid or surface (K)
T_2	Temperature of hot fluid or surface (K)
T	Torque (N m)
$\dot{Q}$	Power or heat rate (W)
$\dot{W}$	Work (W)

Greek Symbols

γ Shear strain (dimensionless)
η Efficiency (dimensionless)
θ Twist angle (rad)
τ Shear stress (N/m^2)
ω Angular speed (rad/s)
Ω Angular acceleration (rad/s^2)

Subscripts

ap Aperture
c Convection
com Compressor
e Electric/electricity
is Isentropic
s Solar
tu Turbine

References

Anderson J (1991) Fundamentals of aerodynamic. McGraw-Hill Inc., New York, NY

Barigozzi G, Bonetti G, Franchini G, Perdich A, Perdichizzi A, Ravelli S (2012) Thermal performance prediction of a solar hybrid gas turbine. Sol Energy 86:2116–2127

Bejan A, Tsatsaronis G, Moran M (1996) Thermal design and optimization. Wiley, New York, NY

Bellos E, Tzivanidis C, Antonopoulos KA (2017) Parametric analysis and optimization of a solar assisted gas turbine. Energy Convers Manag 139:151–165

Benaboud A, Rufer A (2016) Gas turbine: optimization of energy production and high efficiency by using power electronics. Proc Eng 138:337–346

Boyce MP (2002) Gas turbine engineering handbook, 2nd edn. Gulf Professional Publishing, Boston

Budynas RG, Nisbett JK (2011) Shingley's mechanical engineering design, 9th edn. McGraw Hill, Singapore

Cohen H, Rogers GF, Saravanamuttoo HI (1996) Gas turbine theory, 4th edn. Longman Group Limited, Essex

Cui Y, Geng Z, Zhu Q, Han Y (2017) Review: multi-objective optimization methods and application in energy saving. Energy 125:681–704

e Silva VVR, Khatib W, Fleming PJ (2017) Control system design for a gas turbine engine using evolutionary computing for multidisciplinary optimization. Rev Controle Autom 18(4):471–478

Fekete T (2023) On fundamental issues of the structural integrity of large-scale pressure systems. Proc Struct Integr 46:189–196

Frazier AN, Yang H (2023) Utilizing a CRISPR-Cas9 targeted foodborne pathogen antimicrobial within livestock feed: a market research perspective. J Agric Food Res 14:100758

Grove DM, Davis TP (1992) Engineering, quality and experimental design. Longman Scientific and Technical, Essex

Hearn E (1997) Mechanics of materials I: an Introduction to the mechanics of elastic and plastic deformation of solids and structural materials. Butterworth-Heinemann, Oxford

Huston R, Josephs H (2009) Practical stress snalysis in engineering design. Taylor & Francis Group, Boca Raton

Kuravi S, Trahan J, Goswami DY, Rahman MM, Stefanakos EK (2013) Thermal energy storage technologies and systems for concentrating solar power plants. Prog Energy Combust Sci 39(4):285–319

Ligrani P, Potts G, Fatemi A (2017) Endwall aerodynamic losses from turbine components within gas turbine engines. Propul Pow Res 6(1):1–14

Lilliestam J, Barradi T, Caldés N, Gomez M, Hanger S, Kern J, Komendantova N, Mehos M, Hong WM, Wang Z, Patt A (2018) Policies to keep and expand the option of concentrating solar power for dispatchable renewable electricity. Energy Policy 116:193–197

Madhavan S, Jain R, Sujatha C, Sekhar AS (2014) Vibration based damage detection of rotorblades in a gas turbine engine. Eng Fail Anal 46:26–39

Maia TA, Faria OA, Barros JE, Porto MP, Filho BJ (2017) Test and simulation of an electric generator driven by a micro-turbine. Electr Pow Syst Res 147:224–232

Mehos M, Turchi C, Jorgenson J, Denholm P, Ho C, Armijo K (2016) On the path to SunShot: advancing concentrating solar power technology, performance, and dispatchability. National Renewable Energy Laboratory, Golden

National Renewable Energy Laboaratory (2021) Concentrating Solar Power Projects. https://solarpaces.nrel.gov

Pfenninger S, Gauché P, Lilliestam J, Damerau K, Wagner F, Patt A (2014) Potential for concentrating solar power to provide baseload and dispatchable power. Nat Clim Change 4(8):689–692

Schwarzbözl P, Buck R, Sugarmen C, Ring A, Crespo MJ, Altwegg P, Enrile J (2006) Solar gasturbine systems: design, cost and perspectives. Sol Energy 80:1231–1240

Stemmler H (1994) High-power industrial drives. Institute of Electrical and Electronics Engineers, New Jersey

Wagner MJ, Hamilton WT, Newman A, Dent J, Diep C, Braun R (2018) Optimizing dispatch for a concentrated solar power tower. Sol Energy 174:1198–1211

Zhang G, Zhang B, Halang WA (2018) Power electronics converters: past, present and future. Renew Sustain Energy Rev 81:2028–2044

Chapter 8
Simulation of Solar Gas Turbines

8.1 Introduction

A simulation is the imitation of a process or product that could be present in the real world. In this context, engineering simulation uses computer software to analyse, predict, and visualize the effect of processes and the behaviour of product designs while those products and systems are being developed. Simulation is a powerful tool in engineering, enabling the virtual modelling and analysis of complex systems like solar gas turbines. It involves creating mathematical representations of physical products and processes to predict system behaviour under various conditions. This approach allows engineers and scientists to optimize designs, test scenarios, and identify potential challenges without the cost and time associated with physical prototypes. For solar gas turbines, simulations are particularly crucial in understanding the interactions among components such as compressors, turbines, and solar receivers. By leveraging advanced computing techniques, simulations enhance efficiency, reliability, and innovation in renewable energy technologies, supporting sustainable development goals.

Various components of a solar gas turbine (SGT) are presented in Chap. 4. In this chapter, we provide an overview of the importance and objectives of simulation in the context of solar gas turbines. A basic solar gas turbine is presented in Fig. 8.1 for ease of reference. The complexity of an SGT technology increases with the number of components. These components (such as solar receiver, combustor, heat exchanger, and thermal storage units) are added to the basic solar gas turbine to improve its performance. The role of simulation is crucial in optimizing the design and performance of SGT systems. Simulation of solar gas turbines is a profound field of study, encapsulating dynamic principles, effective methodologies, and robust technical interventions to achieve an optimal efficiency in electricity generation. This chapter presents and discusses principles, applications, challenges, and prospects of simulation of solar gas turbines. The overarching objectives of the chapter include equipping readers with the knowledge to comprehend the complex mechanisms

A. Madhlopa and R. Nkhoma, *Principles of Solar Gas Turbines for Generating Electricity*, Green Energy and Technology,
https://doi.org/10.1007/978-3-031-84096-8_8

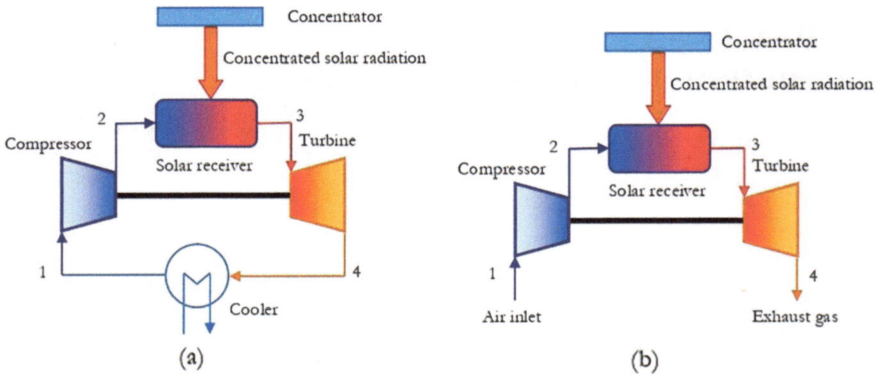

Fig. 8.1 A basic solar gas turbine: **a** closed cycle and **b** open cycle

involved, understand the benefits of these types of turbines, and envisage the transformative potential of solar energy in our pursuit of sustainable development (see Chaps. 10 and 11).

Simulation is a fundamental tool in the development and optimization of technological systems. By simulating different scenarios, engineers and scientists can conduct thorough analyses of solar gas turbines, identify potential issues, enhance design elements, and ultimately improve the overall efficiency and reliability. The use of advanced computing techniques for simulation does not only accelerate the development cycle and reduce costs but also ensures optimized design and performance of SGT systems across a variety of real-world conditions. One of the primary advantages of simulation is its ability to expedite the design process. Traditional methods, which heavily rely on physical prototypes and extensive field testing, can be both time-consuming and expensive. In contrast, simulation allows for virtual testing of design concepts, thereby accelerating the development cycle and reducing costs. As highlighted by Zhang et al. (2015), simulation facilitates the exploration of a broad range of design variables and configurations, leading to more efficient and effective turbine designs.

Each component of a solar gas turbine (including the solar concentrator, solar receiver, compressor, combustor, turbine, and generator) can be independently modelled and optimized. For instance, detailed optical and thermal simulations can maximize the efficiency of the solar field (solar concentrator and receiver) in capturing and converting solar radiation into thermal energy. Simulations also provide a comprehensive understanding of the interactions between various components within the gas turbine system. This holistic approach ensures optimal performance of the entire system. System-level simulations are crucial for identifying and resolving integration issues that may not be apparent when analysing individual components. Simulating thermodynamic processes enables engineers and scientists to optimize the energy conversion efficiency of the SGT system. This includes modelling the heat input from the solar receiver, the compression and expansion processes, and the overall thermal efficiency of the cycle (Bejan 2016). Accurate performance prediction is essential for the reliable operation of solar gas turbines. Simulation tools allow engineers and scientists to forecast system performance under diverse environmental and operating conditions, a predictive capability that is vital for designing gas turbines capable of efficient and reliable real operation.

Given the dependence of solar gas turbines on solar radiation, which may fluctuate due to weather conditions and time of day, simulations can model these variations and predict their impact on turbine performance. This enables the design of systems capable of handling such fluctuations effectively (Kalogirou 2014). Similarly, simulations help in understanding the effects of ambient temperature, pressure, and humidity on turbine performance. For example, models can simulate the impact of high ambient temperatures on compressor efficiency and turbine output (Horlock 2007). To maintain optimal performance across varying loads, solar gas turbines must be adaptable. Simulating different load scenarios allows engineers and scientists to ensure that the turbine can adjust its operations to meet the varying power requirements without significant efficiency losses (Cengel and Boles 2023). Improving the efficiency of solar gas turbines remains the primary objective of simulation. Detailed simulations can reveal system inefficiencies and assist in recommending improvements. Heat transfer process simulations within the gas turbine can identify areas of thermal loss and suggest measures to minimize them, such as optimizing heat exchanger designs and improving thermal insulation (Incropera et al. 2013).

Fluid dynamics simulations optimize the aerodynamic properties of turbine components, reducing drag and enhancing overall efficiency. For example, computational fluid dynamics (CFD) models aid in designing turbine blades with optimal aerodynamic profiles (Versteeg and Malalasekera 2007). Additionally, modelling combustion processes helps in the optimization of fuel–air mixtures and combustion chamber design, leading to more complete and efficient combustion, which not only improves efficiency but also reduces emissions (Turns 2012). Simulation provides a powerful platform for comprehensive risk analysis, enabling engineers and scientists to model various failure scenarios and assess their impact on turbine performance and safety. By predicting potential points of failure within the system, simulations allow engineers and scientists to design for greater reliability; stress analysis models, for instance, can identify areas prone to mechanical failure (Dowling 2013). Understanding these risks enables engineers and scientists to develop strategies for mitigating them, such as incorporating redundant systems, enhancing component durability, or implementing advanced control strategies (Modarres et al. 2016). Virtual safety tests ensure that turbines meet all necessary safety regulations and standards, especially under extreme conditions such as high temperatures and pressures (Lees 2012).

The use of simulation significantly reduces costs associated with the development and operation of solar gas turbines. Instead of building multiple physical prototypes, engineers and scientists can virtually test and refine designs, saving both time and money. This approach reduces development costs by up to 30% (Thomke 2003). Extensive field testing can be minimized, as many performance and reliability assessments can be conducted through simulation, further reducing costs and accelerating the development process (Pahl et al. 2007). Simulation helps in the optimization of operational parameters, leading to reduced operational costs and extended gas turbine lifespans. For instance, optimizing maintenance schedules based on simulation data can prevent costly downtime (Mobley 2002).

Assessing and mitigating the environmental impact of solar gas turbines is increasingly important, and simulation plays a key role in this area by enabling detailed environmental impact assessments. Simulations can predict the emissions produced by turbines under different operating conditions, aiding in the design of systems that minimize environmental impact. For example, models can simulate the formation of nitrogen oxides (NO_x) and carbon dioxide (CO_2) emissions, allowing for the optimization of combustion processes to reduce these pollutants (Heywood 2018). Acoustic simulations help in understanding and mitigating noise pollution caused by gas turbines, which is particularly important for installations near populated areas (Munjal 2013). By modelling the lifecycle impact of gas turbines, engineers and scientists can design more sustainable systems that align with environmental regulations and standards, including analysis of the energy payback period and the carbon footprint of the gas turbine (Hertwich 2005).

Simulation is also a powerful tool for fostering innovation and conducting research in solar gas turbine technology. Researchers can test new ideas and concepts in a virtual environment before moving to physical testing, accelerating the development of innovative technologies (Schrage 1999). Simulation provides detailed insights into the physical and chemical processes within turbines, contributing to the broader understanding of solar gas turbine technology. This knowledge is essential for advancing the field and developing more efficient and effective systems. Furthermore, simulation tools can be used in educational settings to train the next generation of engineers and scientists in the principles and practices of the SGT technology. This hands-on experience is invaluable for developing the skills and knowledge required to drive future innovations (ABET 2016).

8.2 Fundamentals of Simulation

Simulation is a cornerstone in the design and optimization of solar gas turbines, offering a virtual environment to analyse, test, and enhance various aspects of these systems. Understanding the fundamentals of simulation is crucial for engineers and scientists working in this field. In this section, we outline the principles of modelling and simulation, the development and application of mathematical models, and the computational methods that govern these simulations.

8.2.1 *Principles of Modelling and Simulation*

Modelling and simulation are fundamental techniques used to represent and analyse the behaviour of complex systems. In the context of solar gas turbines, these techniques enable the examination of various physical phenomena and operational scenarios without extensive physical testing. At the core of simulation lies the conceptual framework that involves defining the system's structure, behaviour, and interactions, guiding the development of models that accurately represent the physical and functional aspects of solar gas turbines. According to Zeigler et al. (2000), a well-defined conceptual model is essential for ensuring the validity and reliability of simulation results. Simulations can be conducted at various levels of abstraction, from high-level system simulations that capture overall performance metrics to detailed component-level simulations that analyse specific physical processes. The appropriate level of abstraction depends on the simulation objectives and available computational resources (Fishwick 2007). The accuracy of simulation models is of paramount importance, and it is achieved through validation and verification processes. Validation involves comparing simulation outputs with empirical data to ensure the model accurately represents the physical system, while verification focuses on ensuring that the simulation is implemented correctly and is free of errors. Both processes are critical for building confidence in simulation results (Sargent 2013).

8.2.2 *Mathematical Models in Solar Gas Turbine Simulation*

Mathematical modelling involves developing mathematical representations of physical systems and serves as the backbone of simulation by providing the equations and relationships necessary to predict system behaviour. Optical, thermodynamic, hydrodynamic, heat transfer, and combustion models are commonly applied in the simulation of solar gas turbines. Analysis of the thermodynamic performance of an engine layout is the first major design step. Thermodynamic models offer a framework to analyse and predict the behaviour of these systems under various operating conditions. These models facilitate the evaluation of energy transformations, efficiency, and overall system performance.

8.2.2.1 Optical Models

Solar radiation is the most important environmental factor that influences the performance of solar energy systems (Nafey et al. 2000). In this context, optical performance significantly affects the overall performance of a solar gas turbine. It is therefore necessary to model the optical performance. Optical performance models can be classified into three major groups (Blanc et al. 2014): (a) parameterization models, (b) analytical models, and (c) ray tracing models. A simplified way for computing the optical performance is by using parameters or look-up tables that describe the behaviour of these properties based on the position of the sun. Usually, optical parameters are determined through experimentation by using standard methods. The effect of solar concentration is considered via the concentration ratio (C_r). This ratio is defined by a geometrical concentration parameter ($C_{r,g}$) or flux concentration parameter ($C_{r,flux}$). The geometrical concentration ratio is the ratio of the concentrator aperture area (A_{ap}) and solar receiver outer surface (A_{rec}):

$$C_{r,area} = \frac{A_{ap}}{A_{rec}} \tag{8.1}$$

The flux concentration ratio ($C_{r,flux}$) is the ratio of the intensity of radiation on the aperture (G_{ap}) to that of the solar receiver (G_{rec}):

$$C_{r,flux} = \frac{G_{ap}}{G_{rec}} \tag{8.2}$$

where G_{ap} is the intensity of radiation on the aperture and G_{rec} is the intensity of solar radiation on the receiver. Solar radiation flux is assumed but there is non-uniformity in the distribution of flux on the receiver. For instance, most of the concentrated radiation is incident on half of the outer circumference of a tubular receiver for a parabolic trough concentrator or linear Fresnel reflector.

Some analytical formulations have been developed to model the non-uniformity of solar flux distribution. Various assumptions have been made to include the effect of optical imperfections and the shape of the sun in each study. Procedures for obtaining the solar flux distribution on the receiver can be categorized into three main groups: cone optics, semi-finite integral formulation, and explicit formulation. In the cone optics-based method, a ray path is traced from the sun to the reflector and from the reflector back to the sun with optical cones. Then, the solar flux in the receiver is integrated including different functions to model the sun-shape and optical errors.

Incoming solar radiation can be described as a collection of solar rays transmitted from the sun to the concentrators. Eventually, these rays are reflected onto the solar receiver. Ray tracing is the commonest technique for optical performance modelling of concentrating solar power systems (Yılmaz and Mwesigye 2018). This approach consists of tracing the paths of photons or rays from the sun and modelling their interaction with the elements in the medium according to their optical properties. Two global approaches for modelling the ray path from a source light to an image and vice versa are: forward ray tracing (FRT) and backward ray tracing (BRT). The FRT method describes the behaviour of a real world (rays of light propagate from a source to a location where an image is formed). It is therefore not surprising that engineers and scientist most commonly use this approach around the world. All the emitted photons do not however interact with the optical elements in FRT method while all the photons from the elements reach the source in the BRT method which is computationally cheaper (Glassner 1993). The BRT method was therefore more widely applied before the invention and advancement of digital computational capabilities. Available computational tools for ray tracing include STRAL, SolTrace, MIRVAL, and SPRAY (Blanc et al. 2014). These tools compute the path of the sun's rays from the sun shape to the solar receiver by applying physical laws. Usually, Monte Carlo techniques are applied to permit good computational times.

8.2.2.2 Thermodynamic Models

Thermodynamic models describe the energy transformations within solar gas turbines. These models are based on the principles of thermodynamics, including the laws of thermodynamic cycles. For instance, the Brayton cycle, which is the basis for solar gas turbine operation, can be modelled to analyse the efficiency and performance of the turbine under various conditions (Bejan 2016).

(a) Brayton Cycle Model

The Brayton cycle, also known as the Joule cycle, is the fundamental thermodynamic cycle for gas turbine operations. Theoretical and actual cycle analysis involves energy balance and efficiency equations for components and the full gas turbine system. Details about equations for various configurations of the system are presented in Chap. 5. They include solar-only closed cycle gas turbine, hybrid closed cycle solar gas turbine, solar-only open cycle gas turbine, hybrid open cycle solar gas turbine, combined cycle solar gas turbine with closed topping Brayton cycle, and combined cycle solar gas turbine with open topping Brayton cycle.

(b) Exergy Analysis

Exergy analysis is a thermodynamic approach that evaluates the quality and usability of energy within a system. Unlike traditional energy analysis, which considers only the quantity of energy, exergy analysis accounts for the irreversibilities associated with real processes, providing a more accurate measure of system efficiency and performance. In any real process, exergy is destroyed due to irreversibilities such as friction, heat transfer across finite temperature differences, and mixing. The exergy destruction can be quantified and minimized to improve system efficiency. Exergy efficiency (η_E) is defined as the ratio of useful exergy output to the total exergy input. For a solar gas turbine, it can be expressed as (Bellos et al. 2017):

$$\eta_E = \frac{\dot{W}_{net}}{E_{in}} \tag{8.3}$$

where $\left(\dot{W}_{net}\right)$ is the net work output and (E_{in}) is the exergy input from the fuel and solar energy. By identifying and reducing exergy destruction, engineers and scientists can pinpoint areas for improvement and optimize the design and operation of solar gas turbines (Bejan 2016). More details about exergy analysis are reported in Sect. 5.6.

(c) Heat and Mass Balance Models

Heat and mass balance models are essential for analysing the energy and material flows within a solar gas turbine system, ensuring that all energy inputs and outputs, as well as mass flows, are accounted for. These models enable a comprehensive evaluation of system performance. The energy balance equation guarantees that the total energy entering the system equals the total energy leaving the system plus any changes in internal energy. For a control volume, the energy balance is expressed as:

$$\dot{Q} - \dot{W} = \Delta H + \Delta E_{\mathrm{k}} + \Delta E_{\mathrm{p}} \tag{8.4}$$

where $\dot{Q}$ is the heat transfer rate, $\dot{W}$ is the work transfer rate, ΔH is the change in enthalpy, ΔE_{k} is the change in kinetic energy, and ΔE_{p} is the change in potential energy. The mass balance equation ensures that the mass entering the system equals the mass leaving the system plus any accumulation within the system. For a steady-state process, the mass balance is given by:

$$\sum \dot{m}_{\mathrm{in}} = \sum \dot{m}_{\mathrm{out}} \tag{8.5}$$

where $\dot{m}_{\mathrm{in}}$ and $\dot{m}_{\mathrm{out}}$ are the mass flow rates entering and leaving the control volume, respectively. Heat and mass balance models are critical for designing and optimizing solar gas turbines, as they provide insights into the efficiency and feasibility of different configurations and operating conditions.

8.2.2.3 Hydrodynamic Models

Hydrodynamic models (also referred to as fluid dynamics models) are essential for analysing and optimizing the performance of solar gas turbines. These models provide detailed insights into the behaviour of fluids (such as air, carbon dioxide, fuel, and combustion gases) as they flow through the various components of the gas turbine system, such as compressors, solar receivers, combustors, and turbines. Understanding fluid dynamics is crucial for improving efficiency, reducing losses, and ensuring the reliable operation of solar gas turbines. This section discusses several key fluid dynamics models, including computational fluid dynamics (CFD), Reynolds-averaged Navier–Stokes (RANS) equations, large eddy simulation (LES), and direct numerical simulation (DNS). Fluid dynamics models are essential for understanding and optimizing the performance of solar gas turbines. CFD provides a powerful framework for simulating fluid flows and associated phenomena. The RANS equations offer a practical approach for modelling turbulent flows, while LES provides more detailed and accurate predictions by resolving large eddies. DNS offers the highest accuracy by resolving all scales of turbulence but is limited to low Reynolds number flows due to its computational intensity. Together, these models form a comprehensive toolkit for analysing and improving the fluid dynamics of solar gas turbines and other systems.

(a) Mass, Momentum and Energy Conservation Equations

Heat and mass transfer processes are involved in the operation of a solar gas turbine. Heat is commonly transferred between solid surfaces and heat transfer or working fluids of the SGT system. For instance, there is heat transfer from an absorber surface to the heat transfer fluid of a solar receiver. This fluid transports heat from the receiver to other components of the solar gas turbine. The most complete model to represent fluid flows with heat transfer from a solid structure to fluid is a system of partial differential equations constituted by mass, momentum, and energy conservations (Sandá et al. 2019). These conservation equations must be satisfied at each time step in the entire spatial domain. Assuming the heat transfer fluid is single-phase, viscous, and compressible, then the fundamental equations that represent the laws of conservation of mass (continuity equation), momentum, and energy are:

(i) Continuity Equation

$$\frac{\partial \rho}{\partial \mathbf{t}} + \nabla \cdot (\rho \mathbf{u}) = \mathbf{0} \tag{8.6}$$

where u is the velocity vector and ρ is the fluid density.

(ii) Momentum Equation

$$\frac{\partial (\rho u)}{\partial t} + \nabla \cdot (\rho u u) = -\nabla p + \nabla \cdot \tau + \rho f \tag{8.7}$$

where f represents body forces and τ is the stress tensor.

(iii) Energy Equation

$$\frac{\partial (\rho E)}{\partial t} + \nabla \cdot (\rho E u) = \nabla \cdot (k \nabla T) + \Phi + f \cdot u \tag{8.8}$$

where E is the total energy, k is the thermal conductivity, T is the temperature, and Φ represents viscous dissipation.

Equations (8.6–8.8) are coupled and complex, and their complexity and numerical solution algorithms depend on the assumptions made for the independent variables, time and space. For instance, the complexity of conservation equations increases with the number of flow dimensions. Due to the high degree of complexity, the partial differential equations require computer programs to organize them into a suitable form and to find solutions. The set of methods utilized by computers to solve hydrodynamic equations are referred to as computational fluid dynamics (CFD). These methods involve solving the governing equations of fluid dynamics, including the continuity, momentum, and energy equations, using numerical methods. CFD simulations require the discretization of the computational domain into small control volumes or elements. This process, known as meshing, is critical for capturing the complex geometries and flow features within solar gas turbines. The quality and resolution of the mesh directly impact the accuracy and convergence of the solutions.

Turbulence is a common feature in fluid flows through gas turbines, characterized by chaotic and irregular fluid motion. Accurate turbulence modelling is essential for predicting flow behaviour and performance. Several turbulence models are used in CFD, including RANS, LES, and DNS, which are discussed in more detail below.

(b) Reynolds-Averaged Navier–Stokes Equations

Reynolds-averaged Navier–Stokes (RANS) equations are a set of time-averaged equations used to describe turbulent flow. By averaging the Navier–Stokes equations over time, RANS models simplify the analysis of turbulent flows while retaining essential flow characteristics. The RANS equations introduce additional terms called Reynolds stresses that account for the effects of turbulence. The equations are given by (Abubaker et al. 2024):

$$\frac{\partial(\rho\overline{u_i})}{\partial t} + \frac{\partial\left(\rho\overline{u_i}\overline{u_j}\right)}{\partial x_j} = \frac{\partial\overline{p}}{\partial x_i} + \frac{\partial}{\partial x_j}\left[\mu\left(\frac{\partial\overline{u_i}}{\partial x_j} + \frac{\partial\overline{u_j}}{\partial x_i}\right) - \rho\overline{u_i'\rho u_j'}\right] + \rho\overline{f_i} \tag{8.9}$$

where u_i and u_j are the mean velocity components, $\overline{p}$ is the mean pressure, μ is the dynamic viscosity, and $\overline{u_i'\rho u_j'}$ are the Reynolds stresses. To close the RANS equations, turbulence models are used to represent the Reynolds stresses. One of the most widely used models include two transport equations for the turbulence kinetic energy (k) and its dissipation rate (ε). This model is robust and suitable for a wide range of flows. Similar to the k–ε model, the k–ω model includes equations for the turbulence kinetic energy (k) and the specific dissipation rate (ω). The k–ω model is particularly effective for near-wall treatment and flows with strong adverse pressure gradients (Wilcox 1993).

(c) Large Eddy Simulation

Large eddy simulation (LES) is an advanced turbulence modelling approach that resolves the large, energy-containing eddies in the flow while modelling the smaller scales. LES provides more accurate and detailed predictions of turbulent flows compared to RANS. In LES, the Navier–Stokes equations are filtered to separate the large eddies from the small-scale turbulent motions. The filtered equations are given by (Cordova et al. 2023):

$$\frac{\partial\overline{u_i}}{\partial t} + \overline{u_j}\frac{\partial\overline{u_i}}{\partial x_j} = -\frac{\partial\overline{p}}{\partial x_i} + \nu\frac{\partial^2\overline{u_i}}{\partial x_j^2} + \frac{\partial\tau_{ij}}{\partial x_j} \tag{8.10}$$

where τ_{ij} represents the subgrid-scale (SGS) stress tensor that models the effects of the unresolved scales on the resolved scales. The accuracy of LES depends on the SGS model used to represent the unresolved scales. A simple and widely used SGS model that relates the stress tensor to the strain-rate tensor is the Smagorinsky model. This model is considered as the pioneer subgrid-scale model for LES. SGS models adjust the eddy viscosity dynamically based on the resolved flow field, providing more accurate and adaptive representations of SGS stresses (Germano et al. 1991).

(d) Direct Numerical Simulation

Direct numerical simulation (DNS) is the most detailed and accurate approach to simulating turbulent flows, as it resolves all scales of turbulence directly without any modelling approximations. However, DNS is computationally intensive and feasible only for low Reynolds number flows and simple geometries. DNS solves the Navier–Stokes equations with very fine spatial and temporal resolution, capturing all turbulent motions from the largest eddies to the smallest dissipative scales. This requires extremely fine meshes and small-time steps. Due to its high computational cost, DNS is primarily used for fundamental turbulence research and benchmarking other turbulence models. It provides valuable insights into turbulence mechanics and serves as a reference for validating LES and RANS models (Moin and Mahesh 1998).

8.2.2.4 Heat Transfer Models

Heat transfer models are essential for analysing thermal processes within solar gas turbines, by considering conduction, convection, and radiation heat transfer mechanisms to predict temperature distributions and heat fluxes within components. Accurate heat transfer modelling is crucial for optimizing thermal efficiency and preventing overheating (Incropera et al. 2013). Proper modelling of heat transfer within gas turbine components—such as the solar receiver, compressor, combustor, turbine, and generator—is vital for optimizing system design and operation. This section discusses various heat transfer models, including conduction, convection, radiation, and combined heat transfer models, and their application in the simulation of solar gas turbines.

(a) Conduction Heat Transfer

Conduction is the transfer of heat through a solid material or between contacting solids, occurring due to energy transfer between molecules or atoms. The governing equation for conduction heat transfer is Fourier's law as follows:

$$q = -k\nabla T \tag{8.11}$$

where q is the heat flux, k is the thermal conductivity of the material, and ∇T is the temperature gradient. In solar gas turbines, heat conduction is significant through the walls of the solar receiver, combustor, and turbine blades. The heat conduction equation in a solid is described by the heat diffusion equation:

$$\frac{\partial T}{\partial t} = \alpha \nabla^2 T \tag{8.12}$$

where $\alpha = k/(\rho C_p)$ is the thermal diffusivity, ρ is the density, and C_p is the specific heat capacity of the material.

(b) Convection Heat Transfer

Convection is the heat transfer between a solid surface and a moving fluid, influenced by the fluid's motion. Convection can be either natural (driven by buoyancy forces due to temperature differences) or forced (driven by external means such as fans or pumps). The rate of convective heat transfer is described by Newton's law of cooling:

$$\dot{Q}_{\text{conv}} = Ah(T_s - T_\infty) \tag{8.13}$$

where $\dot{Q}_{\text{conv}}$ is the heat transfer rate, h is the convective heat transfer coefficient, A is the surface area, T_s is the surface temperature, and T_∞ is the fluid temperature far from the surface. In solar gas turbines, convection is crucial in cooling turbine blades, heat transfer in the solar receiver or the combustor, and heat exchange processes. The convective heat transfer coefficient depends on fluid properties, flow conditions, and surface characteristics.

(c) Radiation Heat Transfer

Radiation is the transfer of energy through electromagnetic waves and does not require a medium. In solar gas turbines, radiation is particularly important in the solar collector, where solar energy is absorbed and converted into thermal energy. The rate of radiative heat transfer from a surface is given by the Stefan-Boltzmann law:

$$\dot{Q} = \varepsilon\sigma A\left(T_s^4 - T_\infty^4\right) \tag{8.15}$$

where ε is the emissivity of the surface, σ is the Stefan-Boltzmann constant, A is the surface area, T_s is the surface temperature, and T_∞ is the temperature of the surroundings. In complex geometries, the concept of view factors (or configuration factors) is used to account for the geometric relationship between radiating surfaces, determining the fraction of radiation leaving one surface that directly reaches another surface.

In many practical applications, heat transfer occurs simultaneously through conduction, convection, and radiation. Combined heat transfer models are essential for accurately simulating the thermal behaviour of solar gas turbines. For instance, the useful heat ($\dot{Q}_u$) from a solar receiver is equal to the difference between absorbed solar radiation and heat loss through convection and radiation to the ambient environment:

$$\dot{Q}_u = S - \dot{Q}_{\text{conv}} - \dot{Q}_r \tag{8.15}$$

where S is the absorbed solar radiation, $\dot{Q}_{\text{conv}}$ is the convective heat transfer, and $\dot{Q}_r$ is the radiative heat transfer.

8.2.2.5 Combustion Models

Combustion models describe the chemical reactions and energy release during fuel combustion, and they are essential for understanding and optimizing the performance of hybrid solar gas turbines. These models are grounded in chemical kinetics and thermodynamic principles, enabling the simulation of combustion processes to achieve efficient and clean fuel–air mixtures (Turns 2012). Let us consider various combustion models used in the analysis and optimization of solar gas turbines.

(a) Chemical Kinetics

Chemical kinetics is the study of the rates of chemical reactions. It involves understanding the steps in the combustion processes and the factors that influence these rates. The basic reaction mechanism for hydrocarbon combustion typically comprises a series of elementary reactions that can be represented as:

$$\text{Fuel} + O_2 \rightarrow CO_2 + H_2O + \text{Heat} \tag{8.16}$$

The rate of these reactions depends on factors such as temperature, pressure, and the concentrations of reactants and products. Detailed chemical kinetic models incorporate these factors to provide accurate predictions of the combustion behaviour.

(b) Thermodynamic Principles

Thermodynamic principles are essential for modelling the energy release and efficiency of combustion processes. These principles include the conservation of mass, energy, and momentum, which are used to derive the governing equations for combustion. The first law of thermodynamics is particularly important for combustion modelling. The energy equation for a reacting system can be written as:

$$\dot{Q} = \dot{U} + \dot{W} \tag{8.17}$$

where $\dot{Q}$ is the rate of heat addition, $\dot{U}$ is the rate of change of internal energy, and $\dot{W}$ is the rate of work done by the system.

(c) Flame Models

Flame models describe the structure and propagation of flames in combustion processes. There are several types of flame models, including laminar and turbulent flame models. Laminar flame models assume a smooth and steady flame front, while turbulent flame models account for the chaotic and irregular nature of real flames. The flame speed and thickness are key parameters in these models and can be predicted using the following relation for laminar flame speed:

$$S_L = \sqrt{\frac{2k}{\rho C_p} \frac{dT}{dx}} \tag{8.18}$$

where s_L is the laminar flame speed, k is the thermal conductivity, ρ is the density, C_p is the specific heat capacity, and dT/dx is the temperature gradient across the flame front.

(d) Pollutant Formation Models

Pollutant formation models are crucial for designing combustion systems that minimize emissions of harmful pollutants such as NO_x, CO, and unburned hydrocarbons. These models predict the formation and destruction of pollutants based on the combustion chemistry and operating conditions. For example, the formation of NO_x can be described by the Zeldovich mechanism:

$$N_2 + O \rightarrow NO + N \tag{8.19}$$

$$N + O_2 \rightarrow NO + O \tag{8.20}$$

The rate of NO_x formation depends on the temperature, pressure, and the concentrations of nitrogen and oxygen in the combustion zone.

(e) Computational Fluid Dynamics in Combustion

CFD is widely used to simulate combustion processes in solar gas turbines (Matyushenko and Kolmogorov 2021). CFD models solve the governing equations of fluid flow, heat transfer, and chemical reactions to predict the performance of combustion systems. These simulations provide detailed insights into the combustion behaviour, including flame structure, temperature distribution, and pollutant formation. Advanced turbulence models, such as LES and RANS models, are often coupled with combustion models to capture the complex interactions between turbulence and combustion.

8.2.3 Computational Methods for Simulation

The complexity of solar gas turbines necessitates the use of advanced computational methods to solve mathematical models and conduct simulations. These methods include numerical techniques and algorithms that facilitate accurate and efficient analysis of complex systems.

8.2.3.1 Finite Element Method

The finite element method (FEM) is widely used for structural analysis and heat transfer simulations. FEM divides the system into a mesh of smaller, simpler elements, allowing for the approximation of complex geometries and boundary conditions. This method is particularly useful for analysing stress distributions and thermal gradients in turbine components (Bathe 2007).

8.2.3.2 Finite Volume Method

The finite volume method (FVM) is commonly used in fluid dynamics and heat transfer simulations. FVM divides the system into control volumes and applies conservation laws (mass, momentum, energy) to each volume, providing a robust framework for solving fluid flow and heat transfer problems. This method is the foundation of many CFD codes (Versteeg and Malalasekera 2007).

8.2.3.3 Monte Carlo Methods

Monte Carlo methods are statistical techniques used to model probabilistic systems and conduct uncertainty analysis. These methods involve random sampling and statistical analysis to predict the behaviour of systems with inherent uncertainties, such as variations in solar irradiance or material properties (Metropolis and Ulam 1949).

8.2.3.4 Parallel Computing

Given the computational intensity of detailed simulations, parallel computing techniques are employed to enhance simulation efficiency. By distributing computations across multiple processors, parallel computing significantly reduces simulation time, enabling the analysis of large-scale systems and high-resolution models (Gropp et al. 1999).

8.2.3.5 Optimization Algorithms

Optimization algorithms are used to find the best design or operational parameters for solar gas turbines. These algorithms, such as genetic algorithms and gradient-based methods, iteratively adjust parameters to maximize performance metrics such as efficiency and power output while minimizing costs and emissions (Deb 2001).

8.3 Simulation of Components and Subsystems

The performance of solar gas turbines depends significantly on the design and operation of their individual components and subsystems. Each component must be meticulously modelled and simulated to ensure optimal functionality and integration within the overall system. The key components of a basic solar gas turbine are the concentrator, solar receiver, compressor, combustor, turbine, and generator (Fig. 8.1). Details about these components are presented and discussed in Chap. 4. This section focuses on the simulation of these components and subsystems. Accurate simulation of these components is essential for optimizing the system design, enhancing efficiency, and ensuring reliable operation. Advanced simulation tools and techniques, such as computational fluid dynamics, finite element analysis, and thermodynamic analysis, are employed to understand and improve the performance of these components.

8.3.1 Solar Concentrators

A concentrator is an optical component that focuses solar radiation onto the receiver, and there are different types of concentrators (see Sect. 1.3). At present, solar gas turbine systems are based on reflecting concentrators. Generally, this type of reflectors is made from highly reflective materials, such as silvered glass or aluminium, to maximize the concentration of sunlight onto the receiver.

Modelling of solar concentrators is mainly concerned with their optical performance. The angular distribution of incoming rays of the sun and the effects of atmospheric transmission on this distribution is an important consideration. Solar radiation undergoes attenuation as it propagates through the atmosphere. This attenuation includes absorption and scattering of the radiation, resulting in beam and diffuse components of radiation that reach the earth surface (Lu et al. 2023). However, concentrating solar power systems harness only beam radiation. This component of solar radiation travels in straight lines, and therefore, it is traceable. Astronomical angles are used to determine the angle of beam radiation incidence on a given surface of known orientation and inclination. This provides the basis of geometric design and operation of solar concentrators. Four major designs of concentrators (parabolic trough concentrator, linear Fresnel reflector, solar tower, and parabolic dish) are presented and discussed in Sect. 1.3. Due to changes in the position of the sun during the day, concentrators are mounted on solar tracking systems to maintain an optimal focus of solar radiation on the receiver. Details about solar radiation and trackers are reported in Chap. 3.

The optical performance of a concentrator depends on its geometry and surface properties. For instance, the parabolic trough concentrator has a lower ratio of concentration than the parabolic dish. Perfectly smooth surfaces reflect solar radiation specularly, which can enable tracing of the rays of the sun. Specular reflection is the phenomenon in which a surface reflects incident light at equal angles of incidence and reflectance (Boynton 2003). Mirrors are an example of specular surfaces, and so they are often used in the construction of concentrators. Imperfections such as reflecting-surface slope errors, sun-tracking and alignment errors, and mirror-focusing errors affect the optical performance of the concentrator. Optical performance modelling takes into account the angular distribution of incoming rays of the sun, and the effects of atmospheric transmission on this distribution. Imperfections are also considered in some models, depending on the robustness of the model.

8.3.2 *Solar Receivers*

A solar receiver absorbs concentrated solar energy and transfers it to the working fluid. This component operates at elevated pressure and temperature, and it is the primary source of heat for the solar gas turbine. Basic simulation of the solar receiver involves modelling of: Optical processes (reflection, transmission, and absorption of solar radiation), thermal (heat transfer and energy balance), and hydrodynamic (fluid flows) processes (Sandá et al. 2019). Ray tracing and heat transfer equations are commonly used to simulate the absorption of solar radiation and its conversion to thermal energy (Yılmaz and Mwesigye 2018).

The optical efficiency of solar gas turbines is one of the key performance parameters in addition to the thermal efficiency. It is the ratio of solar energy absorbed by the absorber element to the solar radiation intercepted by the aperture of the concentrator. The optical efficiency affects the amount (S) of absorbed radiation per unit area of unshaded aperture given by (Duffie and Beckman 2013):

$$S = \rho(\alpha\tau\gamma)_n K_\theta G_{\mathrm{ap}} = \eta_{\mathrm{opt}} G_{\mathrm{ap}}, \tag{8.21}$$

where G_{ap} is the effective incident radiation measured on the plane of the aperture that includes only beam radiation for all concentrators except those of low concentration ratio ($C_{\mathrm{r}} \leq 10$), K_θ is the incidence angle modifier (determined from experimental data), α is the absorptance of the absorber on the solar receiver, ρ is the specular reflectance (ρ is replaced by transmittance when the concentrator is a refractor), τ is the transmittance of the transparent cover on the solar receiver ($\tau = 1.0$ for unglazed receivers), γ is the proportion of the reflected solar radiation that is incident on the absorbing surface of the receiver (values of γ may be greater than 0.9), and η_{opt} is the optical efficiency of the system.

8.3.3 *Compressors*

The compressor is a critical component of a solar gas turbine system, responsible for increasing the pressure of the working fluid before it enters the solar receiver or any other heater. Accurate simulation of the compressor is essential for optimizing the performance of the entire system.

8.3.3.1 Principles of Compressor Simulation

The primary function of a compressor is to increase the pressure of the incoming working fluid. This function involves thermodynamic and hydrodynamic principles. The compression process is ideally isentropic (Bellos et al. 2017). Figure 8.2 shows the temperature-entropy diagram of the gas turbine cycle. Actual compressors and

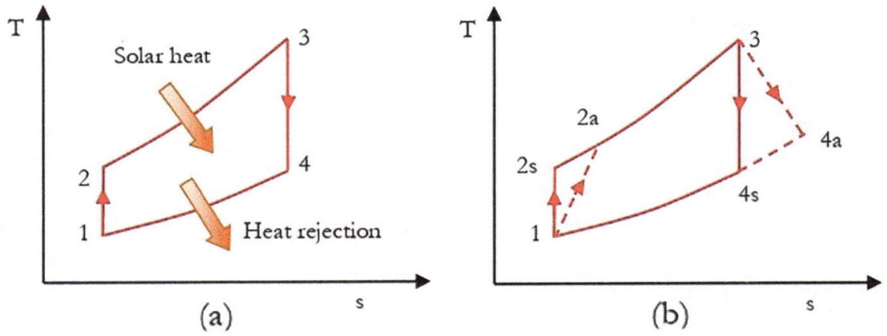

Fig. 8.2 Temperature–entropy (T–S) diagram for: **a** an ideal solar gas turbine cycle and **b** ideal and actual gas turbine cycles. Heat addition and rejection proceed at constant pressures $P_2 = P_3$ and $P_1 = P_4$ in the ideal cycle

turbines deviate from the ideal isentropic processes due to inefficiencies and heat losses. In this context, 2a and 4a are respectively actual exit states of the compressor and turbine, with corresponding isentropic states 2s and 4s. So, the thermodynamic processes of the actual cycle are 1-2a, 2a-3, 3-4a, and 4a-1. The isentropic outlet temperature (T_{2s}) of the compressor can be calculated from (Çengel and Boles 2006):

$$T_{2s} = T_1 \left(\frac{P_2}{P_1} \right)^{(\gamma - 1)/\gamma} \tag{8.22}$$

where P_1 and P_2 are the values of gas pressure at the specified states, and γ is the ratio of the specific heat capacity of a gas at constant pressure to its specific heat capacity at constant volume.

The behaviour of the airflow through the compressor involves complex fluid dynamics, including variations in velocity, pressure, and temperature. In this vein, conservation equations for mass, momentum, and energy are numerically solved to understand these variations. The formation of boundary layers and the occurrence of turbulence within the compressor stages significantly affect the performance. Turbulence models, such as k–ε or k–ω models, are often employed in simulations to capture these effects (Li et al. 2023).

8.3.3.2 Methods for Compressor Simulation

(a) One-Dimensional Models

One-dimensional (1D) models use mean line analysis, providing a simplified approach by averaging properties along a mean line through the compressor stages. These models are useful for preliminary design and performance estimation. Key equations include conservation laws of mass, momentum, and energy that are solved iteratively.

(b) Two-Dimensional and Three-Dimensional Models

Two-dimensional models consider variations in properties in a plane perpendicular to the axis of the compressor, capturing the effects of blade shape and spacing more accurately than 1D models. Three-dimensional simulations provide the most detailed analysis by solving the hydrodynamic equations for the entire compressor geometry, allowing the examination of complex flow phenomena, including secondary flows, blade tip vortices, and flow separation.

(c) Performance Maps

Performance maps, also known as compressor characteristic maps, depict the relationship between key performance parameters such as pressure ratio, mass flow rate, and efficiency. Generated through simulations and experimental data, these maps are used to understand the operational limits and performance of the compressor under different conditions.

8.3.3.3 Performance Analysis

The performance of the compressor is often evaluated in terms of isentropic and polytropic efficiencies. These efficiencies are calculated using the temperature and pressure measurements from simulations and experiments. The pressure ratio and mass flow rate are key indicators of the compressor's capability to handle different operating conditions. Simulation tools help in plotting these parameters across different operating points to identify the optimal performance range. Compressors are prone to instability phenomena such as surge and stall, that can drastically reduce performance and cause mechanical damage. Simulations are used to predict these phenomena and design control strategies to mitigate their effects.

8.3.4 Modelling Combustors

The combustor is an important component of hybrid SGT systems. It converts the chemical energy of the fuel into thermal energy through combustion. Accurate modelling of the combustor involves integrating thermodynamics, fluid dynamics, chemical kinetics, and advanced simulation tools with sophisticated software packages and experimental validation. This comprehensive approach enables accurate prediction and optimization of combustor performance, addressing the unique challenges posed by solar integration to enhance the efficiency, reliability, and environmental performance of solar gas turbines.

8.3.4.1 Principles of Combustor Modelling

The primary reactions in a combustor involve the oxidation of fuel (usually natural gas or other hydrocarbons) with air to produce heat, carbon dioxide, and water. The general reaction for a hydrocarbon fuel can be represented as:

$$C_xH_y + \left(x + \frac{y}{4}\right)O_2 \rightarrow xCO_2 + \frac{y}{2}H_2O + \text{Heat} \tag{8.23}$$

The stoichiometric ratio is the ideal proportion of fuel to air required for complete combustion, but combustors often operate with excess air (lean combustion) to ensure complete fuel oxidation and to control temperatures. The heat generated in the combustion process is transferred to the working fluid (air) flowing through the combustor, involving complex interactions between conduction, convection, and radiation. Understanding the flow dynamics, which include significant changes in velocity, pressure, and temperature, is crucial for efficient combustor design.

8.3.4.2 Methods for Combustor Modelling

(a) One-Dimensional Models

One-dimensional models provide a basic understanding of the combustion process by averaging properties along the flow path. They are useful for preliminary design and performance estimation, using energy balance equations to calculate temperature and pressure distributions along the combustor length.

(b) Two-Dimensional and Three-Dimensional Models

Two- and three-dimensional models offer more detailed insights into the combustor's behaviour by considering variations in properties across different planes and volumes. CFD simulations solve the Navier–Stokes equations along with species conservation and energy equations to capture complex flow and combustion phenomena, essential for optimizing combustor geometry and performance.

(c) Chemical Kinetics Models

Detailed chemical kinetics models include a series of elementary reactions and their rates to simulate the combustion process accurately. These models, coupled with CFD, predict species concentrations, temperature fields, and reaction rates. Simplified reaction mechanisms, or reduced mechanisms, are often used to reduce computational load while maintaining reasonable accuracy.

8.3.5 Unique Challenges in Solar Gas Turbine Combustors

In solar gas turbines, the combustor must operate efficiently with variable solar heat input, requiring flexible design and control strategies to handle transitions between solar and fuel sources of heat. The integration of solar heat can lead to uneven temperature distributions, necessitating advanced modelling to ensure stable operation under varying thermal conditions. Solar integration often results in higher operating temperatures, requiring the use of advanced materials and cooling techniques. Temperature gradients induce thermal stresses in structural members of the SGT system. Frequent temperature fluctuations due to varying solar input can lead to fatigue damage while extreme operating temperatures and pressures can result in creep damage. Both of these modes of damage can occur in the SGT system. Consequently, thermomechanical modelling is necessary to predict them during the design phase of combustors.

8.3.6 Turbine Simulation

The turbine is a vital component in a solar gas turbine system. It converts the thermal energy from hot gases (from combustors and solar receivers) into mechanical energy to drive an electric generator. Accurate simulation of this component is crucial for optimizing the overall efficiency and performance of the gas turbine system.

8.3.6.1 Principles of Turbine Simulation

The primary function of the turbine is to expand gases at high temperature and pressure from the combustor to lower levels of temperature and pressure, thereby converting thermal energy into mechanical work. An ideal expansion process is isentropic (Fig. 8.2), and the relationship between the inlet and outlet states can be expressed as:

$$T_{4s} = T_3 \left(\frac{P_4}{P_3} \right)^{(\gamma-1)/\gamma} \tag{8.24}$$

where T_3 is the turbine inlet temperature, T_{4s} is isentropic outlet temperature, P_3 and P_4 are the corresponding pressures, and γ is the specific heat ratio. Real gas turbines deviate from the ideal Brayton cycle (see Sect. 5.4).

The behaviour of the working fluid through the turbine involves complex fluid dynamics. The conservation equations for mass, momentum, and energy are solved to understand these variations. The design and performance of turbine blades are crucial for efficient energy conversion, involving aerodynamic modelling to understand flow patterns, pressure distribution, and potential flow separation.

8.3.7 Generator Modelling

The generator in a solar gas turbine system converts mechanical energy from the turbine into electrical energy. Accurate modelling of the generator is essential for optimizing system performance, ensuring reliability, and effective integration with the electrical grid. This process involves a comprehensive approach that integrates electromagnetic principles, mechanical dynamics, and advanced simulation tools. By using steady-state, transient, and electromagnetic field models, along with sophisticated software packages and experimental validation, accurate prediction and optimization of generator performance can be achieved. Addressing the unique challenges posed by solar integration is crucial for enhancing the performance of solar gas turbines.

8.3.7.1 Principles of Generator Modelling

The operation of an electric generator is based on Faraday's law, which states that a changing magnetic field within a coil of wire induces an electromotive force (EMF) in the coil. The induced voltage (V) is given by:

$$V = -N\frac{\mathrm{d}\Phi}{\mathrm{d}t} \tag{8.25}$$

where N is the number of turns in the coil, and Φ is the magnetic flux. In an AC generator, a rotating magnetic field produced by the rotor induces an alternating voltage in the stator windings.

Synchronous generators, the most common type used in solar gas turbines, synchronize the rotor's speed with the frequency of the output AC power. Generator efficiency is influenced by various losses, including copper losses (I^2R losses in the windings), core losses (hysteresis and eddy currents in the magnetic core), and mechanical losses (friction and windage).

8.3.7.2 Methodologies for Generator Modelling

Steady-state models use equivalent circuits to represent the generator's electrical behaviour. The synchronous generator can be modelled using a simplified circuit comprising resistances, inductances, and a voltage source representing the induced EMF. This relationship can be expressed as:

$$V = E - (R_s + jX_s) \tag{8.26}$$

where V is the terminal voltage, E is the induced electromotive force, R_s is the stator resistance, X_s is the synchronous reactance, j is the imaginary unit used to represent the phase difference between the voltage and the current due to reactance, and I is the current.

Transient models capture the dynamic behaviour of the generator during start-up, load changes, and fault conditions. These models are essential for understanding the generator's response to varying operating conditions. The dynamic performance is modelled using differential equations that describe the generator's electrical and mechanical dynamics. The swing equation, which describes the rotor dynamics, is given by:

$$J\frac{\mathrm{d}^2\theta}{\mathrm{d}t^2} = T_\mathrm{m} - T_\mathrm{e} - D\frac{\mathrm{d}\theta}{\mathrm{d}t} \tag{8.27}$$

where J is the rotor inertia, θ is the rotor angle, T_m is the mechanical torque, T_e is the electrical torque, and D is the damping coefficient. Electromagnetic field models use finite element analysis methods to solve Maxwell's equations and accurately predict the magnetic field distribution, induced currents, and core losses. These models provide detailed insights into the generator's electromagnetic performance and are crucial for optimizing the design.

8.3.8 Modelling Heat Exchangers

Heat exchangers are integral to solar gas turbine systems, facilitating efficient thermal energy transfer between fluids to optimize performance and reduce energy losses. Their design and operation are very important for achieving high system efficiencies, particularly under the demanding conditions of solar-driven cycles. The accurate modelling of heat exchangers involves capturing the interplay of thermal, fluid, and structural dynamics to predict and enhance performance. The fundamental principle governing heat exchangers is the transfer of heat between two or more fluid streams at different temperatures without mixing them directly. The heat transfer rate is primarily dictated by the temperature difference between the fluids, the overall heat transfer coefficient, and the available heat transfer area. The general expression for heat transfer in a heat exchanger is given by:

$$Q = UA(\Delta T_\mathrm{lm}) \tag{8.28}$$

where Q is the heat transfer rate, U is the overall heat transfer coefficient, A is the effective heat transfer area, and ΔT_lm is the logarithmic mean temperature difference (LMTD). The factor ΔT_lm accounts for the temperature variation along the length of the heat exchanger, ensuring a more accurate representation of the thermal gradient.

In solar gas turbine systems, heat exchangers typically operate as recuperators or regenerators. Recuperative heat exchangers recover waste heat from turbine exhaust gases to preheat the compressed air entering the combustor or solar receiver, improving the thermal efficiency of the cycle. Regenerative heat exchangers, on the other hand, intermittently store heat from the exhaust stream and transfer it to the incoming stream in cycles. The effectiveness-NTU method is often employed to evaluate the thermal performance of heat exchangers, where effectiveness (ε) is defined as the ratio of the actual heat transfer rate (Q_{actual}) to the maximum possible heat transfer rate (Q_{max}). It provides a measure of the heat exchanger's performance relative to its theoretical maximum capacity, given as:

$$\varepsilon = \frac{Q_{\text{actual}}}{Q_{\text{max}}} \tag{8.29}$$

The flow dynamics within the heat exchanger are equally important and are governed by conservation equations for mass, momentum, and energy. Pressure drops and flow maldistribution, which significantly impact system performance, are analysed using computational fluid dynamics. CFD simulations solve the Navier–Stokes equations, often incorporating turbulence models such as k–ε or k–w to capture the high Reynolds number flows typical of gas turbine heat exchangers. Additionally, the thermal boundary layer development along heat transfer surfaces is considered to ensure accurate predictions of local heat transfer rates.

Heat exchanger materials are subjected to intense thermal and mechanical stresses due to high operating temperatures and pressure gradients. Structural modelling using finite element analysis (FEA) evaluates the mechanical integrity of these components under such conditions. Parameters such as thermal conductivity, thermal expansion coefficients, and material creep behaviour are integrated into the model to ensure reliability and durability. Furthermore, transient simulations are employed to capture the dynamic behaviour of heat exchangers under varying solar irradiance and load conditions. One of the significant challenges in modelling heat exchangers is the accurate representation of thermal losses to the surroundings. Boundary conditions must account for radiation, conduction, and convection losses, particularly in outdoor installations exposed to environmental fluctuations. Fouling, which reduces thermal performance and increases pressure drops, is another critical factor. Modelling must incorporate fouling resistance, which evolves over time based on operating conditions and the composition of the working fluids.

Optimization of heat exchanger performance involves balancing heat transfer efficiency with pressure drop minimization. Geometric parameters such as fin density, tube length, and flow channel configuration are adjusted to achieve this balance. Advanced materials with superior thermal properties and resistance to high-temperature degradation are increasingly used to enhance performance. Multiphysics simulation tools, such as COMSOL Multiphysics, enable the integration of thermal, fluid, and structural analyses, providing a comprehensive approach to heat exchanger optimization. Accurate modelling of heat exchangers is very important for the efficient operation of solar gas turbines. When we take advantage of the advanced computational tools and material innovations, we can design heat exchangers that meet the demanding requirements of solar-driven energy systems while contributing to overall cycle efficiency and sustainability.

8.4 Software Tools

Software tools are essential for the design, simulation, and optimization of SGT systems. They enable engineers and scientists to create detailed models, perform complex simulations, and analyse system performance under different conditions. The choice of software depends on project requirements, system complexity, and user expertise. Popular Simulation Software for solar gas turbines are MATLAB/Simulink, ANSYS Fluent, COMSOL Multiphysics, GT-SUITE, Aspen Plus, TRNSYS, ANSYS CFX, and OpenFOAM.

8.4.1 MATLAB/Simulink

MATLAB/Simulink is a high-level programming language and interactive environment extensively used in academia and industry for numerical computation, visualization, and programming. It provides a robust platform for modelling, simulation, and analysis of dynamic systems, making it ideal for applications ranging from engineering and science to economics and data analysis. MATLAB's extensive library of built-in functions and toolboxes supports various specialized tasks such as control system design, signal processing, optimization, and machine learning. Its integration with Simulink allows for graphical simulation and model-based design, enhancing the capability to develop, test, and optimize systems in a user-friendly environment.

8.4.2 ANSYS Fluent

ANSYS Fluent is a leading CFD software known for its comprehensive suite of modelling tools and advanced solvers that allow for detailed analysis of fluid flow, heat transfer, and chemical reactions. Widely used in both academia and industry, ANSYS Fluent provides powerful capabilities for simulating and optimizing solar gas turbines. ANSYS Fluent stands out as a comprehensive and powerful CFD tool for the simulation and optimization of solar gas turbines. Its advanced solvers, extensive libraries, and robust capabilities make it an indispensable resource for engineers and scientists aiming to enhance the efficiency and performance of these systems. By leveraging the features of ANSYS Fluent, the field of solar gas turbines can achieve significant advancements in design, analysis, and overall system integration.

8.4.3 COMSOL Multiphysics

COMSOL Multiphysics is a comprehensive simulation software environment designed for modelling and simulating physics-based systems across various engineering and scientific disciplines. It excels in handling multiphysics problems where multiple physical phenomena interact, making it particularly suited for complex systems such as solar gas turbines. The software provides a user-friendly interface for setting up and solving models, and post-processing the results. It includes an extensive library of predefined physics interfaces, covering areas such as fluid dynamics, heat transfer, structural mechanics, and electromagnetics. With features like custom physics definitions, parametric studies, optimization capabilities, and integration with CAD and MATLAB, COMSOL Multiphysics offers a flexible and powerful platform for detailed analysis and optimization of engineering systems. Its robust post-processing tools and extensive material library further enhance its ability to provide accurate and insightful simulations.

8.4.4 GT-SUITE

GT-SUITE is a comprehensive simulation tool widely used for modelling and analysing multidisciplinary systems in various industries such as automotive, aerospace, and energy. Developed by Gamma Technologies, it integrates multiple physics domains including fluid dynamics, thermodynamics, mechanical systems, electrical systems, and controls, allowing engineers to simulate complex interactions between these domains. GT-SUITE enables the virtual prototyping of powertrains, vehicle dynamics, engine systems, and other complex systems, facilitating the optimization of performance, efficiency, and emissions. It supports detailed modelling of components such as engines, transmissions, cooling systems, and hybrid powertrains, with capabilities for simulating real-world operating conditions and scenarios. The tool also includes advanced features for model-based control design, optimization studies, and integration with other engineering tools, making it a powerful choice for engineers and scientists involved in system-level modelling, simulation, and development.

8.4.5 Aspen Plus

Aspen Plus is a powerful process simulation software used extensively in chemical engineering and related fields for designing, optimizing, and analysing chemical processes. Developed by Aspen Technology, it enables engineers and scientists to model complex chemical processes involving reactions, separation units, heat exchangers, and other pieces of equipment. Aspen Plus uses a rigorous approach based on thermodynamic principles to simulate steady-state and dynamic behaviour of process flowsheets, allowing for detailed analysis of mass and energy balances, as well as process economics. Its extensive library of predefined chemical components, thermodynamic models, and unit operations facilitates the creation of detailed process models that accurately represent actual systems. Engineers and scientists can perform sensitivity analyses, design optimization, and troubleshoot process issues using Aspen Plus, making it an invaluable tool in process development, scale-up, and operational planning. The software's integration with other AspenTech tools and its ability to handle complex simulations make it a preferred choice for industries such as petrochemicals, pharmaceuticals, and special chemicals where process efficiency and product quality are critical.

8.4.6 Transient System Simulation Tool

The Transient System Simulation Tool (TRNSYS) is a software package designed for simulating the dynamic behaviour of systems in the fields of renewable energy, building energy performance, and environmental impact assessment. Developed by the Solar Energy Laboratory at the University of Wisconsin-Madison, TRNSYS facilitates the simulation of complex thermal and energy systems by allowing users to model interconnected components such as buildings, solar thermal collectors, HVAC systems, and renewable energy technologies such as photovoltaics and wind turbines. It employs a modular approach where users can select and connect different components to build a customized simulation model, enabling detailed analysis of system performance under varying environmental and operational conditions. TRNSYS is widely used for evaluating energy efficiency measures, conducting feasibility studies for renewable energy projects, and assessing the impact of design decisions on system performance and energy consumption. Its flexibility, extensive library of components, and ability to simulate transient behaviour make it a valuable tool for researchers, and policymakers involved in sustainable energy and building design.

8.4.7 ANSYS CFX

ANSYS CFX is a computational fluid dynamics software package known for its robust capabilities in simulating fluid flow, heat transfer, and related phenomena across a wide range of engineering applications. CFX was developed by ANSYS Inc. and it utilizes the finite volume method to solve the governing equations of fluid dynamics, offering accurate predictions of flow behaviour, turbulence, and heat transfer in complex geometries. Engineers and scientists use CFX to analyse and optimize designs of turbomachinery (such as gas turbines), HVAC systems, aerospace components, and various industrial processes where fluid dynamics play a crucial role. The software supports multiphysics simulations, allowing for coupling with structural mechanics, electromagnetics, and other disciplines to capture interactions between different physical phenomena. CFX features a user-friendly interface, advanced turbulence modelling options, and powerful post-processing tools that enable detailed visualization and analysis of simulation results. Its ability to handle transient and steady-state simulations, coupled with parallel processing capabilities, makes ANSYS CFX a preferred choice for finding accurate and efficient CFD solutions to complex engineering problems.

8.4.8 OpenFOAM

OpenFOAM, or Open Field Operation and Manipulation, is an open-source CFD software package renowned for its flexibility and extensibility in simulating fluid flow and other related phenomena. Developed by the OpenFOAM Foundation, it employs the finite volume method to solve governing equations of fluid dynamics, offering a versatile platform for academic research, industrial applications, and advanced engineering simulations. OpenFOAM supports a wide range of physical models and turbulence models, enabling users to simulate complex flow behaviours, heat transfer, multiphase flows, and chemical reactions in diverse geometries and boundary conditions. It provides a comprehensive suite of solvers, utilities, and libraries that facilitate the customization and development of tailored CFD simulations, making it suitable for specialized applications in automotive, aerospace, renewable energy, and process industries. With its open-source nature, OpenFOAM promotes collaboration and community-driven development, allowing users to contribute new models, solvers, and improvements to the software. Its scalability on high-performance computing clusters and compatibility with various operating systems enhance its appeal to researchers, engineers, scientists, and developers seeking robust and cost-effective CFD solutions.

8.5 Boundary Conditions

Boundary conditions are essential for accurately simulating the behaviour of solar gas turbines, as they define the external influences on the system, such as environmental and operating factors. These factors significantly impact the simulation results. Proper specification of boundary conditions ensures the reliability and accuracy of the model, leading to meaningful insights. By incorporating solar radiation, ambient temperature, and pressure effects, modelling experts can ensure that their models accurately reflect real-world operating conditions. This leads to more reliable predictions of the system performance and helps in optimizing solar gas turbine designs for enhanced efficiency and effectiveness.

8.5.1 *Incorporating Solar Radiation in Simulation*

Incorporating solar radiation in the simulation of solar gas turbines is a critical aspect that directly impacts the accuracy and reliability of the simulation results. Solar radiation serves as the primary energy source in solar gas turbines, and its accurate representation is essential for predicting thermal performance, efficiency, and overall system behaviour. This involves a comprehensive approach that considers the intensity, distribution, and variability of solar energy. By using advanced solar position algorithms, irradiance models, and heat flux boundary conditions, engineers and scientists can ensure that their simulations accurately reflect real operating conditions. This leads to more reliable predictions of system performance and helps in optimizing the design and operation of solar gas turbines for enhanced performance.

Solar radiation is converted to heat in the solar receiver of a solar gas turbine. This heat is then used to drive the gas turbine cycle, making the precise modelling of solar radiation crucial for accurate simulations. Understanding the variations in solar radiation due to geographic location, time of day, and atmospheric conditions helps in predicting the performance of solar gas turbines under different operating scenarios.

8.5.2 *Modelling Ambient Temperature and Pressure Effects*

Ambient temperature and pressure significantly influence the performance and efficiency of solar gas turbines. Accurate modelling of these environmental factors is essential to simulate actual operating conditions and predict system behaviour under various scenarios because these parameters affect different components.

8.5.2.1 Importance of Ambient Conditions

Ambient conditions, including temperature and pressure, directly affect the thermodynamic properties of the working fluid and the heat transfer rates within the system. Changes in these parameters can lead to variations in the performance of the compressor, combustor, turbine, and overall system efficiency. For instance, ambient temperature and pressure determine the density of air entering the compressor. Higher temperatures and lower pressures result in lower air density, impacting the mass flow rate and efficiency of the compressor. Additionally, properties of the working fluids (gases) vary with temperature and pressure, influencing the heat transfer rates in different components (Benoit et al. 2016).

8.5.2.2 Performance Metrics

Inlet conditions affect the compressor's pressure ratio and efficiency, making accurate modelling crucial for realistic predictions of compressor performance under varying ambient conditions. For hybrid solar gas turbines, the efficiency and stability of the combustion process also depend on the inlet pressure and temperature, which affect the air–fuel mixture and flame characteristics. Similarly, the turbine's performance is influenced by the pressure and temperature of the exhaust gases, which are in turn affected by the ambient conditions.

8.5.2.3 Temperature Effects

Ambient temperature profoundly impacts the thermodynamic cycle of solar gas turbines. The temperature of air entering the compressor affects the density and, consequently, the mass flow rate. Higher ambient temperatures reduce air density, decreasing the mass flow rate and compressor efficiency. The temperature at the combustor inlet influences the combustion process; higher inlet temperatures can improve combustion efficiency but may require adjustments in fuel input to maintain optimal performance. Temperature variations cause materials to expand or contract, potentially affecting the alignment and structural integrity of components, and can create thermal gradients within components, leading to thermal stresses that may impact the longevity and reliability of the system. Additionally, ambient temperature influences the heat transfer rates in heat exchangers and the efficiency of cooling systems, such as intercoolers and regenerators.

8.5.2.4 Pressure Effects

Ambient pressure is another critical factor that influences the performance of solar gas turbines. Lower ambient pressure reduces the pressure ratio across the compressor, decreasing its efficiency. This is particularly relevant at higher altitudes where atmospheric pressure is lower, leading to reduced mass flow rates and power output. Accurate modelling of altitude effects is essential for predicting system performance in different geographic locations. The pressure at the combustor inlet affects the air–fuel mixture and flame stability, making accurate pressure modelling necessary for realistic simulation of combustion efficiency and emissions. The pressure at the turbine exhaust influences the overall pressure ratio and efficiency of the gas turbine cycle.

8.5.2.5 Boundary Condition Implementation

Implementing ambient temperature and pressure as boundary conditions in simulations involves several steps to ensure accuracy and realism. Ambient temperature is applied as a boundary condition at the compressor inlet and turbine outlet of an open cycle gas turbine (OCGT), ensuring accurate simulation of the thermodynamic cycle. For components such as solar receivers and heat exchangers, the ambient temperature is used to model heat exchange with the environment. Ambient pressure is set as a boundary condition at the inlet and outlet of an OCGT system, influencing the flow dynamics and performance of the system. Additionally, pressure losses due to flow resistance in ducts and pipes are considered to model realistic pressure drops in gas turbine systems.

8.5.2.6 Practical Considerations in Simulation

Several practical considerations ensure the accurate representation of ambient temperature and pressure in simulations. High temporal resolution is essential to capture rapid changes in ambient conditions, particularly in dynamic simulations. Simulations often use hourly or minute-by-minute time steps to accurately model the diurnal variation of ambient temperature and pressure. Detailed spatial resolution helps in accurately modelling the distribution of temperature and pressure within the system, ensuring that localized effects, such as temperature gradients and pressure drops, are accurately captured. Validation with experimental data ensures the accuracy of the simulation models by comparing simulation results with experimental measurements. This involves using sensors and measurement devices to record ambient temperature and pressure under real conditions, setting up field experiments, and collecting data to validate the simulation results.

8.6 Performance Analysis

Performance analysis is a crucial aspect of evaluating the effectiveness and efficiency of solar gas turbines. It involves assessing various metrics and parameters derived from simulations or experimental data to gauge the system's operational characteristics and identify areas for improvement. By evaluating metrics such as overall efficiency, component-level efficiencies, power output, and thermal losses, engineers can optimize system design, enhance energy conversion efficiency, and ensure sustainable operation. These analyses help in identifying potential improvements and guiding decision-making processes towards achieving an optimal performance. Key elements involved in performance analysis include efficiency metrics, power output evaluation, and thermal loss assessment.

8.6.1 *Efficiency Analysis Through Simulation*

Efficiency analysis plays a pivotal role in the evaluation and optimization of solar gas turbine systems. By quantifying how effectively solar energy is converted into useful work or electricity, engineers can identify areas for improvement and ensure the system operates at peak performance.

8.6.1.1 Metrics for Efficiency Analysis

(a) Overall Efficiency

The overall efficiency of a solar gas turbine system (η_{overall}) represents the ratio of useful output energy (electricity generated) to the total input energy (solar energy intercepted by the concentrator). Mathematically, it is expressed as:

$$\eta_{\text{overall}} = \frac{\dot{E}_{\text{out}}}{\dot{E}_{\text{in}}} \tag{8.30}$$

where $\dot{E}_{\text{out}}$ is the electrical energy generated by the gas turbine, and $\dot{E}_{\text{in}}$ is the input energy. For solar-only systems, $\dot{E}_{\text{in}}$ is amount of solar energy intercepted by the concentrator. For hybrid systems, $\dot{E}_{\text{in}}$ is equal to the sum of solar energy and backup heat.

(b) Compressor Efficiency

The compressor efficiency measures the effectiveness of the compressor in increasing the pressure of the working fluid. Isentropic and actual processes are involved in the compressor. Consequently, both isentropic and actual efficiencies of the compressor are considered during cycle analysis (see Sect. 5.4).

$$\eta_{\text{com,is}} = \frac{\dot{W}_{\text{com,is}}}{\dot{W}_{\text{com,a}}} = \frac{\dot{W}_{\text{com,is}}}{\dot{m}\Delta h_{\text{com}}} = \frac{h_{2s} - h_1}{h_2 - h_1} = \frac{T_{2s} - T_1}{T_2 - T_1} \tag{8.31}$$

where $\dot{m}$ is the mass flow rate of the working fluid, $\dot{W}_{\text{com,a}}$ is the actual work input to the compressor, $\dot{W}_{\text{com,is}}$ is the isentropic work input to the compressor, h_1 is the specific enthalpy at state1, h_2 is the specific enthalpy at state 2, h_{2s} is the specific enthalpy at the isentropic state corresponding to state2, and Δh_{com} is the specific enthalpy change across the compressor.

(c) Turbine Efficiency

The turbine converts thermal energy into mechanical work. Isentropic and actual processes are involved in the turbine. So, both isentropic and actual efficiencies of the turbine are considered during cycle analysis (see Sect. 5.4).

$$\eta_{\mathrm{tu,is}} = \frac{\dot{W}_{\mathrm{tu,a}}}{\dot{W}_{\mathrm{tu,is}}} = \frac{\dot{m}h_{\mathrm{tu}}}{\dot{m}\Delta h_{\mathrm{tu}}} = \frac{h_3 - h_4}{h_3 - h_{4s}} = \frac{T_3 - T_4}{T_3 - T_{4s}} \tag{8.32}$$

where $\dot{W}_{\mathrm{tu,a}}$ is the actual work output from the turbine, $\dot{W}_{\mathrm{tu,is}}$ is the isentropic work output from the turbine, and $\Delta h_{\mathrm{tu}} = (h_3 - \mathrm{h}_{4\mathrm{s}})$ is the specific enthalpy change across the turbine.

(d) Generator Efficiency

A generator converts mechanical power into electrical power. Part of the work extracted by the turbine is used by the compressor. Therefore, the input work to the generator is equal to the difference between the work extracted by the turbine and work input to the compressor (see Sect. 5.4).

8.6.2 Simulated Power Output

Simulated power output analysis is a critical aspect of evaluating the performance of solar gas turbine systems. It involves calculating and analysing the electrical power generated by the turbine under simulated operational conditions, providing insights into the system's capacity to generate electricity, and aiding in performance assessment and design optimization.

The electrical power generated by the solar gas turbine system, as predicted through computational simulations, can be expressed as:

$$\dot{E}_{\mathrm{gen}} = \eta_{\mathrm{gen}} \dot{W}_{\mathrm{in}} \tag{8.33}$$

where $\dot{W}_{\mathrm{in}}$ is the input work to the generator and η_{gen} is the efficiency of the generator in converting mechanical power into electrical power (see Sect. 5.4).

Simulated power output is important because it offers key insights into the system's capacity to generate electricity under different operating conditions, thereby helping engineers and scientists assess performance, predict energy production, and optimize design for maximum efficiency.

8.6.3 Evaluating Thermal Losses in Solar Gas Turbine

Evaluating thermal losses is crucial in understanding the efficiency and performance of solar gas turbines. Thermal losses occur due to various inefficiencies within the system, including heat dissipation, mechanical losses, and thermal radiation. Thermal losses ($\dot{Q}_{\mathrm{loss}}$) can be represented as:

$$\dot{Q}_{\mathrm{loss}} = S - \dot{E}_e \tag{8.34}$$

where S is the total solar power absorbed by the system, and $\dot{E}_e$ the electrical power generated by the gas turbine system. Sources of thermal losses include the combustion process (incomplete combustion, excess air, and combustion chamber losses), heat transfer (losses through convection, conduction, and radiation from the solar gas turbine components and exhaust gases), and mechanical friction (frictional losses in bearings, seals, and moving parts of the turbine).

To evaluate thermal losses, methodologies such as energy balance analysis, simulation and modelling, and component-level analysis are employed. Energy balance analysis involves performing detailed energy balance calculations to quantify the amount of energy lost as heat within the system by calculating the energy input (incident solar radiation and, for hybrid systems, fuel) and assessing the electrical power output generated by the turbine. Simulation tools such as MATLAB/Simulink, ANSYS Fluent, or COMSOL Multiphysics simulate heat transfer processes and assess thermal losses by modelling heat transfer mechanisms, combustion dynamics, and fluid flow (Abubaker et al. 2024). Component-level analysis evaluates specific components, such as the combustion chamber, turbine, and heat exchangers, to identify sources and magnitudes of thermal losses through detailed simulations or experimental tests measuring heat transfer rates, temperature profiles, and thermal efficiencies of individual components.

Considerations in evaluating thermal losses include efficiency metrics that assess the impact of thermal losses on the overall system performance, and identify opportunities to minimize thermal losses through improved design, material selection, or operational adjustments. Environmental factors, such as ambient temperature, and atmospheric conditions that influence heat transfer and thermal losses, must also be considered to optimize system operation and mitigate the effects of these factors on thermal performance. Evaluating thermal losses in solar gas turbines enables the optimization of optical and thermal efficiencies to enhance overall system efficiency, maximize electricity generation from solar energy, and improve the economic and environmental sustainability of solar gas turbine installations.

8.7 Sensitivity Analysis

Sensitivity analysis is a systematic approach used to assess the sensitivity of system outputs to changes in input parameters or conditions, playing a crucial role in understanding how variations in key factors affect the performance, efficiency, and reliability of solar gas turbines (Cajahuaringa et al. 2024). By systematically evaluating the sensitivity of system outputs to input parameters, analysts can make informed decisions, improve system design, and enhance the economic and environmental sustainability of solar energy technologies. Methods for sensitivity analysis include parameter variation, where input parameters such as solar irradiance, ambient temperature, inlet pressure, turbine efficiency, and component dimensions are systematically varied using computational models or experimental setups to simulate their impact on system outputs such as power output and efficiency. The response surface methodology (RSM) constructs mathematical models describing the relationship between input parameters and system responses, collecting data across a range of parameter values, fitting response surfaces, and analysing the sensitivity to optimize system performance. Monte Carlo simulation stochastically varies input parameters according to their probability distributions to quantify the uncertainty and sensitivity of system outputs, generating multiple simulations with randomized input values to analyse the statistical distribution of outputs and identify critical parameters affecting performance.

Applications of sensitivity analysis in solar gas turbines include optimizing system design by identifying optimal configurations of components to maximize efficiency and power output, assessing the robustness of system operation against variations in environmental conditions and design parameters, and informing decision-making processes regarding investment decisions based on sensitivity analysis outcomes. Key considerations in sensitivity analysis include defining realistic ranges and distributions for input parameters, assessing interactions and dependencies between input parameters, and understanding their combined effects on system behaviour. Practical benefits of sensitivity analysis include performance optimization by fine-tuning system design and operational parameters, risk management by identifying critical factors affecting system reliability and resilience against environmental variability and providing decision support to stakeholders regarding the potential impacts of design choices and operational strategies on renewable energy deployment. Sensitivity analysis is crucial for enhancing system understanding by identifying input parameters that significantly influence system outputs and navigating the trade-offs and synergies between different design choices and operational strategies. It aids in optimizing design and performance by guiding modifications to achieve the desired system objectives and continuously improving operational efficiency and energy conversion rates. Managing risks and uncertainties through sensitivity analysis helps in: (a) assessing the impact of uncertainties on system reliability and economic viability, (b) quantifying risks associated with different scenarios, and (c) implementing robust mitigation strategies and contingency plans. Sensitivity analysis supports decision-making and policy development by providing clear and quantifiable insights into the consequences of different decisions and scenarios, informing evidence-based decision-making, and evaluating the effectiveness of policies promoting renewable energy adoption.

8.7.1 Importance of Sensitivity Analysis

Sensitivity analysis is vital in science and engineering, particularly for optimizing the design, operation, and decision-making processes related to complex systems such as solar gas turbines. It enhances system understanding by identifying input parameters that have the most significant influence on output variables such as power and efficiency. This understanding is crucial for focusing design efforts on optimizing the most influential parameters. Sensitivity analysis also provides insights into complex system dynamics, revealing how interactions between multiple components and environmental factors affect the overall performance, helping engineers and scientists navigate trade-offs and synergies between different design choices and operational strategies. For design optimization, sensitivity analysis guides modifications to achieve the desired system objectives by identifying critical parameters affecting performance metrics such as efficiency, reliability, and cost-effectiveness. It aids in performance improvement by fine-tuning operational parameters and optimizing control strategies to enhance system performance under varying conditions. Additionally, sensitivity analysis plays a crucial role in the management of risks and uncertainties, thereby helping in the assessment of the impacts of weather variations, component degradation, and operational fluctuations on system reliability and economic viability. By quantifying risks associated with different scenarios, engineers and scientists can implement robust mitigation strategies and contingency plans. Uncertainty quantification through probabilistic approaches such as Monte Carlo simulations provides insights into the range and distribution of possible outcomes, assisting stakeholders in making informed decisions about investments, policy development, and technology adoption. Furthermore, sensitivity analysis supports decision-making and policy development by providing clear, quantifiable insights into the consequences of different decisions and scenarios, facilitating evidence-based decision-making, and informing the formulation and implementation of effective energy and other policies.

8.7.2 Varying Input Parameters and Their Impact

Varying input parameters and assessing their impact through sensitivity analysis is a fundamental practice in optimizing the design and operation of systems. This involves identifying key input parameters (such as those affecting power output, efficiency, and operational reliability) by conducting a comprehensive review of system components, environmental factors, and operational variables. Systematic variation of these parameters across defined ranges or distributions is then carried out using computational tools like MATLAB/Simulink, ANSYS Fluent, and COMSOL Multiphysics, or through controlled physical experiments to observe real responses. The resulting changes in system performance metrics are quantified and analysed to understand their impact on the overall system behaviour and operational outcomes. This process reveals optimization opportunities, allowing the identification of optimal configurations and operating conditions that maximize system efficiency and performance, and the refinement of design parameters to improve the component integration and system reliability. Additionally, assessing parameter sensitivity aids in understanding the system's vulnerability to external factors and operational uncertainties, enabling the development of mitigation strategies to ensure robust system performance. Strategically, this analysis: (a) supports design optimization by adjusting parameters related to turbine efficiency, heat transfer characteristics, and operational settings, (b) enhances operational efficiency by fine-tuning parameters to achieve optimal energy output under varying conditions, and (c) provides stakeholders with actionable insights for investment decisions and policy formulation in renewable energy development.

8.8 Validation and Verification

Validation and verification are critical processes in ensuring the accuracy, reliability, and credibility of simulations used to study solar gas turbine systems. Validation aims at demonstrating that the simulation accurately represents the real behaviour of the system by comparing simulation and experimental data, thereby ensuring that the model captures the essential physics. Verification, on the other hand, confirms that the computational model is correctly implemented and performs as intended, focusing on the accuracy of numerical algorithms, boundary conditions, and implementation details.

Methods for these processes include comparing simulation results with experimental data under controlled conditions, using benchmark cases with well-established data, and performing sensitivity and robustness analyses. The importance of validation and verification lies in ensuring the reliability and credibility of simulation results, which supports informed decision-making, reduces risks associated with erroneous predictions, and enhances the overall confidence in the predictive capabilities of computational models. However, challenges such as the complexity of solar gas turbine systems and the limited availability of high-quality experimental data must be addressed. Future directions involve integrating advanced computational tools and data-driven validation techniques to improve the accuracy and predictive capabilities of simulations. Ensuring accuracy in simulations through continuous validation and verification is essential for optimizing system performance, supporting decision-making, and mitigating operational risks. Some researchers have conducted simulation and experimental studies. In this context, two case studies are briefly described in this section: (a) micro gas turbine combined with concentrated solar power system and (b) use of computational fluid dynamics on solar dish in concentrated solar power.

8.8.1 Micro Gas Turbine Combined with Concentrating Solar Power System

Matching the demand for power and reduction of distribution losses are some of the strategies for sustainable energy production and consumption. One way of achieving these strategies is through distributed energy systems. Basically, a distributed energy system is characterized by installing the energy production systems near consumers. Micro gas turbines are suitable for installation close to points of electricity use, and so, they are becoming popular in distributed generation. A micro gas turbine (MGT) is a class of gas turbines that has generally a power output of less than 300 kW (Basrawi et al. 2017).

MGTs driven by concentrated solar radiation are amenable to distributed energy systems. Shuai et al. (2024) examined the simulation and experimental validation of a micro gas turbine combined with a concentrated solar power system. The major components of their system were the heliostat field, the solar air receiver, and the Turbec T100 micro gas turbine. The objectives of their study were in threefold: (i) to investigate the impact of coupling CSP systems with MGTs on system performance, (ii) to analyse steady-state and dynamic characteristics of solar micro gas turbine (SMGT) under varying conditions and (iii) to provide insights into its application for grid stability and renewable energy integration.

The study emphasizes use of experiments and dynamic simulations to analyse the performance of SMGTs under varying ambient and load conditions, with a focus on grid-connected operation and peak-shaving capabilities. MATLAB software was used to model MGT control equations and integrate REFerence fluid PROPerties (REFPROP) for air property data and Heliostat Field Layout and Calibration (HFLCAL) software was used for modelling the solar air receiver energy flux distribution. Their experimental data validated the simulation model, yielding relatively high accuracy in the performance predictions. For instance, the solar receiver model yielded a mean absolute percentage error of 2.0 % between the empirical and simulated data for the receiver outlet temperature.

8.8.2 Use of Computational Fluid Dynamics in a Solar Dish Technology

Solar gas turbines involve fluid flows in different components including solar receivers, compressors, combustors, turbines, heat exchangers, and pipes. So, the performance of a solar gas turbine is affected by hydrodynamics. In this context, Ubando et al. (2022) studied the use of CFD in simulating SGT systems. Their study provides a bibliometric review of the application of CFD in solar dish technology for concentrating solar power. It examined 83 publications, of which 23 focused on CFD modelling of solar dish systems, analysing their contributions to receiver design, thermal efficiency, and optical simulation. The software used in the study included ANSYS Fluent and OpenFOAM for CFD modelling and tools like SolTrace and OptisWorks for ray tracing and optical efficiency studies. Critical equations include the Navier–Stokes equations for fluid dynamics and radiation models like the Monte Carlo Ray Tracing (MCRT) method for flux distribution. The software facilitated evaluation of design variables, such as porosity, tube pitch, and receiver inclination, to enhance performance.

The review shows significant advances in integrating CFD and optical simulations to optimize solar dish CSP systems, indicating substantial efficiency improvements through receiver design innovations. Volumetric receivers with porous absorbers emerged as the most promising due to their ability to operate at elevated temperatures. Despite these developments, challenges like convective losses, material durability, and system integration remain areas for future research, particularly in developing cost-effective and efficient configurations. The study accentuates the importance of numerical tools in promoting CSP technology, bridging the gap between theoretical analysis and practical applications.

8.9 Worked Examples

Example 8.1 An electric generator has an efficiency of 80% and receives an input mechanical power of 10 kW. Calculate the electrical power output of the generator.

Solution

The electrical power output E_{out} is given by:

$$E_{\text{out}} = \eta E_{\text{in}},$$

where η is the efficiency E_{in} is the input energy.

Therefore,

$$\begin{aligned} E_{\text{out}} &= 0.8 \times 10 \\ &= 8\ \text{kW} \end{aligned}$$

Example 8.2 An AC generator has 6 poles and operates at a rotational speed of 1800 revolutions per minute (rpm). Calculate the frequency of the output voltage in Hertz (Hz).

Solution

The frequency f of the output voltage from an AC generator is given by:

$$f = \frac{N_{\text{rpm}} N_{\text{po}}}{120}.$$

But,

$N_{\text{rpm}} = 1800$ rpm.

$N_{\text{po}} = 6$ poles.

Substituting these values yields:

$f = \frac{1800*6}{120} = 90\,\text{Hz}$.

Therefore, the frequency of the output voltage is 90 Hz.

Example 8.3 A solar gas turbine receives 1000 MJ of thermal energy input from the solar receiver and delivers 300 MJ of useful mechanical energy output. Calculate the thermal efficiency of the gas turbine.

Solution

The thermal efficiency η of a gas turbine is given by:

$$\eta = \frac{\dot{W}_u}{\dot{Q}_{in}},$$

where $\dot{W}_u$ is the useful work and $\dot{Q}_{in}$ is the input thermal energy

$$\eta = \frac{300}{1000} = 0.30.$$

Therefore, the thermal efficiency of the gas turbine is 0.30.

Example 8.4 A gas turbine generates a mechanical output power of 5 MW. The turbine operates with an isentropic efficiency of 85% and a specific enthalpy drop across the turbine of 1200 kJ kg^{-1}. Calculate the mass flow rate of the working fluid required to achieve this power output.

Solution

The mass flow rate $\dot{m}$ can be calculated using the formula:

$$\dot{m} = \frac{\dot{W}_{tu,is}}{\eta_{tu,is}\Delta h_{tu}} = \frac{5000}{0.85 \times 1200} \approx 4.90\, kg/s.$$

Therefore, the required mass flow rate is approximately 4.90 kg/s.

Example 8.5 A solar-only open cycle gas turbine comprises a compressor, pressurized solar receiver, and a turbine. Cold air enters the compressor at a pressure of 100 kN/m^2 and a temperature of 300 K, and it exits the compressor at a pressure of 1.2 MN/m^2. If the isentropic efficiency of the compressor is 0.80 and the ratio of specific heat capacities of air is 1.4, calculate the outlet temperature of the compressor.

Solution

$$T_2 = \frac{T_1\left(\eta_{com,is} - 1\right) + T_{2s}}{\eta_{com,is}} = \frac{T_1\left(\eta_{com,is} - 1\right) + T_1\left(\frac{P_2}{P_1}\right)^{(\gamma-1)/\gamma}}{\eta_{com,is}}$$
$$= \frac{300 \times (0.80 - 1) + 300 \times \left(\frac{1200}{100}\right)^{(1.4-1)/1.4}}{0.80}$$
$$= 687.7\text{ K}$$

Nomenclature

A	Area (m^2)
C	Concentration ratio (dimensionless) or specific heat capacity (J/kg K)
D	Damping coefficient (N s/m)
$\dot{E}$	Power (W)
E	Exergy flow (W)
G	Intensity of radiation (W/ m^2)
h	Specific enthalpy (J/kg) or coefficient of heat transfer (W/m^2.K)
H	Enthalpy (J)
k	Thermal conductivity (W/m K) or turbulent kinetic energy (J/kg)
$\dot{m}$	Mass flow rate (kg/s)
N	Number of turns in the coil (dimensionless)
P	Pressure (N/m^2)
$\dot{Q}$	Power (W)
S	Absorbed solar radiation (W)
t	Time (s)
T	Temperature (K) or torque (Nm)
u	Velocity vector (m/s)
$\dot{U}$	Rate of change of internal energy (W)
V	Voltage (v)
$\dot{W}$	Rate of work (W)

Greek Symbols

α	Thermal diffusivity (m^2/s)
γ	Ratio of the specific heat capacity of a gas at constant pressure to its specific heat capacity at constant volume (dimensionless)
ε	Dissipation rate (m^2/s^2)
Φ	Magnetic flux (T)
η	Efficiency (dimensionless)
μ	Dynamic viscosity (N.s/m^2)
θ	Rotor angle (radian)
ρ	Density (kg/m^3) or reflectance (dimensionless)
σ	Stefan-Boltzmann constant (W/m^2.K^4)
τ	Stress tensor (N/m^2)

Subscripts

conv	Convective

e	Electric
m	Mechanical
p	Constant pressure
r	Radiative
ref	Reflected
tu	Turbine
u	Useful

References

ABET (2016) Criteria for accrediting engineering programs. ABET Inc

Abubaker AM, Darwish AA, Singh BB, Manaserh YM, Al-Ghussain L, Akafuah NK, Saito K (2024) Energy and environmental analyses of a solar-gas turbine combined cycle with inlet air cooling. Sustainability 16(14):6229

Bathe KJ (2007) Finite element procedures. Prentice Hall, Watertown

Bejan A (2016) Advanced engineering thermodynamics. John Wiley & Sons, Hoboken, NJ

Bellos E, Tzivanidi C, Antonopoulos KA (2017) Parametric analysis and optimization of a solar assisted gas turbine. Energy Convers Manag 139:151–165

Benoit H, Spreafico L, Gauthier D, Flamant G (2016) Review of heat transfer fluids in tube-receivers used in concentrating solar thermal systems: properties and heat transfer coefficients. Renew Sustain Energy Rev 55:298–315

Blanc P, Espinar B, Geuder N, Gueymard C, Meyer R, Pitz-Paal R, Reinhardt B, Renné D, Sengupta M, Wald L, Wilbert S (2014) Direct normal irradiance related definitions and applications: the circumsolar issue. Sol Energy 110:561–577

Boynton RM (2003) Colour Science. In: Meyers RA (ed) Encyclopedia of physical science and technology. Elsevier, pp 289–313

Çengel YA, Boles MA (2006) Thermodynamics: an engineering approach. McGraw Hill, New York, NY

Basrawi F, Ibrahim TK, Habib K, Yamada T, Daing IDMN (2017) Techno-economic performance of biogas-fueled micro gas turbine cogeneration systems in sewage treatment plants: effect of prime mover generation capacity. Energy 124:238–248

Cajahuaringa A, Palacios RA, Mauricio Villanueva JM, Morales-Villanueva A, Machuca J, Contreras J, Rodríguez Bautista K (2024) Uncertainty evaluation of a gas turbine model based on a nonlinear autoregressive exogenous model and Monte Carlo dropout. Sensors 24(2):465

Cengel YA, Boles MA (2023) Thermodynamics: an engineering approach, 10th edn. McGraw-Hill Education, Boston

Cordova J, Machuca C, Velásquez J, Barrera R (2023) Assessment of large eddy simulation (LES) sub-grid scale models accounting for compressible homogeneous isotropic turbulence. In: 27th International Congress of Mechanical Engineering (COBEM 2023). https://doi.org/10.48550/arXiv.2309.05875

Deb K (2001) Multi-objective optimization using evolutionary algorithms. Wiley, New York

Dowling NE (2013) Mechanical behavior of materials: engineering methods for deformation, fracture, and fatigue, 4th edn. Pearson, Upper Saddle River

Duffie JA, Beckman WA (2013) Solar engineering of thermal processes, 4th edn. John Wiley & Sons Inc., Hoboken, NJ

Fishwick PA (2007) Handbook of dynamic system modelling. CRC Press, Boca Raton

Glassner AS (1993). An overview of ray tracing. In: Glassner AS (editor). An introduction to ray tracing. Academic Press, Eatontown, NJ

Germano M, Piomelli U, Moin P, Cabot WH (1991) A dynamic subgrid-scale eddy viscosity model. Phys Fluids A 3(7):1760–1765

Gropp W, Lusk E, Skjellum A (1999) Using MPI: portable parallel programming with the message-passing interface. MIT Press, Cambridge, MA

Hertwich EG (2005) Life cycle approaches to sustainable consumption: a critical review. Int J Environ Sci Technol 39(13):4673–4684

Heywood JB (2018) Internal combustion engine fundamentals, 2nd edn. McGraw-Hill Education, New York, NY

Horlock JH (2007) Advanced gas turbine cycles. Pergamon Press, Oxford

Incropera FP, DeWitt DP, Bergman TL, Lavine AS (2013) Fundamentals of heat and mass transfer. John Wiley & Sons, Hoboken

Kalogirou SA (2014) Solar energy engineering: processes and systems. Academic Press, Millbrae

Lees FP (2012) Lees' loss prevention in the process industries: hazard identification, assessment, and control, 4th edn. Butterworth-Heinemann, Boston

Li ZL, Lu XG, Wu YF, Han G (2023) Quantitative investigation of the turbulence model effect on high-pressure-ratio centrifugal compressor performance prediction. Int Commun Heat Mass Transf 142:106644

Lu Y-B, Wang L-C, Zhou J-J, Niu Z-G, Zhang M, Qin W-M (2023) Assessment of the high-resolution estimations of global and diffuse solar radiation using WRF-Solar. Adv Clim Chang Res 14(5):720–731

Matyushenko A, Kolmogorov D (2021) An overview of hybrid RANS–LES models developed for industrial CFD. Appl Sci 11(6):2459

Metropolis N, Ulam S (1949) The Monte Carlo method. J Am Stat Assoc 44(247):335–341

Mobley RK (2002) An introduction to predictive maintenance. Butterworth-Heinemann, Woburn

Modarres M, Kaminskiy M, Krivtsov V (2016) Reliability engineering and risk analysis: a practical guide. CRC Press, Boca Raton

Moin P, Mahesh K (1998) Direct numerical simulation: a tool in turbulence research. Annu Rev Fluid Mech 30(1):539–578

Munjal ML (2013) Noise and vibration control. World Scientific, Singapore

Nafey AS, Abdelkader M, Abdelmotalip A, Mabrouk AA (2000) Parameters affecting solar still productivity. Energy Convers Manag 41(16):1797–1809

Pahl G, Beitz W, Feldhusen J, Grote K-H (2007) Engineering design: a systematic approach. Springer, Berlin

Pfenninger S, Gauche P, Lilliestam J, Damerau K, Wagner F, Patt A (2014) Potential for concentrating solar power to provide baseload and dispatchable power. Nat Clim Change 4:689–692

Sandá A, Moya S, Valenzuela L (2019) Modelling and simulation tools for direct steam generation in parabolic trough solar collectors: a review. Renew Sustain Energy Rev 113:109226

Sargent RG (2013) Verification and validation of simulation models. J Simul 7(1):12–24

Schrage M (1999) Serious play: how the world's best companies simulate to innovate. Harvard Business School Press, Havard

Shuai W, Zhang T, Lv H, Ying H, Wang K, Xu H, Chen D, Zhu P, Xiao G (2024) Experiment and dynamic simulation of micro gas turbine combined with concentrated solar power system. Renew Energy 237:121705

Thomke S (2003) Experimentation matters: unlocking the potential of new technologies for innovation. Harvard Business School Press, Havard

Turns SR (2012) An introduction to combustion: concepts and applications. McGraw-Hill Education, Singapore

Ubando AT, Conversion A, Barroca RB, Enano Jr NH, Espina RU (2022) Computational fluid dynamics on solar dish in a concentrated solar power: a bibliometric review. In: Antón JM, Werner JH (eds) Solar, vol 2 (2), pp 251–273

Versteeg HK, Malalasekera W (2007) An introduction to computational fluid dynamics: the finite volume method. Longman Scientific and Technical

Wilcox DC (1993) Turbulence modeling for CFD. La Cañada, DCW Industries
Yılmaz İH, Mwesigye A (2018) Modeling, simulation and performance analysis of parabolic trough solar collectors: a comprehensive review. Appl Energy 225:1351–2174
Zhang HL, Baeyens J, Degrève J, Cacères G (2015) Concentrated solar power plants: review and design methodology. Renew Sustain Energy Rev 22:466–481
Zeigler BP, Praehofer H, Kim TG (2000) Theory of modelling and simulation: integrating discrete event and continuous complex dynamic systems. Academic Press, San Diego

Chapter 9
Testing of Solar Gas Turbines

9.1 Introduction

In the engineering design life cycle, testing is an essential stage, bridging the transition from theoretical models to real applications. It serves as a critical phase for verifying whether a new product or process meets the design specifications and performs as expected under various operating conditions. This iterative process (design, test, modify, and retest) enables engineers and scientists to optimize performance and reliability until the final product is fully refined and satisfactory (Grove and Davis 1992). Testing is also used to observe how changes in independent variables affect dependent variables, providing valuable data for system optimization. Optimization within engineering design aims to maximize or minimize specific performance criteria, often requiring careful consideration of trade-offs. For solar gas turbine (SGT) systems, for example, improving efficiency could potentially increase system costs. To manage these competing objectives, multi-objective optimization techniques are widely employed. Following the completion of the optimal design, prototyping and testing are necessary to ensure the technology performs as anticipated.

To maintain consistency and reliability in testing, the application of industry standards and codes is essential. Traditional gas turbines are tested following well-defined standards; however, adapting these turbines to solar applications introduces unique challenges, because SGTs utilize concentrated solar radiation as a primary heat source. Despite significant advancements in concentrating solar power (CSP) technology, the industry lacks specific standards for testing, which hampers consistent evaluation and development the technology. Key projects discussed in this chapter showcase the integration of existing gas turbine models with CSP systems, revealing technical challenges that stem from modifying conventional turbines to accommodate solar inputs.

The complex design of SGTs incorporates a wide range of materials and components with varied properties, making rigorous testing indispensable. Testing ensures that the final product is efficient, durable, flexible, and sustainable across social,

A. Madhlopa and R. Nkhoma, *Principles of Solar Gas Turbines for Generating Electricity*, Green Energy and Technology,
https://doi.org/10.1007/978-3-031-84096-8_9

economic, and environmental dimensions. Standardized methods are also critical to assessing how well SGTs meet customer requirements because this is essential for market competitiveness. Among the material properties important for SGTs are optical, thermal, and mechanical characteristics. For instance, heliostats require high reflectance to optimize the optical efficiency of a solar field, a property that can be enhanced by recent innovations in material coatings. Likewise, components such as receivers, combustors, and turbines operate under high temperatures and pressure (Heller et al. 2006), requiring materials that can endure these extreme conditions. For instance, the high pressure ratios in modern gas turbines limit operational flexibility, and factors like combustion at elevated temperatures impact component performance and longevity on the turbine's hot side. Accurate measurements of such parameters are essential for evaluating the overall performance of the SGT system.

Since an SGT comprises two primary subsystems (CSP and gas turbine) typically developed and tested independently, system integration testing becomes crucial to determine collective performance. A CSP system generates thermal energy through a process in which solar radiation is concentrated onto a solar receiver that converts a portion of the absorbed solar energy into heat. This heat is then transferred to a heat transfer medium. The thermal output of the solar receiver can be estimated by measuring the mass flow rate, and inlet and outlet temperatures of the heat transfer medium. Accurate measurements of direct normal irradiance on the receiver aperture are also necessary for calculating solar irradiance. Receiver thermal efficiency (η_{rc}) is computed from the following formula:

$$\eta_{rc} = \frac{\dot{m}_{htm} C_{p,htm} \left(T_{out,rc} - T_{in,rc} \right)}{A_{rc} G_{rc}} \tag{9.1}$$

where $C_{p,htm}$ is the specific heat capacity of the heat transfer medium at constant pressure, $\dot{m}_{htm}$ is the mass flow rate of the heat transfer medium, $T_{in,rc}$ and $T_{out,rc}$ are the inlet and outlet temperatures of the medium, A_{rc} is the aperture area of the receiver, and G_{rc} is the solar irradiance on the receiver (Ebert et al. 2016).

For technology performance assessment, adherence to standardized test methods is essential, with major organizations such as the International Organization for Standardization (ISO), International Electrotechnical Commission (IEC), and the American Society of Mechanical Engineers (ASME) providing standards for both CSP and gas turbine technologies. These standards enable uniform testing of protocols and help ensure that technologies meet both operational and regulatory requirements, contributing to their reliability and potential scalability for commercial use.

9.2 Performance Assessment of Concentrating Solar Power Technology

In a CSP system, the solar field, composed primarily of the concentrator and receiver, is the foundation of the CSP technology. The concentrator, typically built with a reflective mirror surface (see Sect. 4.2), plays a critical role in directing sunlight towards the receiver. As mirrors constitute the largest physical component of the CSP system, they significantly influence the system performance. A key characteristic of these reflective surfaces is their solar specular reflectance that measures the mirrors' ability to reflect sunlight precisely towards the target area. High specular reflectance minimizes diffuse reflection that otherwise reduces the optical efficiency of the solar field (Fernández-García et al. 2017).

In solar tower systems, several parameters determine the concentrator's effectiveness. These include the total reflective area, slope and shape deviations across various heliostat orientations, and environmental factors such as temperature and wind loads. Additionally, the tracking accuracy, reflectance, and specularity of the reflective material used in heliostats are essential to maintaining an optimal focus on the receiver, ensuring maximum energy capture (Pfahl et al. 2017; Röger et al. 2017).

The performance of CSP components and systems is often gauged through their energy conversion efficiency that represents the ratio of useful energy output to the total input energy (either solar or a combination of solar and fuel). Efficiency serves as a fundamental metric in evaluating component and system performance, directly influencing the economic and operational viability of the whole system. To accurately assess the performance of the CSP subsystem, several environmental and operational variables must be measured and monitored. These include solar radiation, temperature, flow rate, pressure, and the thermophysical properties of the heat transfer fluid, as well as steam quality in steam-generating CSP configurations. Understanding these parameters is critical for evaluating system performance and ensuring that the components function within their specified operational limits (Janotte et al. 2017).

9.2.1 *Concentrated Solar Radiation*

Solar radiation is the most important environmental factor that influences the performance of CSP systems. Only the direct component of the global solar radiation is exploited by the CSP technology. Concentrators focus the direct solar radiation onto a receiver that absorbs and converts the radiation to heat for driving the CSP system. Details about the measurement of this component on horizontal and tilted surfaces are covered in Chap. 3. Concentrated solar radiation on the receiver may reach the order of megawatts. So, measurement of this quantity requires special radiometric systems. There are different methods for measuring irradiance on the receiver surface, and they fall into direct and indirect methods (Casanova et al. 2022). The commonest direct method for measuring the flux distribution on solar receivers is based on a moving bar with sensors that measure while the bar scans the beam of incident concentrated solar radiation close to the receiver aperture. Another direct method uses cooled radiometers distributed on the surface of the receiver. For instance, the Planta Solar 10 (PS10) CSP plant (in Spain) uses this method to measure the level of irradiance on the receiver. This plant has a capacity of 11 MW, was commissioned in 2007 and operational at the time of writing this book. An example of indirect methods is a system consisting of a white lambertian target that passes in front of the solar receiver, a radiometer and a digital camera. A lambertian surface reflects all the incident radiation equally in all directions. In this context, the white lambertian target reflects the concentrated solar radiation in order to be captured by the digital camera. A radiometer is used to calibrate the system so that it is capable of converting the grey levels of the camera to units of irradiance.

9.2.2 Temperature

Temperature, defined as the degree of hotness or coldness of a body, is a crucial parameter in assessing the thermal efficiency of a component or system. In a CSP subsystem, absorbed solar radiation increases the temperature of the heat transfer medium as it flows through the receiver. In cases where solar heat drives latent processes, the medium's temperature may remain more stable despite the energy input. Precise temperature measurement is essential for understanding the system's performance and is accomplished by using various types of thermometers, each based on different operating principles, including resistance, thermal expansion, thermoelectric effect, and radiation (Nicholas and White 1994).

9.2.2.1 Resistance Thermometers

A resistance thermometer, also known as resistance temperature detector (RTD), offers high resolution (up to 1×10^{-3} K) and a wide operational range (14 to 1233 K). These thermometers, including the commonly used Pt100, use platinum coils with a standard resistance of 100 Ω at 273.15 K. Platinum's specific resistance increases linearly with temperature, making Pt100 thermometers highly accurate and stable across a range of industrial applications. However, these devices require an external voltage or current signal, which can introduce errors due to resistance-measurement inaccuracies.

9.2.2.2 Liquid-in-Glass Thermometers

Liquid-in-glass thermometers measure temperature based on the principle of thermal expansion of a liquid, typically mercury or alcohol within a glass tube. They are effective in the temperature range of 83 to 873 K with resolutions reaching 1×10^{-3} K. Widely applicable in many fields, these thermometers are inexpensive and easy to use but are limited to infrequent or manual readings, making them less suited for continuous monitoring or remote applications.

9.2.2.3 Thermocouples

Thermocouples function on the principle of the thermoelectric effect, where two dissimilar metals form a junction that produces a voltage proportional to the temperature difference between the junction and a reference point. Thermocouples can connect to most data loggers, making them ideal for continuous and automated temperature monitoring in various industrial applications (Delta-T Devices 2000). Despite their versatility, thermocouples may suffer from different sources of errors, including thermal inhomogeneity, reference junction temperature, interference, and wire resistance. Accurate calibration and error correction techniques are necessary to ensure precise readings.

9.2.2.4 Radiation Thermometers

A radiation thermometer is a device that senses radiant flux within a given spectral interval, and it is calibrated to measure the temperature of a body that emits the same flux (Bergman et al. 2011). These sensors can be used to remotely assess temperature without direct contact, which is advantageous when measuring high temperatures of solar receivers. Although ideal for high-temperature applications where direct contact may be impractical or hazardous, radiation thermometers often provide slightly less accuracy than contact-based temperature measurements due to factors such as emissivity variations and environmental interference.

Given the high operating temperatures in solar receivers, remote measurement using radiation thermometers is particularly suitable. However, when high precision is required, contact-based methods such as RTDs or thermocouples may offer superior accuracy. A balanced approach that considers the specific application requirements, measurement accuracy, and safety factors is critical in selecting the appropriate temperature measurement method for CSP applications.

9.2.3 Fluid Flow Rate

The mass flow rate of the heat transfer fluid is a key input for calculating enthalpy in various components of a solar gas turbine system, as discussed in Chap. 5. Accurate flow rate measurement is essential to ensure efficient thermal management and optimal performance. Flow measurement in gases is typically conducted with gas

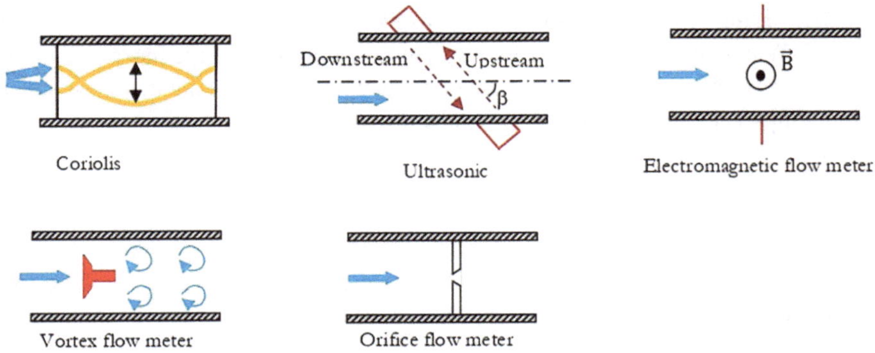

Fig. 9.1 Principles of gas flow measurement. Adapted from (Janotte et al. 2017)

flow meters, which fall into several main types: Coriolis, ultrasonic, vortex, orifice, and electromagnetic meters (Janotte et al. 2017), illustrated in Fig. 9.1.

9.2.3.1 Coriolis Flow Metres

Coriolis flow metres directly measure mass flow by using one or more axially displaced pipes that are induced to oscillate at their resonant frequency. The upstream and downstream sections of the pipes oscillate in phase when there is no flow, but when the fluid flows, the oscillation is disrupted, creating a phase shift proportional to the mass flow rate (Cai 2013). Operating frequencies for Coriolis metres range from 50 Hz to 1 kHz that is influenced by the system's total mass (Henry et al. 2013). They are the only devices that directly measure mass flow rate and can operate up to 673 K. Nevertheless, some components of CSP systems operate at high temperatures. For instance, Heller et al. (2006) report that pressurized volumetric solar receivers can practically heat air to 1273 K. Ultra-high-temperature solar receivers have the theoretical potential to produce heat at temperatures above 1300 K (Abdelsalam et al. 2023). Consequently, the maximum operational temperature of Coriolis flow metres limits their application in high-temperature CSP systems.

9.2.3.2 Ultrasonic Flow Meters

An ultrasonic flow meter (UFM) uses either the transit-time or Doppler-shift principle. In the transit-time type, a pair of transducers inclined at an angle (β) to the flow axis emit sound waves, with the time difference ($\Delta t = t_{up} - t_{down}$) between upstream and downstream transits proportional to the flow rate. The flight times upstream (t_{up}) and downstream (t_{down}) can be computed as follows: (Zhou et al. 2018):

$$t_{up} = \frac{L}{v_{us} - v_{ag}\cos\beta} \tag{9.2}$$

$$t_{\text{down}} = \frac{L}{v_{\text{us}} + v_{\text{ag}} \cos \beta} \tag{9.3}$$

where L is the path length between transducers, v_{us} is the speed of ultrasound, and v_{ag} is the average flow velocity. Ultrasonic flow metres are advantageous because they lack moving parts, have no pressure drop, accommodate large pipe diameters, and allow bidirectional flow measurement. However, when the gas is wet, transit-time UFMs may overestimate flow rates (Xu et al. 2017).

9.2.3.3 Electromagnetic Flow Meters

Based on Faraday's law, electromagnetic flow meters measure the flow rate by detecting the voltage U_e induced in a conductive fluid moving through a magnetic field. The induced voltage is expressed as:

$$U_e = \int \vec{v} \times \left(\vec{B} \times \vec{J}\right) dv \tag{9.4}$$

where $\vec{v}$ is the velocity vector of the fluid, $\vec{B}$ is the magnetic flux, and $\vec{J}$ is the virtual current density (Watral et al. 2015). These meters, with a history of over 180 years, are widely used in both open and closed channels, competing with UFMs for precision in fluid flow measurement. Recent innovations have led to low-energy designs, improving their efficiency and range of applications.

9.2.3.4 Vortex Flow Meters

Vortex meters detect flow rate by sensing the frequency of vortices created as the fluid passes a bluff body in the flow path. Sensors on the channel walls or bluff body detect these vortices, with the frequency being directly proportional to the average flow velocity (Hans and Windorfer 2003). Vortex meters (used for gases, steam, and liquids) are accurate and reliable because they do not use moving parts. However, the bluff body creates a pressure drop, which can affect system efficiency.

9.2.3.5 Orifice Flow Meters

Orifice meters measure flow rate by detecting the pressure drop across a constriction in the flow path. As the fluid velocity increases through the constriction, a pressure gradient is created, allowing for flow rate calculation based on the pressure readings taken upstream and downstream. Orifice meters may have single or multiple holes, with multi-hole configurations improving the discharge coefficient and reducing pressure loss (Singh and Tharakan 2015). They are commonly used due to their compact size, cost-effectiveness, and simplicity, especially in large flow rate applications.

Each type of flow meter offers unique benefits and limitations based on application, accuracy, and operating conditions. Selection should consider the specific requirements of the CSP system, environmental constraints, and temperature ranges to ensure optimal performance and durability in CSP applications.

9.2.4 Pressure

Pressure measurements in CSP systems are crucial for detecting changes across system components, verifying pressure levels, computing thermophysical properties, and measuring volumetric flow rates via devices such as orifices and nozzles (Janotte et al. 2017). Pressure-sensitive properties, such as the specific heat capacity of heat transfer or working fluids, rely on precise pressure data for accurate thermodynamic assessments.

Various pressure measurement methods are available, and selecting the most appropriate one depends on factors such as required measurement precision, operating temperature, fluid type and state, and acceptable uncertainty range. High-temperature fluids, for example, may require thermal separation between the fluid and the pressure sensor to prevent sensor damage. When performing differential pressure measurements, short and straight pressure lines are recommended to reduce measurement error. Additionally, ensuring vertical alignment in liquid-based systems helps to prevent gas buildup, which could interfere with the accuracy of readings.

9.2.5 Thermophysical Properties of Working and Heat Transfer Fluids

CSP plants use various working fluids (such as air, and water) along with heat transfer fluids that include liquid metals and suspended solid particles. The properties of these fluids are essential for thermodynamic calculations and system optimization (see Chap. 5). While air and water are well-documented, data on the thermophysical properties of newer heat transfer media is comparatively limited. Ongoing research is expanding this data set, especially as new fluids are developed to enhance heat engine efficiency, including the CSP technology.

Long-term use of fluids in closed-loop systems leads to gradual fluid degradation, often accelerated by high operational temperatures. Additional factors, such as system design and transient operating conditions, may further impact fluid properties. Routine quality checks are therefore essential to ensure accurate and reliable testing. Key thermophysical properties requiring regular monitoring include density, viscosity, and specific heat capacity, as these significantly influence the performance of CSP systems.

9.2.6 Quality of Steam

At present, most of the operating CSP power plants are based on the steam Ranken cycle. Steam quality is measured by the mass fraction of steam in a saturated steam/water mixture. The presence of a liquid within the steam flow helps wet the inner surfaces of the tubes, thereby enhancing heat transfer and maintaining tube temperatures within safe operational limits. Without this cooling effect, tube overheating could lead to system failures. However, overly high steam quality may cause solid deposits in the tubes, which impede heat transfer and elevate tube temperatures, potentially compromising system reliability (Xie et al. 2013). Conversely, if steam quality is too low, system efficiency is adversely affected.

Steam quality at the vaporizer outlet generally ranges between 0.6 and 0.8, particularly in recirculation mode of CSP systems based on the steam Ranken cycle (Aurousseau et al. 2016). Maintaining a high steam quality reduces the demand on recirculation pumps, thereby optimizing system performance. However, to prevent tube dry-out and potential damage, steam quality should not be excessively high to provide a margin of safety for the system.

9.2.7 Codes and Standards

While Sect. 3.5 discusses solar irradiance measurements, testing solar collection systems extends beyond this, requiring standards such as ISO 9806 for comprehensive guidance. The ISO 9806 standard defines procedures for testing fluid-heating solar collectors across various metrics-performance, reliability, durability, and safety (International Organization of Standardization 2013). These tests can be performed outdoors under either steady-state conditions or over an entire day, covering applications such as tracking solar concentrators used for power generation and industrial heating. This standard allows detailed thermal performance characterization of fluid-heating solar collectors.

At the time of writing this book, no published international standards existed specifically for utility-scale CSP power plant performance assessments. However, the ASME and IEC established committees to create standards for CSP evaluation. The International Committee for Standardization of CSP plants components (IEC TC 117) started to work on a testing standardization for solar power plants (International Electrotechnical Commission 2024). This committee is currently working on eight different projects on the development of standards shown in Table 9.1: As can be seen from this table, all the standards under discussion are expected to be published between 2025 and 2027. In the interim, the National Renewable Energy Laboratory

(NREL) developed performance test guidelines between 2011 and 2013. NREL's guidelines are aimed at performance acceptance tests for large-scale parabolic trough and solar tower power plants, though they are advisory and might require adaptations to suit specific CSP projects, especially for systems with thermal storage, to ensure high accuracy of testing.

The NREL guidelines propose both short- and long-duration performance tests. Short-duration tests, lasting a minimum of 24 h, assess power plant performance under clear-sky conditions for at least four hours with direct normal irradiance exceeding 700 W/m^2 and without any non-solar backup. Long-duration tests span at least a year, focusing on annual production, auxiliary energy use, and performance metrics, direct normal irradiance, net electricity generation, and efficiency.

The Spanish Association for Standardisation and Certification (AENOR) has developed the AENOR UNE 206010 Standard for cylindrical parabolic trough collector (PTC) solar thermal power plants. This national standard applies to PTC plants of any size, with or without auxiliary heaters and optional thermal storage,

Table 9.1 Projects on the development of standards for solar thermal electric plants

Project reference number	Project description	Expected publication date
IEC 62862-1-7 ED1	Solar thermal electric plants—Biphenyl/diphenyl oxide-based heat transfer fluids for use in line-focus concentrated solar power applications	11/2027
IEC 62862-2-2 ED1	Solar thermal electric plants—part 2–2: thermal energy storage systems—technical requirements for molten salt used as heat storage and heat transfer medium	05/2026
IEC 62862-3-4 ED1	Solar thermal electric plants—part 3–4: code of solar field performance test for parabolic trough solar thermal power plant	11/2025
IEC 62862-3-5 ED1	Solar thermal electric plants—part 3–5: laboratory reflectance measurement of solar reflectors	12/2025
IEC 62862-3-6 ED1	Solar thermal electric plants—part 3–6: durability of silvered-glass reflectors—laboratory test methods and assessment	02/2026
IEC 62862-3-7 ED1	Solar thermal electric plants—part 3–7: criteria for design, installation, and performance verification of flexible pipe connectors in parabolic trough collector technology	09/2026
IEC 62862-4-2 ED1	Solar thermal electric plants—part 4–2: heliostat field control system of solar tower plants	09/2025
IEC 62862-4-3 ED1	Solar thermal electric plants—part 4–3: technical requirements and design qualification of heliostats for solar power tower plants	12/2025

Source of data International Electrotechnical Commission (2024)

but does not cover other CSP technologies. Like the NREL guidelines, the AENOR standard includes requirements for both short- and long-duration acceptance testing.

9.3 Performance Assessment of Solar Gas Turbines

In solar gas turbine systems, thermal energy from the solar field is transferred either directly or indirectly to the turbine. In a direct-heat transfer setup, the receiver outlet temperature ($T_{\text{out,rc}}$) needs to closely match the turbine inlet temperature (T_{tin}) for optimal operation. In hybrid solar gas turbines, if $T_{\text{out,rc}}$ falls below T_{tin}, the hot fluid from the receiver is directed to the combustor for additional heating. Alternatively, the fluid can bypass the combustor entirely when $T_{\text{out,rc}} \approx T_{\text{tin}}$.

As observed in Chap. 1, gas turbines operate at elevated temperatures, influencing the performance and longevity of components on the turbine's hot side. Modern gas turbines also use high compressor ratios, yielding a narrow operational margin and rendering the system more susceptible to issues like compressor fouling. In combined cycle or cogeneration systems, backpressure exerted on the system further impacts the overall performance. Other factors, such as pressure drops across air filters, affect turbine efficiency by contributing to pressure losses. Flow resistance from changes in flow cross-section at various points (e.g., compressor inlet, combustor, turbine outlet, recuperator) also impacts the turbine efficiency (Chen et al. 2004).

Vibrations are another critical factor in gas turbine performance evaluation because they waste energy, thereby lowering system efficiency. The main cause of vibration is rotor imbalance that may arise internally due to material non-homogeneity or externally from asymmetries in operation (Boyce 2002). Therefore, assessing gas turbine performance is essential to ensure efficiency and identify any detrimental impacts on component longevity and energy output.

9.3.1 Codes and Standards

Solar gas turbines for electricity generation must adhere to rigorous standards to meet power production requirements. ISO 19859 outlines methods for evaluating the performance of simple cycle, combined cycle, and cogeneration gas turbines, for both onshore and offshore applications (International Organization for Standardization 2016). This standard is based on recognized ISO and IEC standards to ensure consistency in quality and performance. It covers essential requirements related to functional design, safety, and performance, with provisions allowing the use of alternative codes and standards if they meet equivalent standards of reliability, functionality, and safety. This flexibility allows SGT systems to incorporate innovations while still conforming to stringent standards for energy technologies.

9.3.1.1 ISO 2314: Gas Turbines—Acceptance Tests

ISO 2314 is a standard applicable to closed, open, and semi-closed cycle gas turbine power plants that use gaseous and/or liquid fuels (International Organization for Standardization 2009). This standard also includes combined cycle gas turbines integrated with heat recovery systems and can be applied to systems utilizing free-piston gas generators or other specialized heat sources. Emphasizing environmental protection, the standard provides procedures for acceptance tests on gas turbines equipped with emission control or power augmentation technologies, such as fluid injection and inlet air treatment.

ISO 2314 proves valuable for comparing gas turbine performance before and after modifications, upgrades, or maintenance work. It specifies procedures for measuring parameters such as electrical or mechanical output, thermal efficiency, heat rate, and turbine exhaust energy. The standard provides comprehensive methods, including required instruments, calibration protocols, and handling requirements. It also details operating and boundary conditions, ensuring tests are conducted within predefined limits and that the data is recorded and processed appropriately.

9.3.1.2 ISO 3977: Gas Turbine—Procurement

ISO 3977 provides essential criteria for gas turbine procurement, ensuring performance reliability and environmental compliance in various applications. It is dedicated to various aspects of gas turbine procurement, divided into five parts. Part 1 provides definitions and terminology for terms like gas turbine, power plant configurations (e.g., open, closed, and semi-closed cycles), and cycle types (e.g., simple, regenerative, intercooled) (International Organization for Standardization 1997a). Part 2 establishes standard reference conditions to ensure consistency in performance evaluations. Key conditions include air intake parameters (total pressure of 101,300 N/m^2, temperature of 288.15 K, and relative humidity of 60%) and exhaust conditions at the turbine exhaust flange. These reference points allow for global consistency in performance comparisons (International Organization for Standardization 1997b). Part 3 addresses design requirements for procurement, including guidelines for combined cycle gas turbines and ancillary components. It provides technical specifications, assisting buyers in procuring open cycle turbines for power generation. Notably, this part excludes gas turbines for propulsion in vehicles, agricultural equipment, and heavy machinery (International Organization for Standardization 2004). The fourth part focuses on environmental considerations for alignment with the industry's push for reduced emissions. This part provides guidelines for fuel quality and environmental impact data, ensuring that gas turbines meet emission standards for CO_2, NOx, and other pollutants. This aligns with ISO environmental sustainability goals, helping to address concerns over emissions from combustion-based energy sources (International Organization for Standardization 2002). Finally, Part 5 outlines design and testing requirements for gas turbines used in petroleum and natural gas sectors, including compressor-driven applications for extraction and conveyance. This part provides guidance on material quality, fabrication, inspection, and testing, supporting safe and reliable turbine use in these energy-intensive industries.

9.3.1.3 ISO Standard 20816-2: Mechanical Vibration—Measurement and Evaluation of Machine Vibration

ISO 20816-2 provides guidelines for measuring and assessing vibration parameters to ensure reliability, safety, and extended operational life in high-output rotating machines. It focuses on assessing vibration severity in high-power rotating machinery, including land-based gas turbines, steam turbines, and generators with outputs over 40 MW and standard operating speeds of 1500 rpm, 1800 rpm, 3000 rpm, or 3600 rpm (International Organization for Standardization 2017). This standard applies to: (a) Structural vibration at main bearing housings that can impact the alignment and stability of rotating machinery, (b) structural vibration at thrust bearing housings (evaluates vibration severity at thrust bearing points, critical for axial alignment and load handling), and (c) vibration of rotating shafts (monitors and evaluates shaft vibrations, crucial for rotor balancing and reducing wear on the system).

9.3.1.4 ASME PTC 22: Gas Turbines

The ASME PTC 22 standard provides a comprehensive set of guidelines and rules for conducting and accurately reporting thermal performance tests on open cycle gas turbine power plants and engines (American Society of Mechanical Engineers International 2014). It includes detailed procedures for determining corrected power, heat rate, exhaust flow, exhaust energy, and exhaust temperature that are essential metrics for assessing gas turbine efficiency.

The code also outlines procedures for designing test requirements and programs, allowing engineers and scientists to evaluate both absolute and comparative performance metrics of gas turbines. Widely recognized and applied internationally, ASME PTC 22 is instrumental in delivering highly accurate test results, ensuring consistency and reliability across global gas turbine performance assessments.

9.3.1.5 ASME PTC 46: Overall Plant Performance

ASME PTC 46 standard is designed to assess the overall performance of power plants of various sizes under normal operating conditions (American Society of Mechanical Engineers International 2015). This code provides explicit methods and procedures applicable to combined cycle power plants as well as Rankine cycle plants powered by gas, liquid, or solid fuels. A major advantage of ASME PTC 46 is its flexibility—it can be applied to other types of heat engines, provided that the specified methods and equations are used. For instance, the performance equations included in the code can

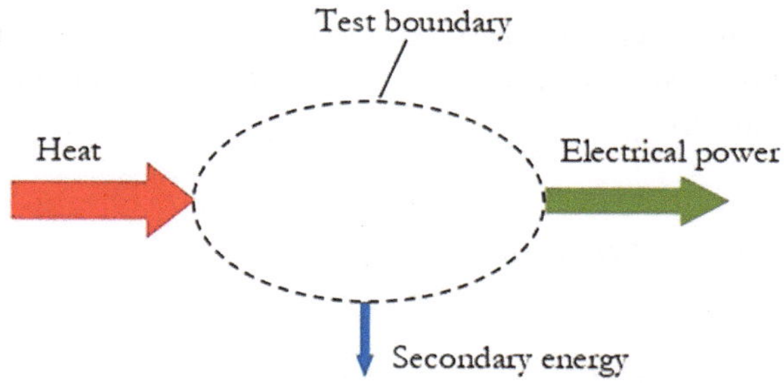

Fig. 9.2 Schematic representation of energy flows into and out of the test boundary

be employed in the steam cycle section of a solar power plant. The code, however, does not apply to testing individual components such as compressors or turbines. It specifies the following conditions that must be satisfied to assess a particular power plant or cogeneration system:

- Availability of methods for direct or indirect measurement of all thermal energy flowing into the test boundary and all electrical power or secondary energy output flowing out (Fig. 9.2).
- Capability to determine all parameters for adjusting test results to the baseline reference conditions.
- Test result uncertainties should not exceed limits specified in Chaps. 1–3 of the code for the applicable plant type.
- For vapour cycles, steam should be the working fluid, as using other fluids may require alternative measurement methods not covered by this code.

As of 2024, there was a noted lack of international standards for testing utility-scale CSP plants. This gap in standardized methods for testing SGTs contrasts with the robust testing methods established for conventional gas turbines (a mature technology with decades of development). However, researchers have conducted some tests on prototype SGTs, as explored in the next section.

9.3.2 Test Rigs

Test rigs are essential tools for assessing the performance, reliability, and safety of SGTs before full-scale deployment. These devices enable engineers and scientists to conduct a variety of tests under controlled conditions (Pan et al. 2022), allowing for detailed analysis of the turbine's thermodynamic and mechanical performance. Test rigs can range from simplified models replicating core aspects of the gas turbine to full-scale prototypes that simulate operating conditions as closely as possible. The following are common test rig configurations used in the development and evaluation of SGTs:

9.3.2.1 Closed Cycle Test Rigs

In closed cycle test rigs, working fluids such as air or other gases circulate in a sealed loop, allowing for controlled examination of the turbine's efficiency and heat transfer characteristics without external environmental influences. This setup is particularly beneficial for studying thermal performance, emissions, and fluid dynamics within the turbine. Closed cycle rigs are often employed to analyse SGTs under specific operating conditions that may be challenging to replicate in open cycles, such as extreme temperatures or pressures. These rigs also facilitate testing of emission control devices and heat recovery systems.

9.3.2.2 Open Cycle Test Rigs

Open cycle test rigs use ambient air intake and exhaust systems, closely replicating real operating conditions for open cycle SGTs. These setups are ideal for assessing power output, exhaust emissions, and fuel efficiency in conditions similar to those encountered in the field. Open cycle rigs are typically equipped with advanced measurement instruments to monitor variables such temperature, pressure, exhaust composition, and vibration. Due to their direct interaction with external air, open cycle test rigs are beneficial for studying the effects of environmental factors, such as humidity and temperature, on turbine performance.

9.3.2.3 Hybrid Test Rigs

Hybrid test rigs offer a combination of closed and open cycle attributes, enabling versatile testing approaches. These rigs can switch between closed-loop and open-loop modes, providing flexibility for engineers and scientists to analyse various aspects of gas turbine (GT) performance in a single setup. Hybrid rigs are particularly useful for studying complex cycle interactions, such as combined cycle operations that integrate additional heat recovery systems.

9.3.2.4 Structural and Vibration Testing Rigs

Given the high rotational speeds and thermal stresses that GTs endure, structural integrity and vibration characteristics are critical for reliable performance. Structural test rigs are designed to evaluate the mechanical stability, material resilience, and creep and fatigue resistance of turbine components under various load conditions. Vibration testing rigs, on the other hand, focus on monitoring and assessing vibration levels to detect potential issues such as rotor imbalance, misalignment, or bearing wear. These rigs are essential for ensuring that the turbine meets ISO 20816-2 standards on machine vibration.

9.3.2.5 Solar Simulator Test Rigs

Solar simulator test rigs are developed to simulate solar irradiance conditions in a controlled laboratory environment because SGTs operate in conjunction with solar concentrators. These rigs typically incorporate high-intensity light sources to mimic sunlight, allowing engineers to test the turbine's responsiveness to varying levels of solar radiation. Solar simulator test rigs are crucial for studying the interaction between solar heat input and turbine operation, as well as for optimizing the efficiency of solar energy conversion in SGT systems.

9.3.2.6 Emission Testing Rigs

A hybrid solar gas turbine possesses a backup source of heat (see Sect. 4.3.1). At present, fossil fuels are commonly used as sources of backup heat. With the increasing focus on environmental impact, emission testing rigs are used to monitor and control pollutants produced by GTs. These rigs assess the effectiveness of emission control technologies in reducing levels of CO_2, NO_x, and other pollutants. Equipped with gas analysers and filters, emission testing rigs can simulate various fuel types and operating conditions to ensure compliance with international environmental standards, such as those outlined in ISO 2314 and ISO 3977.

Each test rig type plays a specific role in the development and validation of solar gas turbines, enabling engineers and scientists to optimize performance, identify potential issues early in the design phase, and ensure compliance with industry standards. The insights gained from these test rigs ultimately contribute to the efficiency, safety, and sustainability of GTs in practical applications.

9.4 Progress in Testing of Solar Gas Turbines

Solar gas turbines rely on high temperatures to achieve optimal efficiency, which necessitates a high outlet temperature of the working fluid from the solar receiver. Solar radiation may be the only source of heat for the solar gas turbine, or the solar radiation may be backed up by another source of heat. Among various CSP technologies, the parabolic dish and solar tower types have proven capable of delivering the required temperature levels. However, while these systems show promise, only a limited number of demonstration projects have been conducted to validate the performance of solar gas turbines.

Several projects have globally tested SGTs in pilot or demonstration-scale facilities. These tests aim at examining critical performance parameters, including thermal efficiency, power output, and stability under solar irradiance fluctuations. Notable examples include research facilities in countries with high solar resources, such as Spain, USA, and Australia. These projects commonly integrate solar receivers with micro gas turbines or small-scale gas turbines to assess the feasibility of SGT systems. Test results highlight the potential for these systems to achieve efficiencies comparable to conventional gas turbines when optimal solar conditions are present.

9.4.1 Project Development

The development of SGT power plants has been driven by several pioneering projects worldwide, aiming to validate the technology and explore its commercial viability. Notable developments in SGT technology have emerged from both small-scale pilot installations and larger demonstration projects, each contributing valuable insights into the operation and optimization of solar-driven gas turbines.

The first operational SGT power plant in the world was constructed by AORA Solar at Kibbutz Samar in southern Israel, and commissioned in 2009. The power plant utilizes the AORA hybrid solar gas turbine system (known as AORA Tulip) produced by AORA Solar. This system is based on the Turbec T100 micro gas turbine, has a nominal thermal power capacity of 330 kW and a nominal pressure of 450 kN/m^2, and it is commercially available for combined heat and power applications, generating 100 kWe (Rovense et al. 2017). In this system, compressed air is heated to around 1273 K and then injected into a 100 kW gas turbine. Following the initial success of this project, AORA Solar installed a second demonstration plant in Almeria, Spain.

Another prominent early initiative is the Solar Hybrid Gas Turbine Electric Power System (SOLGATE) project that began in 2001 to showcase a solar hybrid gas turbine system driven by a solar tower (SOLGATE 2005). The project repurposed an ALLISON 250 helicopter engine, enabling it to accept heat from a solar receiver and drive an electric generator. The modified engine incorporated a combustor that could handle air inlet temperatures of up to 1073 K and produce up to 230 kW of electrical output. Supplementary heat from kerosene ensured stable turbine inlet temperatures. The facility was hosted at the Plataforma Solar de Almería (PSA) in Spain, where simulation tools developed for the project were validated against empirical data, setting a foundation for future solar hybrid gas turbine systems.

In the early 2000s, Germany advanced the REFOS receiver technology through national and international research and development (R&D) projects. This system uses a pressurized volumetric air receiver that was designed to be compatible with fossil-fuel power plants (Schwarzbözl et al. 2006). Building on this, the SOLTREC project (Del Río et al. 2015) explored further receiver enhancements. Utilizing a 1.5 m diameter receiver with a quartz window, this system achieved a thermal output of 1.47 MW_{th} and could heat the air up to 1273 K, achieving up to 80% solar contribution. The addition of the quartz window improved thermal efficiency over earlier models, helping the SOLTREC receiver attain its design temperature more consistently.

Another milestone in solar gas turbine development was achieved through the SOLHYCO project that was initiated in 2006. This project focused on developing solar hybrid gas turbines for power and cogeneration, with an emphasis on reliability and economic viability (SOLHYCO 2011). Unlike previous projects, SOLHYCO employed bio-diesel as a source of backup heat, thereby demonstrating the feasibility of fully renewable gas turbine operation. The project adapted the 250 kW airborne gas turbine setup from SOLGATE, replacing kerosene with bio-diesel and pioneering electricity generation solely from renewable sources. Additionally, a profiled multilayer (PML) tube technology for receivers was developed. This design, with three metallic layers incorporating a copper core, optimized heat conduction, thereby reducing thermal stress and achieving stable receiver temperatures of up to 1073 K. Testing at PSA confirmed the viability of the PML tube that achieved a receiver outlet temperature of 1076 K. In Algeria, economic projections estimated a cost of 0.101 €/kWh for a hybrid solar and bio-diesel system, suggesting potential economic viability in markets with comparable or higher electricity prices.

The SOLUGAS project represents a significant step towards pre-commercial solar hybrid gas turbines. Launched in Seville, Spain, the project constructed a hybrid system including a heliostat field with 69 units, each covering 121.3 m^2, and a 77 m tower with a 3.2 MWth receiver. The system preheated compressed air at 900 kN/m^2 to 1073 K before it entered a Mercury 50 gas turbine that required modification, including bypass pathways and removal of the recuperator, to accommodate the high-temperature solar input. Operated for about 1000 solar test hours, SOLUGAS was the first to demonstrate solar preheating for a turbine with a scale of several megawatts, marking a notable advancement in solar gas turbine technology (Korzynietz et al. 2016).

Further development occurred within the French PEGASE project, which led to several receiver concepts tested at the Thémis Solar Tower in Targasonne, France. Originally constructed in 1979 and rehabilitated in 2004, the Thémis facility has supported experimental efforts for CSP and gas turbine technology. The PEGASE project installed a gas turbine on the tower capable of receiving hot air directly from a solar receiver. Additionally, the Next-CSP project (2017–2021), an initiative involving ten European partners, advanced the use of high-temperature particles as both heat transfer and storage media. Expected to achieve outlet temperatures of up to 1073 K, the Next-CSP project aims to validate a 4 MWth setup at Thémis, reflecting ongoing research to enhance the reliability and efficiency of SGT systems (Next-CSP 2024). In this context, solar receivers for the third-generation CSP systems are expected to utilize air, advanced molten salts, and solid particles as heat transfer media (Alami et al. 2023).

These pioneering projects underscore the potential of solar gas turbines as a sustainable energy solution, although the technology remains at an emerging stage. Continued efforts are required to improve reliability, reduce costs, and achieve commercial scalability, particularly in regions with high direct normal irradiance.

9.4.2 Challenges to Development of Solar Gas Turbines

While there has been progress in demonstrating the SGT technology, the commercialization of SGT power plants remains limited compared to the global deployment of steam-based CSP plants. Several barriers hinder the broader adoption of SGTs, particularly in the solar field, where parabolic dish concentrator (PDC) and solar tower (ST) technologies have shown promising thermodynamic characteristics for powering gas turbines. ST technology, in particular, has been identified as the most suitable option (Madhlopa and Okoroigwe 2017). However, the challenges facing ST development can be grouped into three main categories: (a) technology maturity, (b) financial and policy considerations, and (c) technical limitations (Srilakshmi et al. 2015; Okoroigwe and Madhlopa 2016). The development of solar gas turbines is encountering similar challenges.

9.4.2.1 Maturity of Solar Gas Turbine

Technological development follows a progression from concept to commercialization, with maturity increasing at each stage. Technology maturity indicates the readiness of a technology for commercial application (Selman and Chen 2012). The Technology Readiness Level (TRL) framework has become a widely used metric across industries, allowing developers to assess a technology's maturity, from conceptualization to commercialization (Rybicka et al. 2016). Originally developed by the National Aeronautics and Space Administration (NASA), the TRL framework began with 9 levels but has since expanded to 10, where higher levels indicate increased readiness for market deployment (Straub 2015).

In the development and scaling of new technologies, R&D managers often rely on TRL to make informed decisions about resource allocation and investment. Scientific and engineering research is instrumental in improving a technology's maturity by enhancing its performance and competitiveness. Several studies have evaluated the maturity of ST technology in power production using the Rankine cycle, with findings indicating its relatively low maturity level for power applications. Zhang et al. (2013) and Peterseim et al. (2013) reported that ST technology remains underdeveloped for commercial purposes, particularly in gas turbine applications. In addition, current solar gas turbines make use of modified conventional gas turbine engines that are designed to operate mostly on oil and natural gas. For instance, conventional open cycle gas turbines are widely exploited in the power sector due to their higher thermodynamic efficiency compared with the closed cycle type. Nevertheless, combustors for the open cycle conventional gas turbines accept inlet air at relatively low temperatures. This poses challenges for modifying these engines to accept hot air from solar receivers. The use of supercritical carbon dioxide (s-CO_2) as a working fluid is promising to improve the thermodynamic efficiency of the closed cycle gas turbine (Chacartegui et al. 2011). Nevertheless, the s-CO_2 Brayton cycle is still under active research (see Chap. 5). These gaps demonstrate that the SGT system has a low maturity level, emphasizing the need for intensified research to enhance its competitiveness and readiness for broader adoption.

9.4.2.2 Financial and Policy Instruments

Energy projects typically require substantial capital investment, making access to significant financial resources a critical need (Bhattacharyya 2011). Given the competitive demand for financial resources across various economic sectors, investment decisions aim for profitability and long-term viability. Investment choices must account for uncertainty and risk. Therefore, a history of successful investments in similar projects can help mitigate perceived risks in new projects.

The SGT technology, as an emerging and relatively unproven option, faces unique financial and policy challenges in advancing to commercial scale. A lack of grant facilities or other financial incentives can significantly hinder the commercialization of SGTs. Internationally, instrument instruments, like the Global Environment Facility (GEF), provide financial support for renewable energy initiatives that are essential in promoting new technologies. For instance, in 2000, GEF funded four proposed integrated solar combined cycle (ISCC) plants in India, Egypt, Morocco, and Mexico, aimed at stimulating the development of solar power technologies (Behar et al. 2011). This investment possibly encouraged the establishment of plants like Ain Beni Mathar in Morocco and Kuraymatt in Egypt.

The Ain Beni Mathar ISCC project, for example, received a $43.2 million grant from the GEF, supplemented by two loans totalling $371.8 million from the African Development Bank and a $129 million loan from Spain's Instituto de Crédito Oficial (Amelio et al. 2014). Similarly, the Mexico Hybrid Solar Thermal Power Plant required approximately $180 million, with GEF contributing a grant of $50 million towards the project costs (World Bank 1999). These initiatives helped the establishment of several parabolic trough collector (PTC) ISCC plants in Algeria and the USA, but there remains a lack of equivalent SGT projects at the commercial scale. Table 9.2 provides technical details of some ISCC projects partially funded by GEF, illustrating the significant role that financial and policy instruments play in enabling large-scale renewable energy projects.

9.4.2.3 Technical Factors

Despite the high thermodynamic efficiency of both Parabolic Dish Concentrator (PDC) and Solar Tower (ST) technologies, there were no commercial PDC power plants operational as of 2024. In contrast, the National Renewable Energy Laboratory reported many commercial ST power plants that utilized the Rankine cycle and were operational during the same period. The ST technology can achieve temperatures of up to 1273 K, making it particularly suitable for integration with gas turbines. However, several technical limitations persist, necessitating further research. Key technical factors influencing the deployment of solar gas turbines include solar and

Table 9.2 Technical details of some of integrated solar combined cycle projects partially funded by the global environmental facility

Country	Plant	Solar field (m^2)	Total capacity (MW)	Solar contribution (MW)	Date of operation
Egypt	Kuraymatt plant	130,800	140	20	2011
Mexico	Mexico: hybrid solar thermal power plant	–	300	29	–
Morocco	Ain Beni Mathar ISCC	183,120	470	20	2011

land resources, ambient temperature, tower height, material availability, supply chain constraints, and receiver design (Srilakshmi et al. 2015).

(a) Solar and Land Resources

Solar radiation stands out as the most crucial environmental factor affecting solar energy system performance (Nafey et al. 2000). As discussed in Chap. 3, CSP technology harnesses only direct normal irradiance for electricity generation (Enjavi-Arsanjania et al. 2015). Thus, CSP systems achieve higher optical efficiency in areas with abundant direct normal irradiation. However, there are limited sites that measure direct radiation accurately. Satellite solar radiation data can serve as an alternative but lacks the precision of ground-based sensors (Ali 1990). Consequently, the challenge of obtaining accurate solar resource data may impede the commercialization of SGTs.

The heliostat aperture area is known to correlate linearly with the equivalent power capacity of existing ST plants (Srilakshmi et al. 2015), meaning that an increase in plant capacity necessitates more heliostats and land requirements. The optimization of the solar field is essential to minimize land requirements and optical losses, yet this process can be complicated and time-consuming (Atif and Al-Sulaiman 2015). Although wasteland can be utilized for heliostat fields, data on terrain conditions is vital for estimating potential land availability for CSP installation. Such data remains scarce in many countries that face land limitations which could challenge the widespread deployment of SGTs. Given that a substantial land area is required for heliostat fields, this need may contribute to the low level of commercialization observed in the SGT technology. Thus, addressing these technical factors is critical for advancing SGT projects and enhancing their competitiveness on the renewable energy market.

(b) Ambient Air Temperature

Research indicates that the performance of gas turbines declines as ambient air temperature increases (see Sect. 5.7). CSP technology operates effectively in regions with abundant direct normal irradiance that are often characterized by high ambient temperatures. Additionally, the presence of numerous heliostats may elevate ambient temperatures around solar gas turbines, further impacting their efficiency. Although there is limited information on the influence of heliostat and receiver heat losses on local ambient temperatures, some studies suggest that utilizing low-temperature CSP heat could help chill inlet air, thereby maintaining gas turbine efficiency (Peterseim et al. 2013).

(c) Tower Height

For the ST technology to function optimally, it is essential to maintain an appropriate tower height. In 2024, the operational ST power plants worldwide exhibited tower heights ranging from 30 to 240 metres. The Ashalim Plot B power plant in Israel, commissioned in 2019, has the tallest (National Renewable Energy Laboratory 2024). This plant has a capacity of 121 MW. Future designs may feature even taller towers, but significant tower heights can present maintenance challenges and lead to increased operational costs. While studies specifically examining the effects of tower height on installation, operation, and maintenance costs are scarce, it is plausible that taller structures could complicate these aspects. Further investigation is necessary to assess how tower height influences the levelized cost of electricity for solar gas turbine systems.

(d) Materials and Supply Chain

The choice of heat transfer fluid is critical in CSP technology. Molten salt, a commonly used heat transfer fluid, also serves as an effective thermal storage medium (see Sect. 4.3.2). However, the global supply of molten salt is limited to three suppliers: SQM (Chile), Haifa Chemicals (Israel), and Durferrit Salts and Auxiliary Products (Germany) (Srilakshmi et al. 2015). This limited supplier base may impede the widespread implementation of ST technology and, by extension, solar gas turbines. Research efforts are ongoing to reduce the costs associated with CSP storage systems.

The supply chain for solar technology plays a pivotal role in its advancement. A well-functioning supply chain enables the movement of a product from its point of origin to consumption points (Wang et al. 2015). In the case of ST technology, the supply chain remains underdeveloped, with insufficient local demand for components such as heliostats and receivers to justify domestic manufacturing (Srilakshmi et al. 2015). Currently, operational ST power plants are concentrated in a limited number of countries.

(e) Solar Receivers and Technical Expertise

The solar receiver serves as one of the most crucial components of an ST power plant, acting as the interface between the solar field and the power block. Gas-based receivers are particularly well-suited for integration within the solar gas turbine system. There are three principal designs of gas-based receivers: tubular gas receivers, volumetric air receivers, and microchannel gas receivers (Ho 2017). The advantages of gas-based receivers include their ability to achieve high temperatures (>1273 K), the potential to utilize inert and environmentally friendly gases, and reduced costs associated with heat transfer fluids. However, some challenges accompany each design, as detailed in Table 9.3. Furthermore, existing combustors are primarily designed for conventional gas turbines, presenting additional technical hurdles in the hybridization process of solar gas turbines. Addressing these challenges is essential to enhance the overall performance and viability of solar gas turbines.

Currently, the market for solar receivers is limited, with only a handful of manufacturers specializing in their production. Furthermore, there are very few companies worldwide that focus on the design, construction, and implementation of solar tower technologies (Srilakshmi et al. 2015). This scarcity of expertise can create barriers to the advancement of solar gas turbine systems in many countries.

The limited availability of expertise in SGT technology field poses several challenges such as restricted knowledge base that hampers the ability of many regions to develop, implement, and maintain these systems effectively. The scarcity of pilot projects prevents stakeholders from gaining confidence in the technology's reliability and performance, thereby reducing willingness to invest in SGT systems. Moreover, challenges in developing high-performance solar receivers, a critical component for SGT efficiency, create further obstacles, as the current expertise constraints restrict innovation and slow technological progress. This situation underlines the pressing need for increased research and development investment to deepen knowledge and expand technical capabilities related to solar receivers.

Table 9.3 Summary of gas-based receiver designs

Receiver design	Outlet temperature/ thermal efficiency	Advantages	Challenges/research needs
Tubular gas receivers	973–1073 K/ ~ 90% (modelling), up to 85% (testing)	Ability to use features and geometry (such as cavity) to increase trapping of radiation	Resistance to heat transfer through tube walls, fatigue, and high-cost materials to contain high pressure, high-temperature flow, no direct storage
Volumetric air receivers	> 1073 K/50–90%	Can use ambient air which is cheap, potential for volumetric (interior) heating to minimize heat losses near the aperture	Flow instabilities, high temperatures, and radiative heat losses near the aperture due to limited light penetration, no direct storage
Microchannel gas	973 K/up to 90% (scaled up from bench-scale tests)	Augmented surface area and heat transfer to gas, ability to add up microchannel modules for larger capacities	Potential for increased pressure drop, potential manufacturing complexity, and costs, no direct storage

Source of data Ho (2017)

9.5 Worked Examples

Example 9.1 A solar-only gas turbine is used to generate electricity. During a certain time of the day, the concentrated irradiance on the solar receiver of the solar gas turbine is 2 MW. The receiver has a thermal efficiency of 75% and an area of 1.5 m^2 and uses air as a heat transfer fluid. If air enters the receiver at a temperature of 300 K, find the receiver outlet temperature, assuming the flow rate and specific heat capacity of the air are respectively 5 kg/s and 1020 J/kg.K.

Solution

From Eq. (9.1), outlet temperature $\left(T_{\text{out,rc}}\right)$ is:

$$\begin{aligned} T_{\text{out,rc}} &= \frac{\eta_{\text{rc}} A_{\text{rc}} G_{\text{rc}}}{\dot{m}_{\text{htm}} C_{\mathbf{p,htm}}} + T_{\text{in,rc}} \\ &= \frac{0.75 \times 1.5 \times 2{,}000{,}000}{5 \times 1020} + 300 \\ &= 741 \text{ K} \end{aligned}$$

Example 9.2 A hybrid open cycle solar gas turbine is tested under outdoor conditions. The solar receiver and combustor of the system are connected in series. During a certain time of the test day, the outlet temperature of the receiver is 700 K. The hot air flows into the combustor at a rate of 10 kg/s where fuel (with heating value of 40 MJ/kg) is added at a rate of 0.2 kg/s. The combustion gases exit the combustor at a temperature of 1200 K. Find the combustor efficiency, assuming the combustion is complete, the flow is steady and the specific heat capacity of the combustion gases is1240 J/kg.K.

Solution

Efficiency of the combustor η:

$$\eta = \frac{\dot{m}_{\text{cg}} C_{\mathbf{p,cg}} (T_{\text{out}} - T_{\text{in}})}{\dot{Q}_{\text{f}}}$$

where $\dot{m}_{\text{cg}}$ is the mass flow rate of the combustion gases, $C_{\text{p,cg}}$ is the specific heat capacity of the combustion gases, T_{in}, T_{out} and $\dot{Q}_{\text{f}}$ rate of heat addition

$$\dot{Q}_{\text{f}} = \dot{m}_{\text{f}} H_v,$$

where $\dot{m}_{\text{f}}$ is the mass flow rate of fuel and H_v is the heating value of fuel.

$$\dot{m}_{\text{cg}} = \dot{m}_a \dot{m}_f = 10 + 0.5 = 10.5 \text{ kg}/s$$

Therefore,

$$\eta = \frac{10.5 \times 1240 \times (1200 - 700)}{0.2 \times 40{,}000{,}000}$$
$$= 0.81$$

Example 9.3 A transit-time ultrasonic flow metre of is used to measure the flow rate of air in Example 9.1. The transducers in the metre are inclined at an angle of 10°. In a particular measurement, the velocity of ultrasound is 330 m/s and the transducers emit sound waves with a time difference of 4 min. Calculate the path length between transducers if the average flow velocity is 50 m/s.

Solution

From Eqs. (9.2) and (9.3), path length (L) between the transducers is:

$$L = \Delta t\left(\frac{1}{v_{us} - v_{ag}\cos\beta} - \frac{1}{v_{us} + v_{ag}\cos\beta}\right)$$
$$= 240\left(\frac{1}{330 - 50 \times \cos 10^\circ} - \frac{1}{330 + 50 \times \cos 10^\circ}\right)$$
$$= 0.22\ \text{m}$$

Example 9.4 Describe the significance of technology readiness level in the development of solar gas turbines.

Solution

Technology readiness level (TRL) provides a framework to assess the maturity of a technology from its conceptualization to commercialization. It helps managers in research and development and in making informed decisions about resource allocation and investment. A low TRL indicates that SGT technology is still underdeveloped, necessitating further research and development to enhance its competitiveness.

Example 9.5 What are the primary financial barriers to the commercialization of solar gas turbine (SGT) technology?

SGT projects are capital-intensive and compete for limited financial resources against various economic activities. The absence of adequate financial and policy instruments, such as grants and incentives, can hinder the investment needed for SGT commercialization. Historical investment patterns in similar technologies also play a role in mitigating risk and uncertainty, which can be challenging for emerging technologies like SGTs.

Nomenclatures

A Area (m^2)

C_p	Specific heat capacity at constant pressure (J/kg.K)
H_v	Heating value (J/kg)
L	Distance (m)
$\dot{m}$	Mass flow rate (kg/s)
$\dot{Q}_f$	Rate of heat addition (W)
t	Time (s)
T	Temperature (K)
v	Velocity (m/s)

Greek Symbols

β	Inclination angle (degree)
η	Efficiency (dimensionless)

Subscripts

a	Air
av	Average
cg	Combustion gases
f	Fuel
htm	Heat transfer medium
rc	Receiver
us	Ultrasound.

References

Abdelsalam TI, Tian Z, Robinson A (2023) Directly irradiated liquid metal film in an ultra-high temperature solar cavity receiver. Part 1: concepts and a quasi-steady-state analysis. Sol Energy 255:355–368

Alami AH, Olabi AG, Mdallal A, Rezk A, Radwan A, Rahman SMA, Shah SK, Abdelkareem MA (2023) Concentrating solar power (CSP) technologies: status and analysis. Int J Thermofluids 18:100340

Ali B (1990) Theoretical study of main factors affecting the heliostat field design of tower power plant. Energy Convers Manag 30(2):101–106

Amelio M, Ferraro V, Marinelli V, Summaria A (2014) An evaluation of the performance of an integrated solar combined cycle plant with air-linear parabolic collectors. Energy 69:742–748

American Society of Mechanical Engineers International (2014) Gas turbines. American Society of Mechanical Engineers (ASME) International, Englewood

American Society of Mechanical Engineers International (2015) Standard PTC 46: overall plant performance. American Society of Mechanical Engineers (ASME) International, Englewood

Atif M, Al-Sulaiman F (2015) Optimization of heliostat field layout in solar central receiver systems on annual basis using differential evolution algorithm. Energy Convers Manag 95:1–9
Aurousseau A, Vuillerme V, Bezian J (2016) Control systems for direct steam generation in linear concentrating solar power plants—a review. Renew Sustain Energy Rev 56:611–630
Behar O, Kellaf A, Mohamedi K, Belhamel M (2011) Instantaneous performance of the first integrated solar combined cycle system in Algeria. Energy Procedia 6:185–193
Bergman TL, Lavine AS, Incropera F, DeWitt DP (2011) Fundamentals of heat and mass transfer. Wiley, New Jersey
Bhattacharyya SC (2011) Energy economics: concepts, issues, markets and governance. Springer, London
Boyce MP (2002) Gas turbine engineering handbook, 2nd edn. Gulf Professional Publishing, Boston
Cai Y (2013) Measuring gas flow with MEMS sensors. Electronic Products, http://www.electronicproducts.com
Casanova M, Ballestrín J, Monterreal R, Fernández-Reche J, Enrique J, Ávila-Marín A (2022) Improvements in the measurement of high solar irradiance on a 300 kWth volumetric receiver. Renew Energy 201:441–449
Chacartegui R, Muñoz de Escalona JM, Sánchez D, Monje B, Sánchez T (2011) Alternative cycles based on carbon dioxide for central receiver solar power plants. Appli Therm Eng 31(5):872–879
Chen L, Li Y, Sun F, Wu C (2004) Power optimization of open-cycle regenerator gas-turbine power-plants. Appl Energy 78:199–218
Del Río A, Korzynietz R, Brioso JA, Gallas M, Ordóñez I, Quero M (2015) Soltrec-pressurized volumetric solar air receiver technology. Energy Procedia 69:360–368
Devices D-T (2000) User manual for DL2e data logger. Delta-T Devices, Cambridge
Ebert M, Benitez D, Röger M, Korzynietz R, Brioso JA (2016) Efficiency determination of tubular solar receivers in central receiver systems. Sol Energy 139:179–189
Enjavi-Arsanjania M, Hirbodia K, Yaghoubi M (2015) Solar energy potential and performance assessment of CSP plants in different are as of Iran. Energy Procedia 69:2039–2048
Fernández-García A, Sutter F, Fernández-Reche J, Lüpfert E (2017) Mirrors. In: Heller P (ed) The performance of concentrated solar power (CSP) systems: modelling, measurement and assessment. Woodhead Publishing, pp 67–98
Grove DM, Davis TP (1992) Engineering, quality and experimental design. Longman Scientific & Technical, Essex
Hans V, Windorfer H (2003) Comparison of pressure and ultrasound measurements in vortex flow meters. Meas 33:121–133
Heller P, Pfänder M, Denk T, Tellez F, Valverde A, Fernandez J, Ring A (2006) Test and evaluation of a solar powered gas turbine system. Sol Energy 80(10):1225–1230
Henry M, Tombs M, Zamora M, Zhou F (2013) Coriolis mass flow metering for three-phase flow: a case study. Flow Meas Instrum 30:112–122
Ho CK (2017) Advances in central receivers for concentrating solar applications. Sol Energy 152:38–56
International Electrotechnical Commission (2024) TC 117 solar thermal electric plants. International electrotechnical commission (IEC), Geneva. https://www.iec.ch
International Organization for Standardization (2002) Standard 3977-4: gas turbine—procurement—part 4: fuels and environment. International Organization for Standardization, Geneva
International Organization for Standardization (2004) Standard 3977-3: gas turbine—procurement—part 3: design requirements. International Organization for Standardization, Geneva
International Organization for Standardization (1997a) Standard 3977-1: general introduction and definitions. International Organization for Standardization (ISO), Geneva
International Organization for Standardization (1997b) Standard 3977-2: gas turbine—procurement—part 2: standard reference conditions and ratings. International Organization for Standardization, Geneva
International Organization of Standardization (2009) ISO standard 2314: gas turbines—acceptance tests. International Organization of Standardization, Geneva

International Organization of Standardization (2013) ISO standard 9806: solar energy—solar thermal collectors—test methods. International Organization of Standardization, Geneva
International Organization of Standardization (2016) ISO 19859 satandard: gas turbine applications-requirements for power generation. International Organization for Standardization, Geneva
International Organization of Standardization (2017) ISO standard 20816-2: mechanical vibration—measurement and evaluation of machine vibration—part 2: land-based gas turbines, steam turbines and generators in excess of 40 MW with fluid-film bearings. International Organization of Standardization, Geneva
Janotte N, Wilbert S, Sallaberrt F, Schroedter-Homscheidt M, Ramirez L (2017) Principles of CSP performance assessment. In: Heller P (ed) The performance of concentrated solar power (CSP) systems: modelling. Woodhead Publishing, Measurement and Assessment, pp 31–64
Korzynietz R, Brioso JA, del Río A, Quero M, Gallas M, Uhlig R, Ebert M, Buck R, Teraji D (2016) Solugas—comprehensive analysis of the solar hybrid Brayton plant. Sol Energy 135:578–589
Madhlopa A, Okoroigwe E (2017) Solar gas turbine systems. In: Abraham M (ed) Encyclopedia of sustainable technologies. Elsevier, Amsterdam, pp 377–388
Nafey A, Abdelkader M, Abdelmotalip A, Mabrouk A (2000) Parameters affecting solar still productivity. Energy Convers Manag 41:1797–1809
National Renewable Energy Laboratory (2024) Concentrating solar power projects. National renewable energy laboratory (NREL), Golden. https://solarpaces.nrel.gov/by-technology/power-tower. Accessed 23 Nov 2024
Next-CSP (2024) High temperature concentrated solar thermal power plant with particle receiver and direct thermal storage. http://next-csp.eu. Accessed 23 Nov 2024
Nicholas J, White D (1994) Traceable temperatures: An introduction to temperature measurement and calibration. Wiley, New York
Okoroigwe E, Madhlopa A (2016) An integrated combined cycle system driven by a solar tower: a review. Renew Sustain Energy Rev 57:337–350
Pan T, Shen J, Laumert B, Wang W (2022) A novel gas turbine simulator for testing hybrid solar-Brayton energy systems. Energy Convers Manag 268:116051
Peterseim J, White S, Tadros A, Hellwig U (2013) Concentrated solar power hybrid plants, which technologies are best suited for hybridisation? Renew Energy 57:520–532
Pfahl A, Coventry J, Röger M, Wolfertstetter F, Vásquez-Arango JF, Gross F, Arjomandi M, Schwarzbözl P, Geiger M, Liedke P (2017) Progress in heliostat development. Sol Energy 152:3–37
Röger M, Prahl C, Pernpeintner J, Sutter F (2017) New methods and instruments for performance and durability assessment. In: Heller P (ed) The performance of concentrated solar power (CSP) systems: modelling, measurement and assessment. Woodhead Publishing, pp 205–252
Rovense F, Amelio M, Scornaienchi NM, Ferraro V (2017) Performance analysis of a solar-only gas micro turbine, with mass flow control. Energy Procedia 126:675–682
Rybicka J, Tiwari A, Leeke G (2016) Technology readiness level assessment of composites recycling technologies. J Clean Prod 112:1001–1012
Schwarzbözl P, Buck R, Sugarmen C, Ring A, Crespo MJ, Altwegg P, Enrile J (2006) Solar gas turbine systems: Design, cost and perspectives. Sol Energy 80:1231–1240
Selman JR, Chen C (2012) Scientific and technical maturity of molten carbonate technology. Int J Hydrog Energy 37:19280–19288
Singh V, Tharakan T (2015) Numerical simulations for multi-hole orifice flow meter. Flow Meas Instrum 45:375–383
SOLGATE (2005) Solar hybrid gas turbine electric power system: final publishable report. European Communities, Luxembourg
SOLHYCO (2011) Solar-hybrid power and cogeneration plants: final public report. German Aerospace Centre (Deutsches Zentrum für Luft- und Raumfahrt, DLR), Cologne
Srilakshmi G, Venkatesh V, Thirumalai N, Suresh N (2015) Challenges and opportunities for Solar Tower technology in India. Renew Sustain Energy Rev 45:698–709
Straub J (2015) In search of technology readiness level (TRL) 10. Aerosp Sci Technol 46:312–320

Wang Y, Wallace S, Shen B, Choi T (2015) Service supply chain management: a review of operational models. Eur J Oper Res 247:685–698

Watral Z, Jakubowski J, Michalski A (2015) Electromagnetic flow meters for open channels: current state and development prospects. Flow Meas Instrum 42:16–25

World Bank (1999) Mexico—hybrid solar thermal power plant project: report number PID8128. World Bank, Washington DC

Xie L, Zhao Y, Aziz D, Jin X, Geng L, Goberdhansingh E, Qi F, Huang B (2013) Soft sensors for online steam quality measurements of OTSGs. J Process Contr 23:990–1000

Xu Y, Yu P, Zhu Z, Yuan C, Zhang T (2017) Over-reading modeling of the ultrasonic flow meter in wet gas measurement. Meas 98:17–24

Zhang HL, Baeyens J, Degrève J, Cacères G (2013) Concentrated solar power plants: review and design methodology. Renew Sustain Energy Rev 22:466–481

Zhou H, Ji T, Wang R, Ge X, Tang X, Tang S (2018) Multipath ultrasonic gas flow-meter based on multiple reference waves. Ultrasonics 82:145–152

Chapter 10
Economic Performance of Solar Gas Turbines

10.1 Introduction

Generally, solar technologies have high initial costs but low operating costs, partly due to savings associated with fuel. In case of solar-only energy technologies, there are no operational costs associated with fuel. This is advantageous because fuel costs contribute a significant proportion of the total operational costs. However, solar-only technologies cannot meet baseloads due to the intermittency of the solar resource. This drawback can be circumvented through the inclusion of energy storage, auxiliary source of energy, or a combination of these two components. An auxiliary heater (combustion component) requires fuel during the operational phase. The discussion of fuels for gas turbines has been presented in Chap. 2. Investment in the solar subsystem of an energy technology helps to reduce fuel-related costs in future terms.

Energy projects are usually large and capital-intensive, and so they require a high level of investment (Bhattacharyya 2011). Moreover, various economic activities compete for limited financial resources, and investment in a project is aimed at making some profit. In view of this, choices have to be made carefully among competing opportunities for investment. Even if the thermodynamic efficiency of a technology is satisfactory, cost is an important factor in the consideration of a suitable system design. Cost is the amount of money paid or given up in order to acquire or produce an item. A system design with lower cost may be more preferred for exploitation.

It is usual to base the decision-making on the costs and benefits over the lifetime of a project. So, it becomes necessary to analyse these parameters in order to evaluate the performance of the project. Nevertheless, data may not be available on project benefits in some projects. In such cases, analysis of costs only can be applied to assist with comparison of the economic performance of two or more projects. There are different elements of costs that need to be considered in the economic evaluation of a project. These elements include (Sell 1989; Bhattacharyya 2011): sunk cost, contingency cost, fixed investment, working capital, depreciation, and depletion premium.

A. Madhlopa and R. Nkhoma, *Principles of Solar Gas Turbines for Generating Electricity*, Green Energy and Technology,
https://doi.org/10.1007/978-3-031-84096-8_10

10.2 Project Costs

10.2.1 Sunk Cost

A sunk cost is a cost that has already been incurred and it is non-recoverable. A new project may exploit existing facilities for which investment has already been made. These facilities do not contribute extra financial strain to the new project. Consequently, past costs should be excluded from the inventory of costs of the new project. For instance, two types of sunk costs are considered in auctions for renewable energy projects (Kreiss et al. 2017): entry fees and participation. Every bidder pays an entry fee before participation in the auction, and the fee is non-refundable. Some costs are also incurred by the bidder during the preparation of bids (such as land use plan and feasibility studies).

10.2.2 Contingency Cost

Usually, there are some uncertainties about the exact inventory of the items for estimating the cost of a project. The need for certain items may not be obvious at the inception stage of the project but during project execution. The purpose of a contingency budget is to take care of unforeseen risks during the implementation phase of a project. It is customary to allocate about 5–15% of the total project cost to contingency (Idrus et al. 2011; Shrestha and Shrestha 2016). This traditional method of estimating the cost of contingency is simple but it is inaccurate. Underestimation of the contingency amount may result in a high probability of cost overrun while an overestimated contingency budget may elevate the probability of cost under run with a low magnitude of the feasibility of the project. In view of this, research efforts have been made to develop accurate computational tools for estimating the cost of contingency.

For complex projects, a probabilistic method of estimating contingency costs was proposed by Touran and Liu (2015). Other estimation tools have included the multiple linear regression (MLR), artificial neural network (ANN), and linear regression (LR) approaches. Based on a review of various methods, Shrestha and Shrestha (2016) concluded that regression analysis and ANN methods are more accurate in estimating the cost of contingency in road-maintenance projects.

10.2.3 Fixed and Working Capital

Fixed investment does not change with a rise or fall in the amount of goods or services produced or sold. It includes land and title preparation, building and civil engineering, plant machinery, and industrial property rights (Sell 1989). The solar field occupies a large size of land, and so, land is one of the major constraints in solar gas turbine (SGT) projects (see Chap. 11).

Working capital is the amount of funds required to maintain the operation of a plant before receiving payment for the sale of products. For an SGT power plant, the working capital includes fuels (in case of hybrid systems), spare parts, labour, work in progress, and cash in hand. Costs that represent actual claim on the economic resources of a firm need to be included in the economic analysis. The commonest method of measuring the working capital of a firm is the cash conversion cycle (Ukaegbu 2014). The cash conversion cycle (CCC) is the period (days) between the time the firm spends cash to procure raw materials (or other items) to produce the goods for sale, and the time of cash inflow from the sale of the finished products (say, electricity). A firm can augment its net present value through efficient management of cash. A diminished CCC may result in increased profitability, and therefore viability of a project.

10.2.4 Depreciation and Depletion Premium

Depreciation is the decrease in the value of an asset over a certain period of time, predominantly caused by wear and tear. In economic analysis, the difference between the initial cost of an asset and the residual value (remaining value of an asset after depreciation) is employed. Consequently, the cost of using an asset is taken into consideration in the calculation of project costs.

Economic analysis includes the depletion of rent in order to account for the economic cost to society of exploiting resources (energy and other resources). Economic rent of a natural resource is the extra money earned over the value of the capital, labour, materials, and other production factors employed to utilize the resource (Rothman 2000; Shrestha and Abeygunawardana 2009). When the resource is exhausted, some substitutes have to be made which may entail opportunity costs. Prljić et al. (2018) studied the influence of coal, forest, mineral, natural gas, and oil rents on economic growth. They found that natural resource rents yielded different impacts on economic growth.

Other factors that need to be considered when costing a solar-based project are interest on loan, property and income taxes, resale of equipment, maintenance, insurance, fuel, and other operating costs. In this context, economic analysis aims at finding the cheapest means of meeting an energy demand, bearing in mind the prevailing alternatives that include non-solar options. For solar energy technologies, the challenge is to establish the right share of the solar subsystem that yields a composite system with the lowest cost. Where the backup source of heat is fossil fuel, there is need for trade-off between emissions and cost.

10.2.5 Project Benefits

By and large, the major benefit of a project is the output that has a monetary value. For example, electricity or heat output from a power plant is sold (at a market price) as the main benefit of the project. The market price is not influenced by the project when the output is incremental (Bhattacharyya 2011). In this case, the project is said to be a price taker, which happens when the level of output from the project is smaller than the size of the market and the product is tradable. A price taker has to be willing to accept the prevailing market price (Yucekaya 2013). It is still important for a price-taker power generator to ensure that some profit is made through the project in order to achieve economic viability. For non-tradable goods or services, the supply is inclined to be a non-incremental type and the project output may cause a change in the market price.

Some benefits have non-market value. Among them are social and environmental benefits to society in general. Projects cause environmental impacts that may be negative or positive with varying magnitudes. In view of this, environmental impact assessments are required in project development initiatives. Some of the environmental issues considered in CSP are cooling and water use, materials (for thermal storage, heat transfer, etc.), soiling and atmospheric aerosol loads, and land use and other factors (Lamnatou and Chemisana 2017). A project may be approved or rejected, depending on findings from this kind of assessment. It is possible to approve the implementation of a project with some negative impacts if the forgone benefits are lower than the financial gains of the project. Details about the environmental performance of solar gas turbines are reported in Chap. 11.

10.3 Indicators of Cost–Benefit Comparison

It is necessary to analyse the costs and benefits of a project, once they are identified, to facilitate decision-making. To this end, there are several methods for comparison of costs and benefits that may be brought about by an investment project. These methods can generally be classified into two groups: (a) without time value and (b) with time value of money (Bhattacharyya 2011). Selected methods are presented in this section.

10.3.1 Methods Without Time Value of Money

Methods without time value of money are extensively employed and well understood by many researchers. Two widespread indicators of this group of methods are payback period and average rate of return on investment.

10.3.1.1 Simple Pay-Back Period

The pay-back period indicates the number of years that are required to recoup the project cost of the investment in question (Orioli and Gangi 2017). A shorter pay-back period indicates that the funds of the investor will be recovered quickly. Only cash flows and the initial cost of the project are considered in a simple analysis of the pay-back period. Cash flows are not considered beyond the pay-back period. Usually, this technique is used as a starting point for project screening.

10.3.1.2 Simple Rate of Return on Investment

The difference between total cost and revenue yields profit or loss. When this difference is greater than zero, it is referred to as profit. Otherwise, if the difference is less than zero, there is loss. Profit is always desirable in project investment, and a simple rate of return on an investment is the annual net profit expressed as a ratio of initial investment. For instance, if a project makes an annual profit of US$9000 on an investment of US$100,000 then the rate of return is 9%. This method is simple and generally yields information that can easily be understood. Nevertheless, the method is commonly concerned with the commercial profitability of the project and may therefore be less helpful in projects that mostly give social benefits (Bhattacharyya 2011).

10.3.2 Methods with Time Value of Money

Practically, cash flows take place over a certain period of time. It is therefore necessary to take into account the variation of the economic value of a project with time. Some methods employ the time factor for determining indicators of economic performance. These indicators include net present value (NPV), discounted pay-back (DPB), internal rate of return (IRR), and levelized cost of electricity/energy (LCOE).

10.3.2.1 Net Present Value

The present value of a sum of money is the current worth of a future amount of money at a specified rate of return. So, future costs and benefits of a project have present values, and the net present value (NPV) of an investment is the difference between the present value of the costs and the present value of benefits. It is a scaled measure with units of currency such as US dollar, Euro, and others. This analytical tool yields a robust measure of the net benefits (or costs) and it is commonly employed in decision-making about the economic feasibility of a project. The NPV can be calculated from:

$$\mathrm{NPV} = \sum_{t=1}^{t=N} \frac{R_t - C_t}{(1+r)^t} - C_\mathrm{o} \tag{10.1}$$

where C_t is the cost in year t, C_o is the initial investment cost, R_t is the revenue in year t, r is the discount rate, and N is the project lifespan. The difference (R_t–C_t) is the net cash flow in year t.

A positive NPV indicates that a project is economically feasible. The degree of project feasibility increases with the NPV. Equation (10.1) shows that the NPV of an investment is influenced by the discount rate. Generally, an investor is interested in projects that give returns (profits) on the investments, with an expectation that the return will be equal to (or higher than) the rate of interest earned from similar investments in alternative projects. Often, investors use the market interest rate to estimate the NPV but other values of interest may be applied depending on the nature of the project. The NPV is said to be undiscounted if $r = 0$.

10.3.2.2 Discounted Pay-Back

The discounted pay-back (DPB) period takes into account the time value of money. At the start of the project, the NPV is negative because of the initial costs and absence of revenue. As time (t) increases, cash commences to flow in and out, and eventually, the total costs and benefits balance out (NPV = 0). The DPB period is the value of time ($t = t_0$) that corresponds to NPV = 0, and so:

$$\sum_{t=1}^{t=N} \frac{R_t - C_t}{(1+r)^t} - C_\mathrm{o} = 0 \tag{10.2}$$

The left-hand side of Eq. (10.2) varies from negative through zero to positive values for a given discount rate (r) as t increases. Solving for t in this equation would yield the DPB but the equation cannot be solved analytically. Numerically, the NPV is calculated progressively from the initial year until NPV $\approx$ 0, using discounted cash flows. The value of t that corresponds to a NPV of zero gives the discounted pay-back period.

10.3.2.3 Internal Rate of Return

An internal rate of return (IRR) is a metric for estimating the profitability of an investment. This measure is unscaled with a unit of % per unit time (e.g., IRR = 9% per year), and has been used by economists, engineers, and other experts for many years. It is the discount rate that gives a net present value of cash flows equal to zero at the end of the project. So, the IRR can be computed by setting NPV to zero and then solving for r:

$$\sum_{t=1}^{t=N} \frac{R_t - C_t}{(1+r)^t} = C_o \tag{10.3}$$

The discount rate cannot be computed analytically from Eq. (10.3) but through trial-and-error or use of specialized software for calculation of IRR. This parameter can also be estimated by using linear interpolation (Lucey 2002). Some attractive features of the IRR are (Mellichamp 2017): (a) Optimization of the desirable result is obvious, one just needs to maximize IRR, and (b) a lot of experience on IRR has been acquired by many individual practitioners and design groups over the time.

Projects are usually assessed by using IRR and NPV rules, based on both socio-economic and financial considerations (Percoco and Borgonovo 2012). This blend of metrics is particularly important because some projects may have comparable IRR but different scope or size. In such cases, the IRR and NPV can be blended in assessing the economic performance of alternative projects.

10.3.2.4 Levelized Cost of Electricity/Energy

Levelized cost of electricity/energy (LCOE) is the ratio of the sum of all costs incurred to the sum of all the electricity/energy produced during the lifetime of the project. This parameter gives the average unit cost of electricity generated by a power plant over its lifetime and enables a comparison of different generation technologies of unequal lifespans and capacities. The levelized cost of electricity can be given by (Zhao et al. 2017):

$$\text{LCOE} = \left\{\sum_{t=0}^{t=N} C_t(1+r)^{-t}\right\}\left\{\sum_{t=0}^{t=N} E_t(1+r)^{-t}\right\}^{-1} \tag{10.4}$$

where E_t is the amount of energy produced during the period t.

Some advantages of the LCOE over the net present value are: (a) absence of restrictions on project scale, (b) LCOE is independent of energy technologies, and (c) LCOE is applicable even when energy technologies are different. Consequently, many researchers have used this method to evaluate the feasibility of SGT projects. Theoretical values of the LCOE, ranging from 0.06 US$/kWh for a closed cycle solar gas turbine driven by the solar tower technology (Kribus et al. 1998) to 0.40 US$/kWh for PTC-driven steam injected solar gas turbine (SISGT) (Selwynraj et al. 2015), have been reported in the literature. For a given design, the LCOE decreases with annual insolation (Chahartaghi and Baghaee 2020). In the study of a coal power plant, Cruz et al. (2017) reported a LCOE range of 0.092–0.095 US$/kWh, which shows that the economic performance of an SGT is theoretically approaching parity with conventional thermal power plants. A review of the LCOE worldwide by Larsson et al. (2014) showed that the values of LCOE for a given energy technology vary

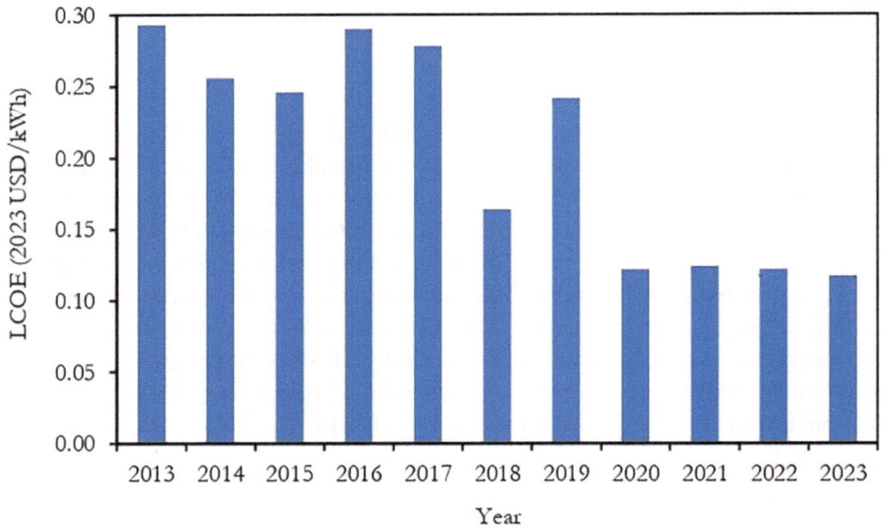

Fig. 10.1 Variation of the global average weighted levelized cost of electricity (LCOE) for concentrating solar power. *Source of data* International Renewable Energy Agency (2024)

with region and study (authors report different values due to variations in methods and assumptions).

In the last decade (2013–2023), the global weighted LCOE for concentrating for power has decreased by about 60% (International Renewable Energy Agency 2024). However, it has remained almost constant at 0.12 USD/kWh over the years 2020–2023 (Fig. 10.1). On a global scale, coal is the largest carbon emitter in the power sector. So, replacement of this fossil fuel is necessary to achieve deep decarbonization. In this context, the International Renewable Energy Agency (2013) argues that carbon–neutral electricity can compete against coal power if its generation costs are lower than approximately 0.08 USD/kWh. Based on this benchmark, the CSP technology is still not competitive on the market. The high costs are attributed to: (a) the low level of maturity of this technology and (b) the modest net electric conversion efficiencies of the power blocks in operation that are mainly based on the steam cycle (Morosini et al. 2023). Most of the operating CSP plants are currently based on the steam Ranken cycle that yields annual solar-to-electricity efficiencies of 10–20% (Alami et al. 2023), which is relatively low. Consequently, one of the key strategies for cost reduction is to exploit the Brayton cycle in the power blocks of the third-generation CSP technology. This development is expected to increase the solar-to-electricity efficiency of the CSP technology to 25–30% and to reduce the levelized cost of electricity. Thus, the solar gas turbine technology is central to the advancement of CSP systems. In this context, the closed cycle solar gas turbine is expected to play an important role in the reduction of electricity generation costs, and the supercritical carbon dioxide Brayton cycle is a very promising power cycle in this regard (Schöffer et al. 2021). This power cycle is expected to contribute to the improvement of the economic performance of CSP systems.

10.4 Worked Examples

Example 10.1 An independent power producer is considering investment in a solar micro gas turbine for generating electricity. The initial investment in the project is USD 250,000. The estimated net cash flow in year 1 is USD (– 9000). Thereafter, the annual net cash flow is USD 60,000 until the power plant is decommissioned. Calculate the net present value if the discount rate is 20% and the lifespan of the project is 25 years. Is the project viable?

Solution

From Eq. (10.1), the net present value (NPV) is:

$$\begin{aligned}\text{NPV} &= \sum_{t=1}^{t=N} \frac{R_t - C_t}{(1+r)^t} - C_o \\ &= -250{,}000 - \frac{9000}{(1+0.2)} + 600{,}000 \sum_{t=2}^{t=25} \frac{1}{(1+0.2)^t} \\ &= \text{USD } (-10645).\end{aligned}$$

Therefore, the project is not economically viable because the net present value is negative.

Example 10.2 In Example 10.1, how is the investment decision affected when the discount rate is 15%?

Solution

We need to find the net present value (NPV) when the discount rate is 15%.

$$\begin{aligned}\text{NPV} &= \sum_{t=1}^{t=N} \frac{R_t - C_t}{(1+r)^t} - C_o \\ &= -250{,}000 - \frac{9000}{(1+0.15)} + 600{,}000 \sum_{t=2}^{t=25} \frac{1}{(1+0.15)^t} \\ &= \text{USD } 77849.\end{aligned}$$

The project is viable at a discount rate of 15% because the net present value is positive.

Example 10.3 A solar gas turbine power plant is expected to operate for 20 years. The initial investment cost is USD 10 million. The annual operating cost (maintenance, labour, etc.) is USD 500,000. Calculate the levelized cost of electricity at a discount rate of 20%, assuming the plant produces 35 GWh/year.

Solution

$$\begin{aligned}\text{LCOE} &= \left\{\sum_{t=0}^{t=N} C_t(1+r)^{-t}\right\}\left\{\sum_{t=0}^{t=N} E_t(1+r)^{-t}\right\}^{-1} \\ &= \left\{10{,}000{,}000 + 500{,}000\sum_{t=1}^{t=20}(1+0.2)^{-t}\right\}\left\{35{,}000{,}000\sum_{t=1}^{t=20}(1+0.2)^{-t}\right\}^{-1}. \\ &= (10{,}000{,}000 + 500{,}000 \times 4.8996)(35{,}000{,}000 \times 4.8996)^{-1} \\ &= 0.073\ \text{USD/kWh}.\end{aligned}$$

Example 10.4 An entrepreneur is considering investment in a solar gas turbine for generating electricity. The initial investment in the project is USD 250,000. The estimated net cash flow in year 1 is USD (−9000). Thereafter, the annual net cash flow is USD 60,000 until the power plant is decommissioned. Calculate the discounted pay-back period if the discount rate is 15% and the lifespan of the project is 25 years.

Solution

Pay-back period occurs when the net present value (NPV) is approximately equal to zero. Therefore,

$$\text{NPV} = \sum_{t=1}^{t=N} \frac{R_t - C_t}{(1+r)^t} - C_o = 0.$$

This implies that:

$$\frac{-9000}{(1+0.15)} + 60{,}000\sum_{t=2}^{t=N}\frac{1}{(1+0.15)^t} - 250{,}000 = 0.$$

After calculating the discounted cash flows progressively, NPV = USD (−8873.88) in year 10, and NPV = USD 4022.71 in year 11.

Therefore, the discounted pay-back period is approximately 11 years.

Example 10.5 In Example 10.2, compute the internal rate of return (IRR).

Solution

IRR can be computed by setting the net present value equal to zero and then solving for r:

$$\sum_{t=1}^{t=N} \frac{R_t - C_t}{(1+r)^t} - C_o = 0.$$

This implies that we need to solve for r in the following equation:

$$\frac{-9000}{(1+r)} + 60{,}000 \sum_{t=2}^{t=25} \frac{1}{(1+r)^t} - 250{,}000 = 0.$$

There are different methods of solving this equation. We can use linear interpolation to find r. In this case, we need to choose two values (r_1 and r_2) of r such that one of them gives a positive NPV while the other value yields a negative NPV. Let $r_1 = 0.15$ and $r_2 = 0.25$.

When $r_1 = 0.15$, NPV = USD 77,848.95, and when $r_2 = 0.25$, NPV = USD (− 66,106.69). This shows that IRR lies between 0.15 and 0.25. Through linear interpolation, we obtain:

$$r = 0.2040.$$

Therefore, the IRR is approximately 20%.

Nomenclature

- C Cost (USD or other currency)
- E Energy/electricity production (kWh)
- r Discount rate (dimensionless)
- R Revenue (USD or other currency)
- t Time (year)

References

Alami AH, Olabi AG, Mdallal A, Rezk A, Radwan A, Rahman SMA, Shah SK, Abdelkareem MA (2023) Concentrating solar power (CSP) technologies: status and analysis. Int J Thermofluids 18:100340

Bhattacharyya SC (2011) Energy economics: concepts, issues, markets and governance. Springer, London

Chahartaghi M, Baghaee A (2020) Technical and economic analyses of a combined cooling, heating and power system based on a hybrid microturbine (solar-gas) for a residential building. Energy Build 217:110005

Cruz MA, Araújo OQF, Medeiros JL, Castro RPV, Ribeiro GT, Oliveira VR (2017) Impact of solid waste treatment from spray dryer absorber on the levelized cost of energy of a coal-fired power plant. J Clean Prod 164:1623–1634

Idrus A, Nuruddin MF, Rohman MA (2011) Development of project cost contingency estimation model using risk analysis and fuzzy expert system. Expert Syst Appl 38:1501–1508
International Renewable Energy Agency (2013). Southern African Power Pool: planning and prospects for renewable energy. International Renewable Energy Agency (IRENA), Abu Dhabi
International Renewable Energy Agency (2024) Renewable power generation costs in 2023. International Renewable Energy Agency (IRENA), Abu Dhabi
Kreiss J, Ehrhart K, Haufeb M (2017) Appropriate design of auctions for renewable energy support—prequalifications and penalties. Energy Policy 101:512–520
Kribus A, Zaibel R, Carey D, Segal A, Karni J (1998) A solar-driven combined cycle power plant. Sol Energy 62:121–129
Lamnatou C, Chemisana D (2017) Concentrating solar systems: life cycle assessment (LCA) and environmental issues. Renew Sustain Energy Rev 78:916–932
Larsson S, Fantazzini D, Davidsson S, Kullander S, Höök M (2014) Reviewing electricity production cost assessments. Renew Sustain Energy Rev 30:170–183
Lucey T (2002) Quantitative techniques, 6th edn. ELST, London
Mellichamp D (2017) Internal rate of return: good and bad features, and a new way of interpreting the historic measure. Comput Chem Eng 106:396–406
Morosini E, Villa E, Quadrio G, Binotti M, Manzolini G (2023) Solar tower CSP plants with transcritical cycles based on CO_2 mixtures: a sensitivity on storage and power block layouts. Sol Energy 262:111777
Orioli A, Gangi A (2017) Six-years-long effects of the Italian policies for photovoltaics on the pay-back period of grid-connected PV systems installed in urban contexts. Energy 122:458–470
Percoco M, Borgonovo E (2012) A note on the sensitivity analysis of the internal rate of return. Int J Prod Econ 135:526–529
Prljić S, Nikitović Z, Stojanović AG, Cogoljević D, Pešić G, Alizamir M (2018) Management of business economic growth as function of resource rents. Physica A 491:325–328
Rothman R (2000) Measuring and apportioning rents from hydroelectric power Developments, World Bank Discussion Paper no. 419. World Bank, Washington
Schöffer SI, Klein SA, Aravind PV, Pecnik R (2021) A solid oxide fuel cell- supercritical carbon dioxide Brayton cycle hybrid system. Appl Energy 283:115748
Sell A (1989) Calculation of working capital for the evaluation of projects in developing countries. Project Appraisal 4:151–156
Selwynraj AI, Iniyan S, Polonsky G, Suganthi L, Kribus A (2015) An economic analysis of solar hybrid steam injected gas turbine (STIG) plant for Indian conditions. Appl Therm Eng 75:1055–1064
Shrestha KK, Shrestha PP (2016) A contingency cost estimation system for road maintenance contracts. Procedia Eng 145:128–135
Shrestha R, Abeygunawardana A (2009) Evaluation of economic rent of hydropower projects. Energy Policy 37:1886–1897
Touran A, Liu J (2015) A method for estimating contingency based on project complexity. Procedia Eng 123:574–580
Ukaegbu B (2014) The significance of working capital management in determining firm profitability: evidence from developing economies in Africa. Res Int Bus Finance 31:1–16
Yucekaya A (2013) Bidding of price taker power generators in the deregulated Turkish power market. Renew Sustain Energy Rev 22:506–514
Zhao Z, Chen Y, Thomson J (2017) Levelized cost of energy modeling for concentrated solar power projects: a China study. Energy 120:117–127

Chapter 11
Environmental Performance of Solar Gas Turbines

11.1 Introduction

Energy drives social and economic development in the world. Nevertheless, there are growing concerns about environmental degradation arising from activities in energy and other systems. An energy system is the chain of processes from the extraction of primary energy to the use of final energy to supply goods and services to a given society or economy (Pfenninger et al. 2014). The energy system consists of two separate subsystems: (a) energy supply and (b) energy demand. An energy supply system is made up of all parts that convert primary energy to final energy. Extraction and treatment of primary energy, conversion technologies, and secondary energy and transmission/distribution technologies fall within the supply system. An energy demand system consists of all parts that convert final energy to end-use services. The demand parts include various end-use conversion technologies such as automobiles, light bulbs, and furnaces. Major commodities required by society include electricity, heating, cooling, water conveyance, water treatment, and energy storage (such as storage of oil and natural gas).

Energy systems interact with the subsystems of the earth system in complex ways (Fig. 11.1). There are four major subsystems of the earth system (also known as spheres), namely: atmosphere, hydrosphere, geosphere, and biosphere. The atmosphere is the gaseous envelop that surrounds the earth (Fig. 11.2a). Within this sphere, there are five major layers arranged from lowest to highest: the troposphere, stratosphere, mesosphere, thermosphere, and exosphere. Atmospheric layers have different characteristics, suited for their various functions. For instance, within the troposphere, the air temperature decreases with an increase in the height above the earth's surface, which promotes convective heat and mass transfer. It contains air that is required by animal and plant life. On the other hand, temperature increases with height in the stratosphere, thereby suppressing convection and dilution of pollutants deposited in this layer. One of the most important characteristics of the stratosphere

A. Madhlopa and R. Nkhoma, *Principles of Solar Gas Turbines for Generating Electricity*, Green Energy and Technology,
https://doi.org/10.1007/978-3-031-84096-8_11

is the possession of an ozone layer that filters off the harmful ultraviolet (UV) radiation from the sun. The hydrosphere is the sum of all the water on earth in the three forms of matter (vapour, liquid, and solid), and dissolved substances (Speight 2020). It includes groundwater, surface water (rivers, lakes, and seas/oceans), and water in the atmosphere. The frozen part of the hydrosphere is known as the cryosphere. This sphere is made up of snow and ice on land, ice caps, glaciers, permafrost, and sea ice. It helps in maintaining the earth's climate by reflecting incoming solar radiation back into the outer space (Kumar 2022). The geosphere is the earth itself that consists of the inner core, outer core, mesosphere, asthenosphere, and lithosphere (Fig. 11.2b). The biosphere comprises parts of earth where organisms live, and included in this sphere are human beings. Except for solar resources (from the sun), major primary energy resources (bioenergy, geothermal, hydro, fossil fuel, and nuclear fuel) are found in the four spheres.

Activities in the earth system can broadly be classified into natural and anthropogenic categories. Natural activities are processes that exist or are produced by natural mechanisms. Anthropogenic activities are caused or influenced by human beings, and they are currently the main cause of environmental degradation (Nwankwegu et al. 2022). The natural characteristics of the environment are being significantly altered by different kinds of anthropogenic activities such as energy consumption and production.

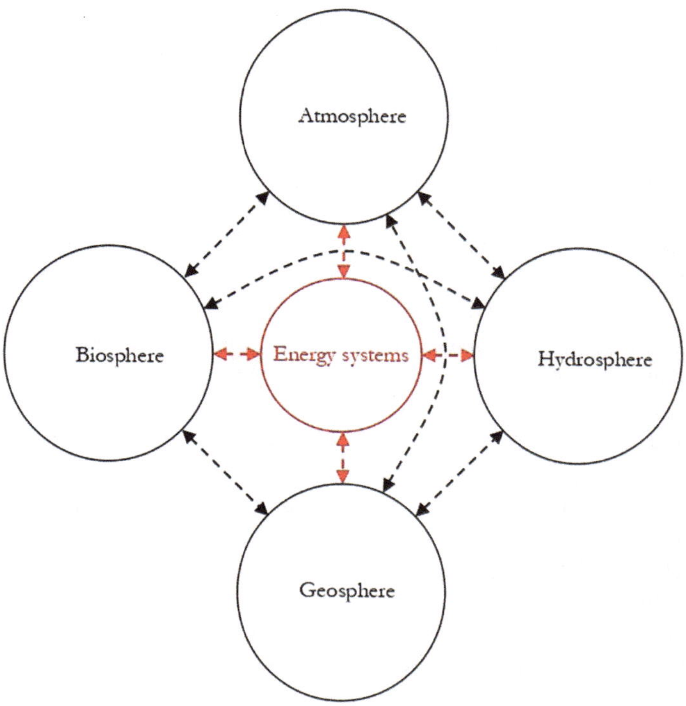

Fig. 11.1 Energy systems and their interactions with the spheres of the earth system

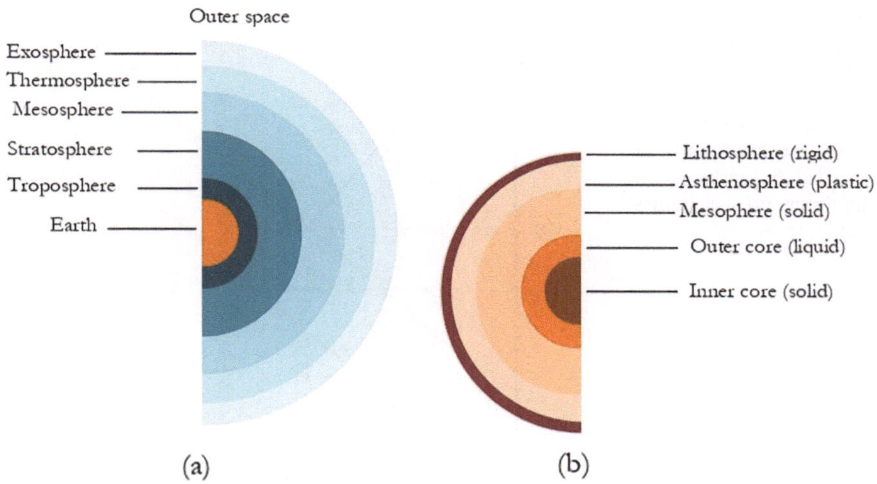

Fig. 11.2 Cross-section of major layers of the: **a** atmosphere and **b** geosphere (earth)

Energy consumption refers to the amount of energy used in a process. It is a critical parameter that affects the environment. The generation of carbon emissions increases with energy consumption (Panagiotopoulou et al. 2024). Thus, it is necessary to reduce the consumption of energy from sources with high levels of pollution (Mahmoudi et al. 2021). Energy consumers can be classified into two broad categories: domestic and industrial consumers. Domestic energy consumption is the total end-use of energy for cooking, lighting, heating, refrigeration, amusement, transportation, and others (Zhou et al. 2008). This category of consumers accounts for about 30% (excluding transportation) of the total global sector (Hu and Man 2023). Industrialization is a process in which a human society is socially and economically transformed from being agrarian to industrial-based (Kyule and Wang 2024). It started around the eighteenth century and is a key factor to power, wealth, and higher standards of living. In this context, industrial energy use is mostly for production, lighting, and other businesses (Jordan 2014). Some industries are energy-intensive, and so they generally contribute a high proportion of industrial energy consumption. This category of industries includes: (a) chemicals, (b) iron and steel, (c) nonmetallic minerals, (d) pulp and paper, and (e) nonferrous metals (Abdelaziz et al. 2011). Industrial energy accounts for about 37% of the total global energy consumption. The global demand for energy is rising due to population and economic growth (International Energy Agency 2023). This energy demand increase requires a corresponding rise in the level of energy supply to reduce energy prices because the supply is driven by the demand side in the current structure of the energy market (Fig. 11.3). About 80% of the global energy supply is currently contributed by fossil fuels: coal, oil, and gas, and coal is the main source of electricity (Malico 2024). Unfortunately, exploitation of fossil fuels is significantly contributing to the degradation of air, water, and soil/land resources. Moreover, there is an increasing shortage of essential

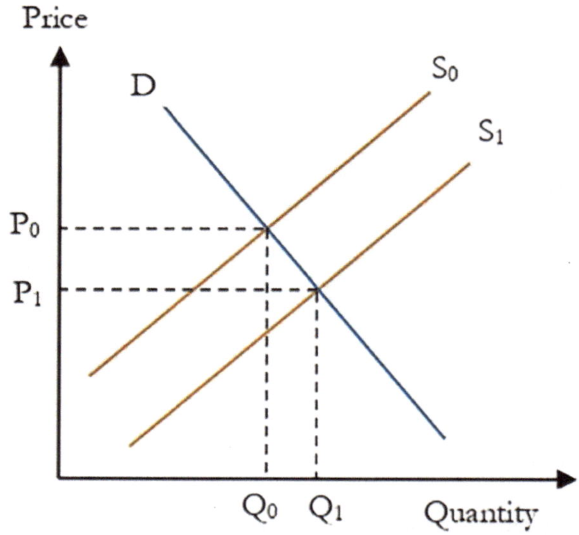

Fig. 11.3 Illustration of energy price variation with energy demand (*D*), supply (*S*), and quantity (*Q*): For a given demand, the supply curve must shift from S_0 to S_1 so that the market equilibrium price decreases from P_0 to P_1

natural resources such as water and land in many parts of the world because of population and industrial growth (Wang and Azam 2024). The major concerns about air pollution are greenhouse emissions, particulate matter, and acid rain. Environmental impacts on water and soil/land include the quality and quantity of these resources.

11.2 Air Pollution

Air pollution is the contamination of the indoor or outdoor environment by any biological, chemical or physical agent that changes the natural characteristics of the atmosphere. These pollutants include gaseous and particulate matter (PM). The gaseous emissions include greenhouse and other gases.

11.2.1 Greenhouse Gas Emissions and Climate Change

Greenhouse gases are gaseous substances that trap heat in the atmospheric environment. In normal concentrations, these gases help in maintaining appropriate levels of ambient air temperature for living things and desirable environmental processes. They raise the temperature of the earth by absorbing energy and reducing the rate at which the energy escapes from the atmosphere into the outer space. Key GHGs include: carbon dioxide (CO_2), chlorofluorocarbons (CFCs), hydrochlorofluorocarbons (HCFCs), methane (CH_4), nitrous oxide (N_2O), hydrofluorocarbons (HFCs), perfluorocarbons (PFCs), and sulphur hexafluoride (SF_6). Other GHG emissions are water vapour (Varanasi 1993) and tropospheric ozone (Lu et al. 2021). Among these gases, carbon dioxide accounts for about 76% of the total annual global GHG emissions, and so it is the major greenhouse gas (Mousavi et al. 2022). Therefore, carbon dioxide equivalent (CO_2eq) is applied as a standard unit for measuring the concentration of GHG emissions. The justification for this is that each GHG should be expressed in terms of the amount of carbon dioxide that would produce the same amount of warming.

GHGs have different abilities to absorb energy (radiative efficiencies). In addition, they also have different times of residence in the atmosphere. So, they have different values of global warming potential (GWP). The GWP enables comparison of the amount of energy absorbed by 1 tonne of a gas compared with the amount of energy absorbed by 1 tonne of carbon dioxide over the same time period (often, 100 years). For gases with shorter time of residence, an averaging time of 20 years is used instead of 100 years. Atmospheric concentrations of carbon dioxide will last thousands of years because this GHG has a long residence time in the atmosphere. The average atmospheric residence time of methane is about a decade but its radiative efficiency is significantly higher than that of carbon dioxide, yielding a GWP of 28–36 (Vallero 2019). Nitrous oxide has an average residence time of 100 years with a GWP of 265–298 times that of carbon dioxide. Chlorofluorocarbons, hydrofluorocarbons, hydrochlorofluorocarbons, perfluorocarbons, and sulphur hexafluoride have high values of GWP that can be up to tens of thousands. Nevertheless, chlorofluorocarbons and hydrochlorofluorocarbons also contribute to the depletion of the stratospheric ozone layer (Chipperfield et al. 2015). The discovery about ozone depletion led to the signing of the Montreal Protocol in 1987, an international agreement that limits the production of chlorine-containing and bromine-containing ozone-depleting substances. Consequently, the global warming contribution from chlorofluorocarbons and hydrochlorofluorocarbons is neglected because they are being phased out under the Montreal Protocol. Therefore, the Kyoto Protocol specifies only six key GHGs of concern: carbon dioxide, methane, nitrous oxide, hydrofluorocarbons, perfluorocarbons, and sulphur hexafluoride (United Nations 1998). The concentration of GHGs is increasing predominantly due to the high dependence on fossil fuels in modern economies. Since the start of the industrial revolution in around 1750, the concentration of carbon dioxide has risen from 280 parts per million (ppm) to 417 ppm in 2021, representing a rise of about 49%, predominantly from the exploitation of fossil fuels. Projections show that the concentration of atmospheric carbon dioxide will reach 1045 ppm by the end of 2100 (Intergovernmental Panel on Climate Change 2023).

It is believed that the rise in the concentration of GHGs contributes to climate change, a phenomenon of long-term swings in temperatures and weather patterns (United Nations 2024). Climate change can be caused by natural mechanisms such as large volcanic eruptions or variation of the activity of the sun. Nevertheless, since the commencement of industrialization, different anthropogenic activities have predominantly been the drivers of climate change. These activities are producing greenhouse gases that account for most of the rise in the ambient temperature (Saikku et al. 2008). Due to the greenhouse effect, there is a gradual increase of the mean global surface air temperature of the earth, predicted to rise by about 2.4–4.8 K by the end of the twenty-first century compared to the temperature levels in 1995–2014. Generally, this warming has progressed at a constant rate of about 0.2 K/decade from around 1970 (Samset et al. 2024). This rise in the surface air temperature is known as global warming, defined as the long-term heating of the earth's surface witnessed since the pre-industrial period because of anthropogenic activities that raise GHG levels in earth's atmosphere. Major impacts of climate change are: higher surface temperatures, increased drought, severe storms, warming-rising oceans, insufficient food, loss of species, higher risks of health, and poverty and displacement.

11.2.2 Particulate Matter

Particulate matter (PM) comprises microscopic solid particles or liquid droplets present in the air. It is produced by primary and secondary sources (Daellenbach et al. 2020). Primary particulate matter is emitted directly from a source such as mining sites, construction sites, fossil-fuelled power plant, and dusty roads. Secondary particulate matter is produced in the atmospheric environment through complex chemical reactions. Almost all forms of mining operations emit particulate matter (Petavratzi et al. 2005), including coal mining. Coal processing, handling, and transportation also produce particulate matter. Some stages (such as refinery processes) of oil and gas operations also release particulate matter into the air (Adebiyi 2022; Dadkhah-Aghdash et al. 2022). The combustion of coal and oil produces relatively higher concentrations of particulate matter than natural gas.

When inhaled, particulate matter can cause serious health problems (Jiang et al. 2024a, b). Particles of diameter below 10 µm can diffuse into lungs while others may even enter the bloodstream. Of these, particles with a diameter of less than 2.5 µm in diameter (also known as fine particles or PM2.5) pose the greatest risk to health (Li et al. 2023). Moreover, particulate matter may contain heavy metals such as arsenic (Ar), cadmium (Cd), lead (Pb), zinc (Zn), and mercury (Hg) (Dehkordi et al. 2024). Heavy metals are metallic elements in the periodic table that possess a density of more than 5000 kg/m^3 and an atomic weight of higher than 65, and they are poisonous even at low concentrations (Ojovan and Lee 2005). So, their presence in the atmospheric environment poses a serious hazard to the living things.

11.2.3 Acid Rain

The combustion of fossil fuels also releases sulphur dioxide (SO_2) and oxides of nitrogen (NO_x) into the atmospheric environment (Pryor et al. 2015). Sulphur dioxide is produced when sulphur-containing fuels are burnt. Oxides of nitrogen are emitted during the combustion of fossil fuels at high temperatures. In addition, they are naturally released from soils and lightning events into the atmospheric environment.

Within the atmospheric environment, sulphur dioxide and oxides of nitrogen react with water to respectively form sulphuric and nitric acids. Acid rain is the term often used to describe precipitation (such as rain, snow, fog, and dew) with a pH value of less than 5.6 (Luo et al. 2024). It should be noted that acid rain with a pH of 5.6 is considered normal because the atmospheric carbon dioxide at a pressure of 101 kN/m^2 and temperature of 293 K reduces the pH of rainwater from 7 to 5.6 (Zhang et al. 2012). However, abnormality sets in when gases such as sulphur dioxide and nitrous oxide contribute to a further reduction in pH below 5.6 because of elevated hydrogen ion concentrations.

Acid rain causes the acidification of water and soils, and loss of biodiversity from these ecosystems. It also reduces agricultural productivity and causes corrosion of metallic and other structures. Therefore, acid rain is a serious environmental problem that must be mitigated.

11.3 Water and Soil/Land Degradation

Water and soil/land perform many critical functions that make them socially, economically, and environmentally very important in the world. Water is the most abundant natural resource on the surface of the earth, and it is used in agricultural, domestic, industrial, commercial, and other activities. Most of the water is stored in the oceans that cover about 71% of the surface area of the earth (Kundzewicz 2008). In addition, only about 2.5% of the earth's water resources are fresh while the rest (97.5%) are salty with limited applications. Consequently, there is limited access to fresh water (especially safe potable water) around the world. Potable water is suitable for human consumption such as drinking and cooking, and it must meet stipulated standards (Brenniman 1999). Raw water is often treated so that it meets relevant standards. In general, water treatment is expensive, and the water treatment costs tend to rise with the level of contamination (Hopland and Kvamsdal 2023). It is therefore necessary to protect water and soil/land resources from degradation.

11.3.1 Water Pollution

Water is required in different phases of a power plant, such as during extraction and processing of fuel, extraction and processing of raw materials, manufacturing of components, and construction, operation and decommissioning of the power plant. So, there are many ways in which water can be contaminated with a wide spectrum of pollutants (ranging from sediment to synthetic chemicals) because of its multiple uses in the energy production chain. Water is adversely affected at all stages of prospecting, exploration, and development of fossil-fuel fields, and production, transportation, and use of these fuels. Prospecting and exploration activities aim at discovering reserves of the natural resource, and drilling is one of the key activities in mineral/ore exploration. The mining/extraction operation commences only if resource recovery is economically feasible. Exploration and mining affect both the quantity and quality of water. For instance, extraction of one tonne of coal can result in the depletion of water resources by 1.32×10^9 m^3, contamination of 8.8×10^8 m^3 of water resources, and impairment of 1.7×10^8 m^3 of ecological environments (Hasii and Gasii 2024). Coal mining exposes groundwater to coal seams, rock formations, and anthropogenic activities, resulting in significant degradation of the water quality and ecological environment (Wang et al. 2021). Coal mining and processing generate a significant amount of solid waste known as gangue. This solid waste contains trace elements such as antimony (Sb), arsenic (As), barium (Ba), beryllium (Be), mercury (Hg), vanadium (V), cadmium (Cd), chromium (Cr), cobalt (Co), copper (Cu), nickel (Ni), molybdenum (Mo), lead (Pb), selenium (Se), strontium (Sr), thorium (Th), uranium (U), vanadium (V), and zinc (Zn). It should be noted that mercury, lead, strontium, and uranium are radioactive elements while arsenic, cadmium, chromium, copper, lead, and zinc are heavy metals. Surface water, groundwater, and soils can be contaminated by these elements. The main pollutants from oil and gas production are formation fluids (oil, gas, formation water) during well drilling and stimulation of oil and gas production (Poberezhna et al. 2024). Minor pollution is also caused by chemicals and materials used in the drilling and stimulation processes. These formation fluids penetrate into the surface water, groundwater, and soil resources, thereby degrading the quality of these resources. Combustion of fossil fuels at power plants causes thermal pollution of water resources, whereby cooling water (in case of wet cooling) absorbs excess heat from the plant and rejects it into streams or lakes, thereby raising the biological and chemical oxygen demand in the receiving water.

11.3.2 Land Degradation

Land is the surface area of the earth that is not covered by surface water (such as rivers, seas, and oceans). It provides the foundation for various ecosystems, natural resources, and human activities (Zhao et al. 2024a, b). The size of the land is fixed, but there are many competing demands for this resource. It is therefore necessary to reduce the wasteful use of land when developing energy projects. In this context, the land area required by an electric power plant depends on the specifications of an individual utility design, cost of land, installed capacity, and other factors.

The pressure on land is aggravated by degradation, such as land subsidence (sudden sinking of land surface due to natural or anthropogenic activities). Some of the activities that cause land subsidence are: coal mining and extraction of oil and gas by pumping or fracking (hydraulic fracturing). Different mining methods are used depending on: (a) the depth and quality of the coal seams, (b) geological factors, and (c) environmental factors. Broadly, these methods fall into surface and underground mining methods and they entail disruption to the land. Underground coal mining involves the initiation and propagation of internal cracks in the overlying rock and soil mass, which results in the movement, deformation, and destruction of the rock strata (Piao et al. 2024). The structural deformation and failure of the overlying strata can result in surface subsidence. Generally, factors of surface subsidence include mechanical properties of the overlying strata, ratio of mining thickness to mining depth, thickness of loose layers, dip angle of coal seams, panel size, and mining and roof-control methods (Gong et al. 2024). Out of these factors, mechanical properties of the overburden are the most influential. Fracking is a method of extracting oil or gas from an underground shale rock (O'Leary 2022). In this method, a fluid is injected at high pressure into the rock formation to create fissures and allow the trapped natural gas or crude oil to flow through a pipe that is linked to a wellhead located at the ground surface. Once the oil/gas is extracted, the pressure is reduced, which can make the land to sink.

Soil is a thin grainy layer that covers the land, and it consists of minerals, weathered rocks, and organic matter. Its functions include: provision of habitat for biological activity, medium for biomass production, water filtration and storage, nutrient recycling and storage, and carbon storage (Wiesmeier et al. 2019). Pollutants in the soil reduce the capacity of the soil to produce food and they yield contaminated crop, resulting in malnutrition and diseases. Thus, soil pollution is risking the health of about 40% of the global population (Münzel et al. 2024). Moreover, pollutants deposited in the soil easily migrate to the surface water and groundwater resources.

11.3.3 Heavy Metals, Acid Mine Drainage, and Ocean Acidification

The long-term stockpiling of the solid wastes (say, from coal and other mining activities) releases heavy metals into the groundwater and soil environments through leaching, sedimentation, percolation, and other mechanisms (Zhao et al. 2024a, b). Oil spills and stormwater runoff containing oil derivatives comprise heavy metals and other pollutants. Similarly, natural gas extraction can generate large volumes of produced water that is usually contaminated with heavy metals and other pollutants. Combustion of fossil fuels also spreads these dangerous metals in the environment (Kanwal et al. 2024). Unfortunately, heavy metals are toxic, non-biodegradable and they accumulate in the soil. Excessive concentrations of soil heavy metals can affect the physicochemical properties of soil and hinder plant growth and reproduction. It has been established that soils around coal mining areas face serious heavy metal contamination problems and should be given more attention.

Acid mine drainage (AMD) is associated with coal and oil extraction (Hasrod et al. 2024; Ling et al. 2024). It is produced from complex chemical and biological reactions of ores (such as pyrite and marcasite) that contain sulphides. When exposed to air, water and microorganisms, these ores undergo oxidation and produce acid mine drainage with an approximate pH value of less than or equal to 3, and it contains a high concentration of heavy and other metals. The reduction in pH makes dissolved metals to be readily available for uptake by animals that live on the sea floor (such as

Table 11.1 Selected impacts of fossil-fuel extraction, processing, and use (combustion) on water and soil/land resources

Process	Impact on water/soil quality
Coal mining	Tailings and drainage can reduce the surface water and groundwater quality, collapses of coal mines can harm water flow and reduce water quality
Coal processing	Coal washing generates a significant amount of coal gangue with heavy metals and other pollutants that can enter the water and soil systems
Oil and gas exploration	Formation fluids can enter surface water, groundwater, and soil resources
Oil and gas production	Formation fluids can enter surface water, groundwater, and soil resources
Oil and gas processing	End use can affect water and soil quality
Combustion of fossil fuels at power plants	Wastewater from power plant may contaminate surface water resources, wet cooling can elevate biological and chemical oxygen demand of the receiving water

fish), thereby providing a major pathway of their introduction into the human food chain. This acid mine drainage may infiltrate into groundwater via fractures in the mine tailing dumps or via shafts of abandoned mines (Singh 1987). So, acid mine drainage directly and indirectly poses a risk to human health.

Ocean acidification is caused by the combustion of coal, oil, and gas. The combustion of fossil fuels releases carbon dioxide and other gases into the atmosphere. Oceans can maintain the balance of the carbon cycle by absorbing about 30% of anthropogenic carbon dioxide emissions (Shangguan et al. 2024). Nevertheless, the oceans are becoming more acidic due to the rise in carbon emissions from different sources. The amount of calcium carbonate decreases with an increase in water acidity. Unfortunately, this substance is utilized by oysters, lobsters, and other marine organisms to form shells. So, the rise in water acidity can retard the growth rates of these organisms, weaken their shells, and disrupt critical food chains, resulting in loss of biodiversity. There are many environmental problems that have been caused by the operation of fossil-fuel fields and use of fossil fuels. Presented in Table 11.1 is a summary of selected impacts on the quality of water.

11.4 Sustainable Production of Electricity

Sustainable development is defined as the development that meets the needs of the present without compromising the ability of future generations to meet their own needs (Mahdi 1987). It can be classified into social, economic, and environmental dimensions that are interrelated (Geissinger et al. 2019; Serpa and Ferreira 2019).

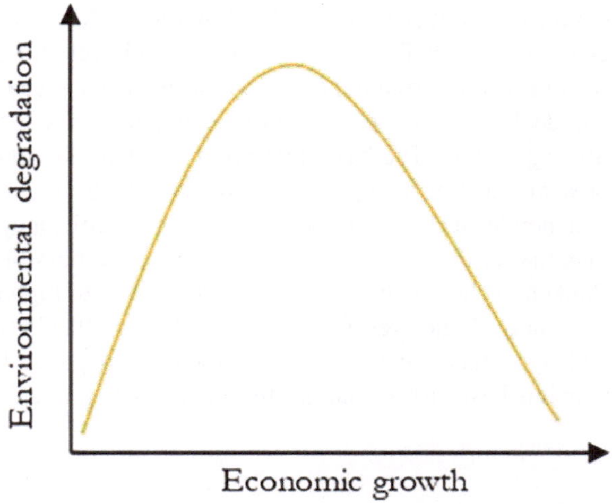

Fig. 11.4 Variation of environmental degradation with economic growth

The social dimension covers issues such as fairness in the distribution of wealth, equal opportunities, and ethical behaviour throughout the life cycle of a project, from inception to demolishing stages. Any investment is aimed at making profit. In this context, the economic dimension involves increasing the profitability of projects, and it includes issues such as efficient exploitation of resources, including energy, water, materials, and labour. Prudent exploitation of natural resources and prevention of detrimental and potentially irreversible effects on the environment are the key issues of the environmental dimension. Theoretically, the level of environmental degradation increases with economic growth until reaching a maximum point where it starts declining with an increase in economic growth (Zambrano-Monserrate et al. 2018; Prasad 2024). So, there is a positive relationship between environmental degradation and economic growth in the first stage of economic growth (Fig. 11.4).

In the energy sector, sustainable development is founded on three key pillars (Kowalska-Pyzalska 2024): (a) reduction in GHG emissions, (b) increase in the energy efficiency on the demand side, and (c) increase of the renewable energy share in the energy mix. Consequently, the goal of increasing access to electricity is also connected to policies of climate change (van Vuuren et al. 2012). These policies aim at achieving net-zero GHG emissions by 2050 at global level (United Nations 2015; Cheng et al. 2023). Net-zero means the balance between the amount of anthropogenic emissions by sources and removals by sinks of greenhouse gases. Attainment of this balance demands remarkable decreases in emissions across all economic sectors. In this context, renewable energy sources are perceived to be clean, replenishable, and low-risk (Amponsah et al. 2014).

Concentrating solar power (CSP) is one of the key renewable energy technologies. It is a scalable technology with enormous potential to supply baseload and dispatchable power (Pfenninger et al. 2014). These characteristics are required to maintain a reliable electric power system as the share of renewable electricity increases in the energy mix. The CSP technology uses solar thermal energy to drive conventional engine power cycles such as the Brayton, Rankine, and Stirling power cycles (Islam et al. 2018). The Brayton cycle is the ideal cycle for gas turbine (GT) engines, and so, it is also referred to as the gas turbine cycle (see Chap. 5). The Ranken cycle is the ideal cycle for the steam turbine engines while the Stirling cycle is ideal for the Stirling engine. The basic principles of a CSP system are reported in Chap. 1). At present, most of the operating commercial CSP plants use the steam Ranken cycle to generate electricity. However, the thermal efficiency of the Rankine cycle is lower than that of the Brayton cycle. Moreover, it is possible to combine the Brayton and Ranken cycles to obtain a combined cycle with higher energy and exergy efficiencies than a single cycle (Khorampoor et al. 2020). Consequently, the third-generation CSP systems are expected to be based on the gas turbine cycle (Khan et al. 2022), which is basically a solar gas turbine for generating electricity.

11.5 Environmental Impacts of Solar Gas Turbines

A solar gas turbine (SGT) consists of the following major components: solar concentrator, solar receiver, compressor, turbine, and cooler. These components are made from metallic (such as aluminium, copper, and steels) and non-metallic (such as glass and ceramic) materials. A thermal energy storage (TES) unit is often added to enhance the capacity factor of the SGT system (Turchi and Ma 2014). Integration of the SGT system with a TES unit enables the system to operate on solar energy only. It is also possible to add a backup heater driven by renewable or non-renewable energy sources (see Sect. 4.3). The use of fossil fuels to back up the SGT system tends to elevate carbon emissions and other pollutants. Consequently, fossil-fuelled backup heaters should be used only in the absence of a more sustainable alternative.

Water and land resources are required and get affected at different stages of the SGT life cycle. So, while aiming to reduce the lifecycle GHG emissions, it is also necessary to consider other potential environmental impacts of SGTs in order to achieve sustainable development. An environmental impact is a change in the natural or built environment, resulting directly from an activity, that can have adverse effects on the air, water, soil/land, and living organisms (Abdallah 2017). Environmental impacts can be classified into two types: direct and indirect: A direct environmental impact refers to an impact that is directly ascribable to an activity or process and it can be controlled within a given project. Indirect impacts are caused through the consumption of goods and services, and the activities of the many producing industries at national and international levels. This category of impacts is infinite and it is not controllable by a given project. SGT projects do consume external goods (such as materials and components) and services (such as transportation of raw materials). Consequently, SGT systems exert direct and indirect environmental impacts.

11.5.1 Greenhouse Gas Emissions

There are direct and indirect GHG emissions from an SGT power plant at different stages of its life cycle. GHGs are emitted during extraction and processing of raw materials (such as metallic ores), manufacturing of components, plant construction, plant operation, and plant decommissioning (dismantling and disposal). The operation phase of solar-only SGT systems produces emissions due to the consumption of goods and services (such as electricity from fossil-fuel combustion, and fossil-fuelled transportation of raw materials and goods). Hybrid SGT systems that exhibit higher GHG emissions during operation are mainly due to fossil-fuelled backup units (Khorampoor et al. 2020). Fossil-fuelled plants directly emit most of the GHGs during combustion (Saikku et al. 2008), and natural gas is often used as a source of backup heat. Thus, deployment of SGT systems avoids a significant proportion GHGs.

11.5.2 Particulate Matter and Heavy Metals

Primary particulate matter is emitted directly at different stages of the life cycle of SGT power plants. Construction particulate matter is discharged from various materials and equipment utilized, and activities in different stages of the construction process (Yan et al. 2023). These particles are transported by air flows and diffusion from the construction site to surrounding areas. Consequently, construction of SGT power plants contributes to air pollution.

The SGT system also consumes various goods and services that are associated with the generation of particulate matter. For instance, particulate matter is emitted during extraction of raw materials for manufacturing plant components. Some of the goods and services (such as transportation) heavily depend on fossil fuels. As seen in Sect. 11.2.2, the fossil-fuel cycle generates significant amounts of particulate matter.

11.5.3 Acid Rain

SGT systems integrated with fossil-fuelled backup heaters directly emit sulphur dioxide and oxides of nitrogen during the operation phase. For such systems, the intensity of these emissions declines with an increase in the solar share. It should also be noted the combustion of biomass releases oxides of nitrogen (Ozgen et al. 2021), and so its use as a source of backup heat still contributes to the emission of oxides of nitrogen. In addition, solar gas turbines consume some goods and services that have relatively high footprints of sulphur dioxide and oxides of nitrogen. Consequently, the SGT technology can directly and indirectly contribute to the formation of acid rain. Nevertheless, considering that most of SO_2 and NO_x is emitted during the combustion of fossil fuels, the contribution to acid rain from the solar-only SGT technology would be remarkably lower than that from fossil-fuelled power plants.

11.5.4 Impact on Water

Construction of an SGT facility exposes large areas of bare land with soil and unattached nutrients or pollutants that (if not properly managed) can be washed from construction sites into nearby streams, rivers, lakes, and wetlands during storm events. The use of best management practices of erosion control can assist in reducing the amount of sediment and pollutants that would otherwise be washed from the site into surface water resources. Some of the contaminated surface water may percolate into the soil and transport pollutants to groundwater resources.

During operation, wastewater is generated by various operational activities including cooling of the engine (wet cooling requires more water than dry cooling). Wastewater from the power plant must be cleaned, filtered, and processed before being discharged into a stream, river or lake. At the discharge point, the wastewater must be relatively neutral with little heat or newly dissolved materials. In addition, SGT systems indirectly cause water pollution through consumption of some external goods and services that significantly pollute water resources. Nevertheless, compared with fossil-fuelled power plants, the SGT contribution to water pollution is lower than that from fossil-fuelled plants.

11.5.5 Land Use and Visual Impact

The quality and quantity of land are affected by the various activities associated with the production of electricity from CSP systems. Collector arrays in CSP facilities affect the speed of wind (Egerer et al. 2024). In this context, the average wind speed is significantly reduced while the kinetic energy of turbulence grows within the collector field. Reduction of wind speed is greatly beneficial because speed is the primary factor for spreading desert. In this vein, the collector field helps in controlling soil erosion. Another benefit is moderation of soil temperature. The collector field affects the soil temperature under it. The temperature of the soil under the collector field may be higher in winter but lower in spring and summer compared with the soil temperature outside the collector field.

The land area required by an electric power plant depends on the specifications of an individual utility design, cost of land, installed capacity, and other factors. The land use of CSP systems decreases with an increase in the level of insolation. This implies that a CSP plant of given capacity occupies more land in a location with low insolation than in a location with high insolation. CSP plants require a significant amount of relatively flat land (Bošnjaković and Tadijanović 2019). Land is classified as flat when its slope is less than 10%. In this context, slopes of less than 1% and 10%

are ideally required for centralized-generation and distributed-generation systems, respectively for public land in the USA (Azoumah et al. 2010).

Landscape is another important aspect of land use. It is a collection of visible features of an area of land that are considered in terms of aesthetics. Notable elements of visual impacts are: (a) visibility of the SGT facility, (b) occurrence of glare, (c) geometric patterns of reflected light, (d) viewer position and viewing geometry, and (e) ancillary facilities and ground clearing. Generally, SGT plants occupy larger areas of land compared with a conventional gas turbine power plant for a given level of power output. This has significant implications for the landscape of the area where SGT facilities are installed. Landscapes can be classified into natural and anthropogenic categories (Reddy et al. 2017). A natural landscape consists of a collection of natural features: landform (slope, hills, mountains, plateaus, and plains), drainage system (wetlands, streams, rivers, and lakes), and vegetation. Anthropogenic landscapes are derived by transforming natural landscapes through human activities on the earth's terrestrial surface. The perception of a landscape is heavily dependent on human vision (Misthos et al. 2023). Human beings actively experience and engage with the surrounding landscape through the sense of vision. So, they may be pleased or displeased with a given landscape. In this context, landscape transformation can produce visual impacts that may be acceptable or unacceptable by viewers. Visual impacts are changes to the scenic characteristics of a given landscape produced by the introduction of visual contrasts. The negative perception of the visual aspects of SGT power plants by the general public can hinder their wider implementation.

Although the engine block of an SGT power plant occupies a small area, the solar field requires a relatively large space. Table 11.2 shows examples of the total land area occupied by selected CSP plants. The combination of large size, strong regular geometry of SGT power plants, and the use of mirrors with metal supporting structures may yield high visual contrast that is visible from distant viewing points in many instances. The power plant may not be recognized as a solar facility when viewed from long distances. In certain cases, it may appear like a natural feature, depending on its layout and contrast. On the other hand, it may appear like an artificial feature with insufficient visual details to be clearly identified as an SGT facility.

The SGT power plant has a strong regular geometry and highly reflective surfaces (in the concentrator). Reflecting surfaces may produce glare that is a possible hazard to human vision (Ho et al. 2015). Glare is strong or dazzling light that causes difficulty of seeing objects. In general, glare can be classified into three types: (a) discomfort glare, (b) disability glare, and (c) dazzling or blinding glare (Krumina et al. 2021). Discomfort glare refers to the glare that causes visual annoyance or irritation without essentially impairing the ability to see objects (Jain et al. 2023). Prolonged exposure to discomfort glare may result in headaches and other physiological sensations. Disability glare is the kind of glare that damages the vision without essentially causing some discomfort (Löfving et al. 2015). It can cause visual problems such as afterimage effects, veiling, and retinal burn. An afterimage is the visual effect that continues after cessation of the stimulation. Scattered light in the human eye produces a veiling luminance superimposed on the retinal image, thereby diminishing the levels of contrast. Retinal burn can be caused by exposure to laser or concentrated solar

Table 11.2 Land area of selected concentrating solar power plants

Name of plants	Location	Capacity (MW)	Technology	Year started	Land area (m^2)
Posadas Córdoba Andalusia	Spain	50	Parabolic trough	2012	2,500,000
Godawari solar project	India	50	Parabolic trough	2013	1,500,000
Dhursar	India	125	Linear Fresnel reflector	2014	3,400,000
Ivanpah solar electric generating system	USA*	377	Solar tower	2014	14,200,000
Lanzhou Dacheng Dunhuang	China	50	Linear Fresnel reflector	2020	3,200,000

* USA = United States of America

radiation. Dazzling glare is so intensive that no visual perception is possible for an appreciable period of time after the stimulation has ceased. The solar tower system makes the highest visual impacts compared with the other CSP technologies such as the parabolic trough concentrator, parabolic dish concentrator, and Fresnel reflector because of the height of the central tower (Bošnjaković and Tadijanović 2019). This system produces intense brightness, causing significant glare. Nevertheless, it is the most promising technology for driving SGT systems (Okoroigwe and Madhlopa 2016), suggesting that the SGT technology is very likely to make higher visual impacts.

Solar radiation may be simultaneously reflected from regularly spaced metal surfaces in the solar field, yielding some geometric patterns of reflected light. Various surfaces may participate in the reflection of light. These surfaces include the ends of heat transfer fluid tubes in parabolic trough and linear Fresnel systems, supporting structures for the tubes in parabolic trough and linear Fresnel systems, and supporting structures for heliostats in solar tower systems. These reflections may appear as straight lines or grids of light across the whole solar field or large parts of it. Although these reflections are not bright enough to cause discomfort, they may significantly change as the viewer moves and generally create remarkable visual effects that attract the visual attention of the viewer.

The degree of visual contrast depends on the viewer position with respect to the power plant, the degree of illumination, and the angular position of the sun. This position can be described by the horizontal distance from the plant and the angle of elevation or depression. Surfaces with high reflectance (such as mirrored surfaces of heliostats) may reflect light from surrounding features such as the sky, clouds, vegetation cover or surrounding soil if viewed from a particular angle at a specific time of day and under certain illumination conditions. The view may be of colour (or no colour), reflective support structures, soil, or vegetation between the rows of the collector arrays, when mirrored surfaces are not facing the viewer.

Power blocks, substations, and other ancillary facilities may significantly contribute to observed visual contrasts although the solar field is generally the major contributor to such contrasts. This is particularly true for low-elevation views of the power plant. Taller ancillary facilities project well above the generally low-profile collector arrays so that they are usually more visible than the arrays, from low-viewing angles. Ancillary structures produce strong contrast with the uniform geometry of the collector arrays regardless of viewing angle. Usually, they are also the source of glinting or even glare in certain instances. Visual impacts of CSP plants (arising from the large size of the solar field, strong regular geometries, and use of reflecting surfaces) can be unavoidable but these plants are often located in arid zones with low population or touristic value (Merchán et al. 2022).

11.5.6 Impact on Flora and Fauna

Other impacts on the local environment may be caused by the construction activities of SGT power plants. Local flora is destroyed during construction of facilities, roads, and car parks (Bošnjaković and Tadijanović 2019). For instance, vegetation is cleared when constructing building and road infrastructure. The magnitude of damage varies with the area affected and the type of land use prior to plant construction. Plant construction can lead to destruction of habitats, resulting in loss of biodiversity. Some natural resources (such as water) are used by power plants, thereby reducing necessary resources for flora and fauna in the locality. In the Mediterranean areas, vegetation that dries up due to the deficiency of water can also contribute to the risk of fire around the premises of power plants. The power plants directly affect the vertebral mortality in two ways: (a) collision with the upper parts of the plant structures (such as the central tower) and (b) heat stroke or burning of the surrounding area because of fire caused by the reflection of concentrated solar radiation (Hernandez et al. 2014). Birds collide with the tower of the plant when the visibility is low, and a reasonable number of them die through such collisions. When visibility is high, collisions are rare. Birds can also visually replace reflective surface mirrors with water and, when convinced of diving into the water, they collide with mirrors and get injured. Plants and animals slowly adapt to the new environment during the operational phase. For instance, some birds may try to build a nest on the heliostat. Other animals may start feeding on the vegetation under the heliostats. Power plants may block migratory pathways to local plant and animal populations, thereby reducing their chances of survival. Invasive species may be introduced to the site of the power plant during transportation of construction equipment and materials for the SGT facility. There may be a rise in the number of invasive species in the area where the SGT facility is located. These species have a significant impact on biodiversity and ecosystem services (Gallardo et al. 2024). They can quickly propagate and pose a huge risk to domestic species.

11.5.7 Noise Pollution

Noise pollution refers to unwanted sound that can cause various adverse effects on health and hearing when living beings (especially people) are exposed to it for a certain period of time. Sound is said to be unwanted if it exceeds a certain threshold. Exposure to noise pollution can cause cardiovascular diseases (such as high blood pressure), compromised childhood learning, loss of hearing, and sleep disorders (Jacuzzi et al. 2024).

Noise may be generated at different of stages electricity production. During construction of an SGT facility, noise can be generated by activities such as excavation, drilling, movements of vehicles, concrete breaking, cutting metallic materials, pumping, and welding. Construction noise be can reduced by implementing best practice work methods such as (Murphy and King 2022): (a) scheduling of very noisy activities during less sensitive periods of the day, (b) using the quietest and most suitable plant and equipment for the project, (c) ensuring that all pieces of equipment are properly maintained and operating within stipulated specifications, (d) making use of mitigation measures where appropriate such as application of acoustic screens, and (e) using work practices that reduce noise activities such as restriction of dropping waste material from high points.

Gas turbines produce significant levels of noise during operation (Cumpsty and Marble 1977; Tupov and Taratorin 2020). Noise is generated in all the sections of the gas turbine engine, including the air intake unit, compressor, combustor, turbine, and the exhaust. The air intake unit and the exhaust are the main producers of noise pollution. The rule is that noise from the exhaust section is the determinant of the acoustic environmental impact of the whole GT system. It should also be noted that noise pollution is affected by climate factors (air temperature, humidity, and barometric pressure) and ground surface effect (Tupov and Taratorin 2013). The sound pressure from the same source significantly varies with ambient air temperature and humidity. Barometric pressure has insignificant influence on the sound pressure. The ground surface effect can cause an increase or a decrease in the sound pressure levels. Other sources of noise during operation are transformers and maintenance activities. Transformers, employed to step up the voltage for transmission, can also generate noise (Tang et al. 2024). Usually, transformer noise is continuous and it can vary in intensity. Routine maintenance activities (such as cutting metallic raw materials, welding, cleaning of concentrators, and vegetation management) can also generate occasional noise.

11.6 Assessment of Lifecycle Environmental Impacts

As shown in Sect. 11.5, many activities directly or indirectly affect the environment over the lifetime of the SGT system. These impacts need to be thoroughly tracked and evaluated to enable policymakers and other stakeholders to make reliable decisions. Sustainable development requires suitable methods and instruments for measuring and comparing the environmental impacts of anthropogenic activities for different products (Dincer 2001). The SGT is an emerging technology under active research. So far, there has been some progress in the development of this technology for demonstration purposes. However, limited work has been done on its lifecycle environmental impacts.

11.6.1 Indicators of Environmental Impact

Fossil fuels are currently the main source of electricity and heat. Nevertheless, the heavy reliance on these fuels is detrimental to the environment. Consequently, it has become necessary to replace them with renewable energy. The goal is to reduce environmental impacts that arise from the exploitation of various energy resources. As seen from the preceding sections, different environmental impacts result from energy production and consumption, and so, they require different indicators.

An environmental indicator is used to evaluate the effect of different activities/processes on the environment (Joglekar and Khadke 2023). There are many indicators of environmental impacts but the selection of a given indicator depends on various factors such as (Blattert et al. 2017): Availability of data, relevance to a given study and to the entire indicator framework, validity of measurements provided by an indicator, sensitivity of the indicator and possibility of forecasting the trend. Analyses of lifecycle environmental performance require extensive inputs of data that may not be practically available. Only relevant indicators must be considered in a given study. The validity of measurements is imperative to avoid misleading policymakers and other stakeholders. So, an indicator must really measure what it theoretically claims to. It is difficult to establish the environmental impact of an activity if the responses before and after the activity are indistinct. Consequently, it is necessary for an indicator to be sensitive enough to enable clear detection of changes in system responses. Forecasting is one of the key instruments used by policymakers. So, an indicator should enable the prediction of future developments and potential trends. For CSP systems, lifecycle analyses are readily available on three indicators in the literature: GHG emission, water withdrawal, and land requirements.

The key driver of renewable energy is to reduce GHG emissions and therefore, mitigate climate change. The steam turbine-based CSP technology is matured with many commercial power plants that have been operating for more than 10 years now. So, a lot of empirical data has been accumulated. Consequently, many studies have analysed the lifecycle GHG emissions from the conventional CSP technology.

On the other hand, the SGT is not yet matured. Therefore, a limited number of studies have analysed the lifecycle GHG emissions from this system. Water is a critical resource for social and economic development of any country, and so, there are many competing demands for this resource. Moreover, the demand for water is increasing due to population and industrialization. Consequently, withdrawal of water by the SGT technology has great potential to strain the limited water resources. Nevertheless, studies that focus on water withdrawal by the SGT system are scarce. In view of this, results obtained from the lifecycle analysis of water withdrawals by the conventional CSP technology can provide a rough idea about potential water withdrawals by the SGT technology. Land is required for installation of SGT power plant. Other phases of the plant life cycle also require land.

11.6.2 Assessment Methods

There are many methods for assessing the environmental impacts of a system. In this book, selected notable approaches are briefly discussed. Methods for environmental assessment include (Gargalo et al. 2024): Eco-indicator 99, environmental risk assessment (ERA), cumulative exergy analysis (CEA), eco-efficiency and lifecycle assessment (LCA). The Eco-indicator 99 method is one of the earliest attempts to incorporate sustainability as a post-hoc analysis. Nevertheless, one of its main drawbacks is lack of clear objectives in defining levels of sustainable targets.

ERA is a systematic method for assessment and management of the risk associated with human health and ecological systems caused by an event that takes place within the environment. The ERA method identifies and evaluates the potential impacts of a proposed activity or project on the environment. Environmental impact assessment (EIA) focuses on predicting and analysing the environmental impacts of anthropogenic activities before they occur on a specific project. It yields qualitative and quantitative information based on checklists on social, economic, and environmental concerns. Possible mitigation measures are also proposed in the EIA report, thereby enabling the reduction of environmental impacts during project implementation.

Exergy analysis quantifies the potential impact of a material or energy stream on the environment. In this context, the environmental impact of a process or material decreases while the potential sustainability of the process or material increases with an increase in the exergy efficiency of the process. This has led to the further development of exergy-based approaches such as the: (a) cumulative exergy demand/exergy energy footprint analysis and (b) water exergy footprint methods.

Eco-efficiency is a sustainability indicator that combines economic and environmental performances. It is based on the idea of producing more goods and services while using fewer resources and minimizing the levels of waste and environmental pollution (Muthu 2020). This approach is a management strategy and it has gained momentum. Four basic variants of eco-efficiency are environmental cost-effectiveness, environmental productivity, environmental intensity of production, and environmental improvement costs (Čuček et al. 2015). Some of the benefits of eco-efficiency are: (a) increased revenue created by developing innovative products and processes, and increasing market share, (b) reduced costs attained through more efficient utilization of energy and input materials, (c) diminished risk and liability by designing out best techniques for reducing or eliminating toxic substances, (d) enhanced image of a brand by applying improvement efforts of marketing and communication, (e) augmented productivity and employee morale through close alignment of company values with personal employee values, and (f) increased environmental performance through the reduction of toxic emissions and increase in the recovery and reuse of waste.

LCA tries to systematically quantify the environmental impacts related to a system over its life cycle in terms of emissions, health impacts, and resource consumption. It can assist in: (a) informing decision-makers in public and private sectors; (b) improving the environmental performance of goods and services at different phases in their life cycle; and (c) marketing of goods and services such as the implementation of an eco-labelling scheme, making an environmental claim, and making an environmental product declaration. During transition to more sustainable systems, it is essential to examine the whole life cycle of the system under study, including all mobile system components such as processes, activities, and supply chains (Čuček et al. 2015). A solar gas turbine power plant comprises many fixed and movable components, and so LCA is naturally a suitable method for environmental assessment of this system.

The SGT technology can be assessed and compared with other energy technologies in order to determine its relative sustainability. Depending on the scope of the LCA study, life stages of the SGT plant may include all or part of the following (Fig. 11.5): (a) extraction of raw materials for plant components, (b) processing of raw materials for plant components, (c) manufacturing of plant components, (d) plant construction, (e) plant operation and maintenance, and (f) plant decommissioning. Some of the industries along the energy production chain are energy-intensive, and they include mining & mineral processing, manufacturing of components, and transportation. Globally, the mining and metal production industry falls into the category of energy-intensive industries. It currently consumes about 38% of the global industrial energy and 11% of global energy, and this energy is derived from fossil fuels (Torrubia et al. 2023). Most of these other industries are also driven by fossil fuels. Water and land resources are also important inputs at different stages of the electricity production chain. In this context, the SGT power plant can also directly or indirectly affect the quality and quantity of these resources.

11.6.3 Greenhouse Gas Emissions

A wide range of lifecycle GHG emissions are reported in the literature, attributable to variation in study scope and assumptions made in the analysis. Lifecycle GHG emissions for the CSP technology fall in the range of about 7–90 gCO_2eq/kWh (Intergovernmental Panel on Climate Change 2011). On the other hand, the range

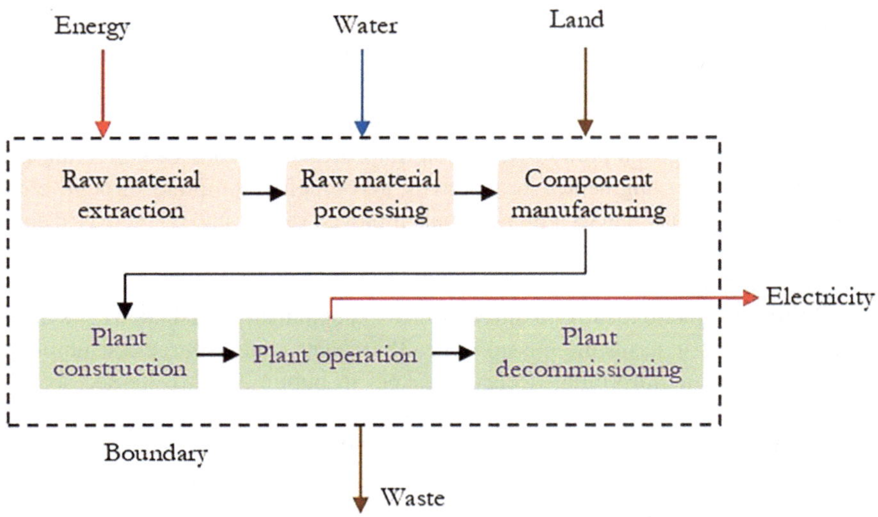

Fig. 11.5 Production chain of electricity from a solar gas turbine power plant

of GHG emissions for natural gas-fuelled power plants is 380–1000 gCO_2eq/kWh (Turconi et al. 2013). Natural gas is the cleanest fuel among the fossil fuels. It is a fuel of choice for conventional single-cycle and combined cycle gas turbine power plants. Single-cycle gas turbines have diminished energy efficiencies (26–35%) while combined cycle turbines have higher energy efficiencies of up to 60%. So, single-cycle gas turbines have generally higher carbon footprint than combined cycle cycles. However, the former technology supplies peak electricity while the latter technology mainly provides baseload power. This operational arrangement reduces the overall carbon footprint of the conventional gas turbines. Figure 11.6 shows emission factors for selected technologies. As can be seen, CSP technology has a significantly lower emission factor than fossil fuels. Coal exhibits the highest emission factor while wind technology has the lowest factor.

11.6.4 Water Requirements

Water is required at various stages of the SGT life cycle (such as extraction of raw materials, manufacturing of plant components, and plant construction, operation (such as engine cooling, and cleaning reflectors) and decommissioning). At the *i*th stage, a certain amount of water is withdrawn (W_i) from a water resource/reservoir, consumed (C_i), recycled (R_i), and discharged back into the reservoir (D_i) as shown in Fig. 11.7. Withdrawal of water from a resource reduces the available water for other applications. The problem is more serious in countries/regions with water scarcity. Water is said to be scarce in a country when the annual available water per capita is

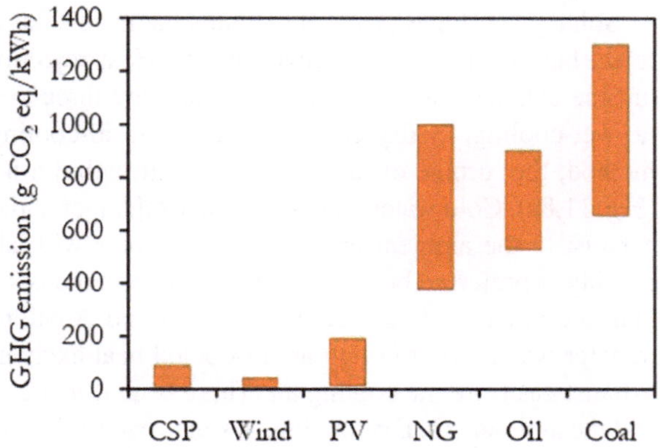

Fig. 11.6 Lifecycle greenhouse gas (GHG) emission factors for electricity generation from selected technologies: concentrating solar power (CSP), wind, photovoltaic (PV), natural gas (NG), oil and coal

less than 1000 m^3 (Rijsberman 2006). Water consumed refers to the quantity of water that is permanently removed from the water resource or so degraded in quality that it is no longer fit for use as a supply of water. Recycling helps in reducing the amount of water withdrawn from a reservoir. The scarcity of water is posing a complex challenge throughout the world, with serious agricultural, social, economic, and ecological implications. Over its lifetime, the SGT system withdraws a total amount of water (W_{total}) given by:

$$W_{\text{total}} = \sum_{i=0}^{i=n} W_i \tag{11.1}$$

where n is the total number of stages considered in the analysis.

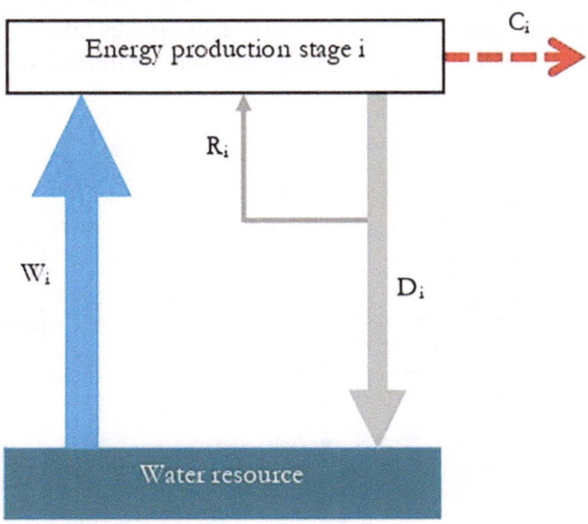

Fig. 11.7 Water use at a given stage of the life cycle of a solar gas turbine

Solar gas turbines primarily obtain heat from solar receivers, do useful work, and reject heat to the ambient environment. Heat rejection is achieved by cooling the turbine exhaust gas (see Chap. 5). There are three key methods of engine cooling: (a) wet-cooling, (b) dry-cooling, and (c) hybrid-cooling methods. In the wet-cooling method, the turbine exhaust gas is cooled within a heat exchanger by using water (Fig. 11.8a). Cold water enters the heat exchanger and transfers heat from the turbine exhaust to the ambient environment. In areas with abundant water resources, wet cooling is preferred because it has lower capital costs and higher thermal efficiency than the other cooling methods (Siddiqi and Anadon 2011). Dry-cooling systems transfer waste heat through an air-cooled heat exchanger that separates the turbine exhaust gas from the cooling air. There is no direct contact between the exhaust gas and the ambient air. Dry-cooling systems can be classified into two basic categories: direct and indirect systems. In the direct dry-cooling system, the turbine exhaust is directly cooled through an air-cooled heat exchanger (Fig. 11.8b). In the indirect dry-cooling system, the turbine exhaust is cooled in a surface heat exchanger where the heat of the exhaust fluid is transmitted to the cold water pumped from an air-cooled heat exchanger. Hybrid-cooling systems combine features of dry and wet cooling to evacuate waste heat from the engine.

Commercial CSP plants are economically feasible in locations where the direct normal irradiance (DNI) is high. Usually, the levels of DNI are favourable in arid and semi-arid regions. Consequently, CSP power plants are often located in such regions. There is however a lot of dust and sand that results in soiling and dust accumulation on the reflectors of CSP plants in these regions. Soiling and accumulation of dust cause the absorption and scattering of direct solar radiation, thereby reducing the reflectance of mirrors and the efficiency of the power plant. In view of this, it is necessary to clean the reflectors in order to maintain good levels of mirror reflectance and plant efficiency. Water is frequently used in different cleaning methods such as (Fernández-García et al. 2014): (a) high pressure demineralized water, (b) demineralized water and a natural horse-hair bristles brush, (c) high pressure demineralized water with detergent, (d) high pressure demineralized water plus 2% water additive,

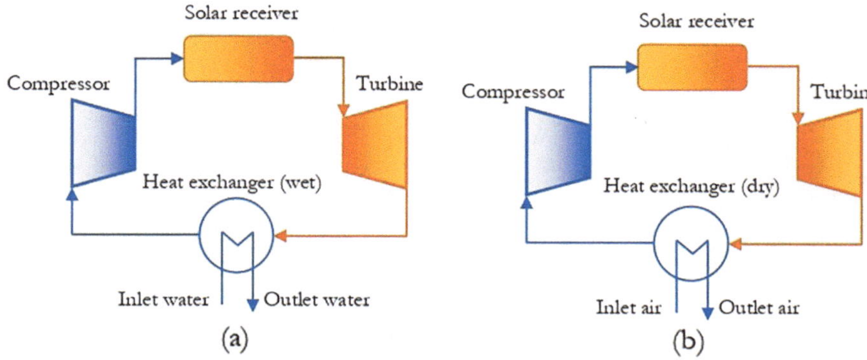

Fig. 11.8 Schematic representation of basic cooling systems: **a** Wet cooling and **b** dry cooling

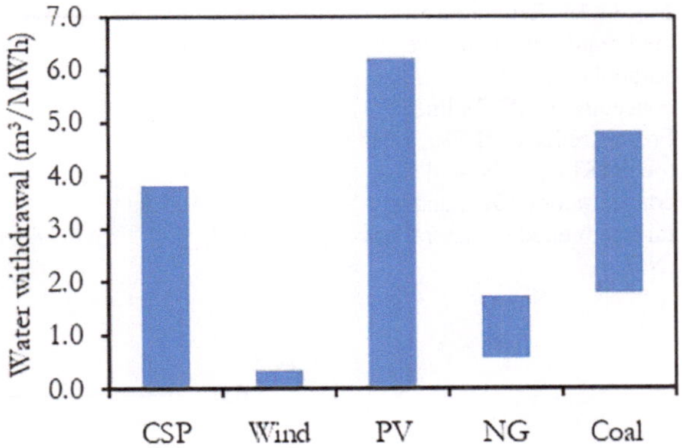

Fig. 11.9 Estimated values of lifecycle water withdrawal based on data reported by Meldrum et al. (2013) and Islam et al. (2018)

(e) demineralized water plus natural horse-hair bristles brush plus 2% water additive, and (f) pressurized steam and soft tissue. A high amount of water may be required because the area of the reflectors is relatively large.

Approximately, water withdrawals are in the range of 0.2–3.8 m^3/MWh for the parabolic, linear Fresnel reflector and solar tower technologies (Fthenakis and Kim 2010; Islam et al. 2018). Water is required only for cleaning in the parabolic dish concentrator, and so it has a very low factor of water withdrawal (0.015 m^3/MWh). Generally, CSP systems with dry cooling exhibit the lower water demand while wet-cooling systems have the highest water demand. Most of the water is withdrawn during the operational phase of the CSP plant. Water withdrawal for selected technologies is presented in Fig. 11.9. Wind technology exhibits the lowest factor of lifecycle water withdrawal while PV technology has the highest range of water withdrawal. It should nevertheless be noted that there is wide variation in the estimated values of water requirements reported by different authors due to differences in study scope (boundary), analytical methods, assumptions, data used, and other factors. Research is still going on to establish the most accurate estimates of water usage factors for different types of power plants.

11.6.5 *Land Requirements*

Different metrics are applied when evaluating the impacts of land use by energy technologies. Two commonest metrics are based on: (a) total surface area and (b) area occupied by individual plant components. The total surface area refers to all land covered by the boundaries of the plant site. These boundaries are usually enclosed in a fence. The area occupied by individual plant components is the area directly occupied by the solar field, power block, substations, service buildings, access roads, and other pieces of infrastructure. Direct area falls within the border of the total area.

Quantification of land requirements is based on capacity (area/MW) or production (area/GWh/yr). Capacity-based requirements are used in decision-making about land

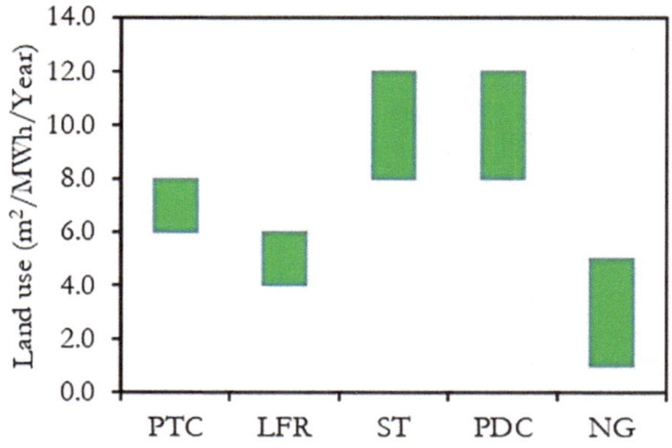

Fig. 11.10 Estimated total land requirements for the parabolic trough concentrator (PTC), linear Fresnel reflector (LFR), solar tower (ST), parabolic dish concentrator (PDC), and gas turbine fuelled by natural gas (NG)

use and new-project costs because these costs are usually based on plant capacity. Generation-based requirements give a more consistent basis for comparing technologies with different capacity factors and allow an analyst to assess land use impacts that change with insolation, storage options, sun-tracking method, and applied technology. Land requirements vary with the type of CSP system, namely parabolic trough concentrator (PTC), linear Fresnel reflector (LFR), solar tower (ST), and parabolic dish concentrator (PDC) as shown in Fig. 11.10. Generally, PTC and LFR systems use less land than ST and PDC systems. In addition, we observe that the maximum land use of all these CSP systems is higher than that of the conventional gas turbine powered by natural gas. This observation is attributed to the presence of the solar field that occupies a relatively large area of land in CSP systems.

11.7 Alignment with Policies of Pollution Control

Concerns about environmental degradation arising from energy production and consumption have led to the formulation and revision of energy and other related policies at national and international scales. It has become necessary to control pollution through energy efficiency and increasing the share of renewable energy in the energy mix. Pollution can be caused by stationary or mobile sources (Bhattacharyya 2011). A stationary source is fixed in one place while a mobile source moves from one point to another. Stationary sources are further classified into point sources and area sources. A point source is a large stationary source that emits remarkable levels of pollutants. For instance, power plants are point sources of pollution. In contrast, area sources are small stationary sources of pollution that are spread over a large area and each source is individually an insignificant emitter but they can form a significant pollution source when combined together. A notable source of area source pollution is domestic exploitation of wood fuel in cooking and space heating, especially in rural areas of developing countries. It should be noted that stationary sources can cause: air pollution, water pollution, land degradation, and noise pollution. Due to the flow of air and surface water, both air and water pollution from stationary sources can produce local, national, and international impacts (Jiang et al. 2024a, b). In view of this, there is need for direct and indirect policies to control pollution from power plants. As seen in Sects. 11.5 and 11.6, SGT systems have some environmental benefits as well as negative environmental impacts, and so, they need to comply with stipulated standards.

11.7.1 Environmental Regulations and Standards

There are two major types of approaches to pollution control (Kartsan et al. 2023; Wang et al. 2024): economic and regulatory approaches. Several economic instruments exist for environmental protection, and they include taxes, emission trading, subsidies, and green bonds. Taxes are forced on organizations that pollute the environment to motivate them to reduce emissions and promote cleaner production. Emission trading is a market-based system in which companies can purchase and sell permits for the right to pollute the environment, thereby producing a financial incentive for companies to cut down emissions because those who emit a smaller quantity of pollutants can sell their unused permits to those who are in need of them. Subsidies are financial incentives given to companies or individuals that apply principles or technologies friendly to the environment. Green bonds are bonds issued to finance projects such as renewable energy or green infrastructure that are environmentally sustainable. Solar gas turbines fall under the category of green technology, and so, SGT projects can benefit from this policy instrument.

Regulatory methods often stipulate a set of emission rules that apply to all polluters of a specific pollutant. Usually, the same regulations apply all over the country. This approach uses the power of the state to influence polluter behaviour and limit emissions below a prescribed quality target. In this approach, there are three major aspects. Firstly, relevant laws and rules specify an acceptable behaviour. Secondly, the regulator checks compliance against the stipulated behaviour. Thirdly, violations result in penalties. This approach is also referred to as a command-and-control approach because of its dependence on state power.

Usually, a state agency is mandated to prescribe the acceptable behaviour in the form of a standard. Various standards are developed and used for this purpose, including ambient air quality standards, water quality standards, industrial emissions standards, and fuel quality standards. Ambient air quality standards specify the maximum level of concentrations of pollutants. For instance, Table 11.3 shows recommended air quality guidelines (AQG) levels of selected pollutant based on the data reported by (World Health Organization 2021). Water quality standards are legal provisions that describe the desired condition of a water body and the means by which that condition will be protected or achieved. Thus, the quality of wastewater discharged into water bodies must meet certain limits. Industrial emissions standards specify the upper limits of pollution from industrial activities. Often, these limits vary with the type of technology, era of the power plant, and power plant

Table 11.3 Recommended air quality guidelines (AQG) levels for selected air pollutants

Pollutant	Averaging time	AQG level
CO (mg/m^3)	24-h	4
N_2O (μg/m^3)	Annual	10
SO_2 (μg/m^3)	24-h	40
$PM_{2.5}$ (μg/m^3)	Annual	5
PM_{10} (μg/m^3)	Annual	15

location. In this vein, older power plants are permitted to emit pollutants at a higher level due to consideration of the difficulty in meeting more stringent targets by them. Fuel quality standards specify the minimum quality of fuels that may be utilized in the energy generation activities.

11.7.2 Compliance Requirements and Implications for Solar Gas Turbines

Greenhouse gas emission from power plants is the main environmental concern within the energy industry. Therefore, national standards for limiting GHG emission are necessary for the electricity industry within a country. For example, Table 11.4 shows limits of carbon dioxide emissions from conventional gas turbine power plants in the USA (Environmental Protection Agency 2024). Even though this standard is for conventional gas turbines, we observe that the lowest limit (201 g CO_2/kWh) is significantly higher than the potential carbon emissions from SGT systems during the operational phase. Alarcón et al. (2024) analysed the carbon footprint of a combined renewable simple and recompressed Brayton system of supercritical carbon dioxide and organic Ranken cycle during three phases: plant construction, operation, and decommissioning. The total carbon emission during these three phases was in the range of 11–12 gCO_2eq/kWh. They found that the construction and operation phases respectively contributed about 95% and 0.1% of the total emissions. This shows that most of the goods and services consumed by the construction phase of the SGT system had relatively high emissions. Consequently, the SGT system would distinctly comply with the carbon emissions standard under discussion.

Noise pollution from different sources is a notable environmental concern worldwide. Gas turbines produce noise of varying frequencies and sound pressure levels.

Table 11.4 Recommended upper limits of carbon dioxide (CO_2) emissions from gas turbine power plants in the USA

Capacity factor (%)	Carbon dioxide emission (g CO_2/kWh)
< 20	201
20–40	635
> 40	726

Often, the level of sound pressure is used as an indicator of the strength of an acoustic wave, expressed in decibels (dB) at a given point in space. The reference sound pressure (P_r) is set to the threshold of human hearing at about 1000 Hz for young people. The sound pressure level (L) is defined as (Blake 2017):

$$L = 10\log\left(\frac{\overline{P^2}}{P_r^2}\right) \tag{11.2}$$

$$\overline{P^2} = \frac{1}{T}\int_{-T/2}^{T/2} P^2(t)\mathrm{d}t \tag{11.3}$$

where p is the sound pressure at time t, $\overline{P^2}$ is the time average of a pressure squared, and p_r is the reference pressure. The reference sound pressure can be taken as 2×10^{-5} N/m^2 and 1×10^{-6} N/m^2 in gases and liquids, respectively. L can be weighted based on frequency (A-weighted) to better reflect the human ear, equivalent to continuous sound pressure level during a specified time interval $t_1 = -T/2$ and $t_2 = T/2$, and it is expressed as dBA.

The impact of exposure to noise also varies with the magnitude of the noise, distance from the source, and duration of exposure. In view of this, noise pollution regulations commonly specify the permissible duration of exposure to sound levels. For instance, the Department of Labour in the USA limits impulse noise to 140 dBA, and continuous exposure to 90 dBA for 8 h exposure and 115 dBA for 15 min exposure (Otieno and Loosen 2016). As can be seen from Fig. 11.11, the maximum permissible duration of daily exposure declines with an increase in the level of sound pressure. The upper limits set by the World Bank noise abatement guidelines range from 45 dBA outside facilities in residential and institutional areas at night to 70 dBA outside facilities in industrial and commercial areas during both day and night. For gas turbines to comply with legislation, noise control equipment (such as silencers and acoustic enclosures) is used to reduce the noise levels to acceptable levels (McAuliffe et al. 1963).

11.8 Innovations for Improved Environmental Sustainability

Research is going on to improve the environmental performance of CSP systems. Some innovations have taken place to reduce environmental impacts on the subsystems of the earth system.

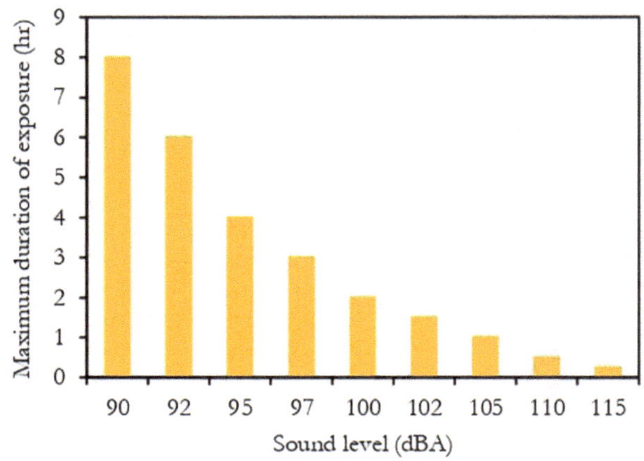

Fig. 11.11 Maximum permissible duration of noise exposure per day

11.8.1 Greenhouse Gas Emissions

Strategies for reducing greenhouse gases include: efficiency improvements (through advanced power cycles, waste heat recovery and other systems), use of clean fuels and carbon capture and storage (CCS). One of the most promising advanced power cycle is the supercritical carbon dioxide (s-CO_2) Brayton cycle. This power cycle is expected to remarkably increase the efficiency of gas turbine power plants. Waste heat can be recovered through the combined cycle, recuperation and other systems (see Chap. 5). Combined cycle systems are significantly more efficient than single cycle systems. Waste heat from gas turbines can also be recovered and exploited through polygeneration systems. A polygeneration system is an energy facility that simultaneously produces multiple outputs (such as electricity, heat, cooling, fresh water, hydrogen and methanol) from a single source of input energy, thereby increasing efficiency. The increase in efficiency results in lower GHG emissions. However, it should be noted that greenhouse gases from solar-only SGTs are associated with the consumption of external goods and services that emit these gases, mainly during construction. Thus, one way of reducing the GHG emissions from solar-only gas turbines is to use goods and services with low carbon footprints wherever possible. Goods with low carbon emissions can be identified by using green labelling. A green label is a symbol printed on a product (or its packaging) to advertise the environmental quality or characteristics of the product (Zhang et al. 2019).

For hybrid SGT systems, the first choice of backup source of heat should have low GHG emissions. Some hybrid SGT systems utilize fossil fuels in the backup heater unit (Bellos et al. 2017), and so, they tend to have higher GHG emissions. In this context, clean fuels are being investigated for possible exploitation in hybrid systems. Hydrogen is one of the most promising clean fuels. It can be blended with other input fuels (such as natural gas) to reduce carbon emissions (Li et al. 2024). Technologies for capturing and storing carbon dioxide from the flue gases of gas turbine plants are also being developed to reduce the overall carbon emissions. Consequently, optimization of the hybrid systems is necessary to establish an optimal design with respect to GHG emission and system cost. Hybrid systems also consume some goods and services. In this vein, the use of goods and services with low GHG levels can contribute to the environmental sustainability of the hybrid SGT system.

11.8.2 Water Requirements

Water is one of the important inputs to the SGT system, mainly for cooling and cleaning. The output power of a gas turbine increases with a decrease in the inlet temperature of the working fluid (a gas, such as air) entering the compressor (Unver et al. 2018). This is attributed to the fact that the density of the gas increases with a decrease in temperature, and the rise in gas density increases the mass flow rate of the working fluid. The efficiency of the gas turbine system increases with the mass flow rate through the turbine section. In case of closed cycle gas turbine systems, the exhaust gas from the turbine section is cooled before entering the compressor. Wet cooling is the most efficient method of cooling but it uses the highest amount of water compared to dry and hybrid methods of cooling. Although dry-cooling methods use a minimal amount of water, they tend to decrease the efficiency of the power plants. An important innovation has been the development of hybrid-cooling systems (the combination of dry- and wet-cooling systems). Hybrid-cooling systems offer the advantages of: (a) lower water requirement and (b) higher efficiency of electricity production (Asfand et al. 2020). Other coolants can be used instead of water in hybrid systems. It should also be noted that gas turbines operate at high temperatures (that may reach 1873 K) to achieve high thermodynamic efficiencies (Al-Hadhrami et al. 2011). Nevertheless, the heat transmitted to the turbine blades increases with the turbine inlet temperature. These operating temperatures are significantly above the permissible service temperature of metals that are limited to approximately less than 1673 K (Ashby 2011). It is therefore necessary to cool the turbine blades for safe operation. Cooling of the blades is achieved by using extracted air from the compressor of the engine (dry cooling). The cooling system must be designed to ensure that the blade surface temperature during operation does not exceed the maximum service temperature of the blade material.

Soiling of reflectors on concentrators is a common problem. It reduces the optical efficiency of the system. Consequently, it is necessary to clean these reflectors regularly to maintain a high level of reflectance. In general, methods for removing dust/soil from the reflectors are based on water (Fernández-García et al. 2014). For instance, the surface area of the parabolic trough reflectors of a 50-MW commercial CSP plant with a 7.5-h molten-salt storage system in South Spain is 500,000 m^2. The reflector area increases with the capacity, which shows that the amount of water required for reflector cleaning would increase with the plant capacity. Therefore, minimizing the water required for cleaning reflectors is important. It has been found that the most effective cleaning method is the one based on demineralised water and a brush, with average efficiency values of 98.8 and 97.2% in rainy and dry seasons, respectively. Another innovative cleaning method is the use of additives such as polymers and natural clay (El Baraka et al. 2022). These substances can increase the spread of water when in contact with the reflective surfaces, thereby reducing the amount of water required to cover a given surface.

11.8.3 Land Use

Generally, SGT systems require larger land compared with the conventional gas turbine. The solar field occupies a significant proportion of the land, and the aperture area increases with the equivalent power capacity of the system. This implies that the required land increases with the plant capacity.

One way of reducing the required land is to increase the optical efficiency of the solar field (concentrator and receiver). The optical quality of the solar field directly affects the thermal efficiency of CSP plants (Arancibia-Bulnes et al. 2017). All of the different types of CSP plants heavily depend on the optical accuracy of the solar collector field to generate competitive electricity. Thus, a constant goal is to reduce collector field costs because they contribute up to 60% of the overall plant investment. This goal can be achieved by improving the optical quality of the systems. In this vein, it should be noted that reflector geometries deviate from their design points, and imperfections exist in real reflecting surfaces, resulting in optical losses. Optimization of the solar field can assist in reducing optical losses (Al-Sulaiman and Atif 2015) and, therefore, raise the optical efficiency of the SGT system. In turn, an increase in the optical efficiency can result in the reduction of the solar field size and the size of required land.

11.9 Environmental Implications of Future Prospects and Developments

At present, research efforts are directed at improving the main characteristics of the CSP technology, namely: receiver outlet temperature, power cycle, peak temperature of the power cycle, efficiency of the power cycle, annual solar-to-electricity efficiency, heat transfer media, and thermal energy storage (He et al. 2020). These improvements generally aim at increasing the system efficiency and the dispatchability of power.

11.9.1 System Efficiency

A solar receiver is the key component that links the heliostat field and the power block. Thus, it has received significant research attention. The research efforts include solar receiver design and optimization. Innovative receivers operating at high temperatures have been proposed to achieve higher receiver efficiency. The receiver outlet temperature ranges achieved by the first and second generations of the CSP technology are respectively 523–723 K and 773–993 K. In contrast, the third generation is expected to have receiver outlet temperatures that exceed 973 K to achieve higher thermal efficiency. Research attention has also been given to the development of ultra-high-temperature (UHT) receivers that can deliver heat at temperatures above 1300 K (Abdelsalam et al. 2023).

UHT receivers are expected to be deployed in the third-generation CSP technology. Deployment of these receivers would unlock the use of advanced power cycles, such as the gas turbine cycle. There is a consensus among researchers that gas turbine-based CSP can yield higher efficiency of the power blocks of the CSP systems. This is expected to result in better overall plant efficiencies, improving the solar-to-electricity efficiency by up to 50% (Abdelsalam et al. 2023). The gas turbine cycle can be combined with other cycles such as the Rankine cycle and the organic Rankine cycle to obtain a combined cycle that exhibits higher efficiency than a single gas turbine cycle. Consequently, both the gas turbine and combined cycles have been tested to work with various fluids such as air, carbon dioxide, or helium under different thermodynamic conditions (subcritical, supercritical, and transcritical) (Dunham and Iverson 2014). The supercritical carbon dioxide (s-CO_2) gas turbine cycle is the most promising cycle. Use of liquid metals and solid particles as heat transfer media in receivers is also promising to raise the receiver outlet temperature. Research efforts aimed at improving efficiency have great potential to reduce some negative environmental impacts of the SGT system. For instance, the carbon, water, and land footprints can decrease due to an increase in the output power from the plant.

11.9.2 Enhancement of Power Dispatchability

Solar radiation is available intermittently. In this vein, thermal energy storage (TES) is aimed at providing a continuous supply of heat under fluctuating weather conditions. TES allows the dispatchability of concentrating solar power (Linares et al. 2023), and therefore it is a key enabler of CSP technology uptake. Details about TES systems are reported in Sect. 4.3. However, there are some challenges associated with this mechanism of energy storage. Most (95.6%) of the CSP plants use sensible heat storage systems, with molten salts being the commonest storage media (Merchán et al. 2022). Molten salts are available on the market and they can store a high energy density for over 20 years and 10,000 cycles. Nevertheless, they have a maximum service temperature of about 833 K, which is lower than the anticipated operational temperatures (>973 K) of the third-generation CSP systems. Moreover, molten salts are corrosive and they have low thermal conductivity. Addition of metallic nanoparticles can increase the thermal conductivity of these salts but nanofluids are not yet commercially available. Further innovation is required before these fluids can attain technological maturity. Other promising TES media include concrete, porous beds, and solid particles. Another important innovation is the development of ultra-high-temperature TES technology. It has been established that TES is viable at ultra-high temperature, and it yields high energy densities and roundtrip efficiencies (Amy et al. 2019). The development of UTH receivers and UTH thermal energy storage will enable the production of dispatchable electricity from solar gas turbines that operate at high temperatures. Improvements in TES can assist in promoting the deployment of solar-only SGT systems, thereby avoiding negative environmental impacts associated with the exploitation backup heaters. Nevertheless, solar radiation may not be available for prolonged periods (longer than the TES capacity of a given plant), resulting in depletion of the stored heat. Thus, backup heating becomes necessary to enhance the dispatchability of power under all weather conditions.

Hybrid SGT systems are a feasible technology but they require more research and development to reduce the cost of electricity production (Merchán et al. 2020). To reduce carbon emissions, fossil fuels in the hybrid SGTs can be replaced by biofuels from biomass (such as forestry residues or wood waste) or by geothermal energy resources to obtain a completely renewable and sustainable system. Exploitation of renewable energy resources in the backup unit offers an additional potential for reducing carbon emissions from SGT systems with relatively low cost (Nathan et al. 2018). Nevertheless, the cost of CSP-based electricity is higher than that from competing renewable technologies such as PV and wind. Consequently, further innovation is required to reduce the cost of SGT systems. More details about hybrid SGTs are covered in Sect. 4.3.

11.10 Worked Examples

Example 11.1 The rate of global warming has progressed at a constant rate of about 0.2 K/decade from 1970. In view of this, the Paris Climate agreement targets net-zero GHG emissions by 2050 to achieve less than 1.5 K average change in surface temperature. Greenhouse gases such as carbon dioxide have long times of residence (about 100 years). This implies that the rise in surface temperature will continue for some decades after cessation of carbon emission. Assuming the rate of global warming remains constant, determine if the average surface temperature of less than 1.5 K is achievable by 2050.

Solution

Let r be the constant rate of global warming, n be the number of decades from 1970, ΔT be the change in surface temperature, and Y be any year after 1970.

Then,

$$Y = 1970 + 10n.$$

But $n = \Delta T / r$.

Therefore,

$$\begin{aligned}\Delta T &= r(Y - 1970)/10 \\ &= 0.2 \times (2050 - 1970)/10 \\ &= 1.6\ \mathrm{K}\end{aligned}$$

The change in surface temperature is slightly greater than the target level. Consequently, the set target will not be achieved under the current assumptions.

Example 11.2 A hybrid solar gas turbine power plant is designed comprising solar and natural gas (NG) heaters with a total capacity of 50 MW. The capacity factors of the solar and NG subsystems are respectively 0.60 and 0.20. Data on environmental and economic parameters is presented in Table 11.5. Determine the minimum total installation costs when the maximum allowable total emissions are 300, 350, 400, and 450 kg CO_2 eq/MWh.

Solution

This is a problem of minimization with a single objective. Let the capacities of the solar and natural gas subsystems be X_s and X_n, respectively. Then, the problem can be set as follows:

Minimize installation cost, $Y = (u_s X_s + u_n X_n)$.

Subject to:

$$\begin{aligned}&X_s + X_n = 50, \\ &(c_s f_s X_s + c_n f_n X_n)/(f_s X_s + f_n X_n) \le L,\end{aligned}$$

where c is carbon emission factor, f is the capacity factor, u is unit installation cost, L is the maximum allowable carbon emission from the plant, and subscripts n and s represent natural gas and solar subsystems, respectively.

The minimization can be performed in any computational software with appropriate capability. In this example, the minimization is done by using the Microsoft Excel solver. The system of inequalities is solved for a given value of L. Results of the analysis are presented in Fig. 11.12. The installation cost decreases from USD 97 million at $L = 300$ kg CO_2 eq/MWh to USD 68 million at $L = 450$ kg CO_2 eq/MWh. This is attributed to differences in the intensity of carbon emission and unit

Table 11.5 Capacity factor, carbon emission factor, and unit installation cost of solar and natural gas (NG) of the plant subsystems

Subsystem	Capacity factor	Carbon emission (kg CO_2 eq/MWh)	Unit cost (million USD/MW)
Solar	0.60	80	4
NG	0.20	600	1

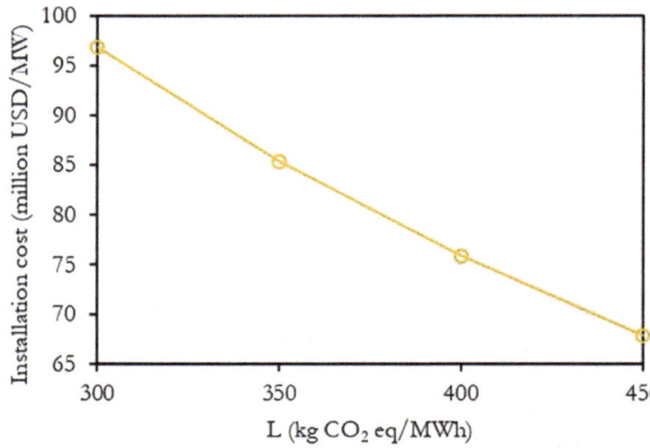

Fig. 11.12 Variation of minimum installation cost with carbon emission limit (*L*)

cost. The solar subsystem has lower carbon emission but higher unit cost. Consequently, the solar contribution to the total capacity decreases with an increase in the emission limit, resulting in higher contribution from the natural gas and diminished installation costs.

Example 11.3 A solar gas turbine plant is used to generate electricity for an isolated community. Technical data shows that it can generate a noise level of 65 dB at 1 m from the facility. Calculate the sound pressure of this noise assuming reference pressure levels of: (a) 2×10^{-5} N/m^2 and (b) 1×10^{-6} N/m^2.

Solution

From Eq. (11.1), sound pressure is given by

$$P = \sqrt{p_r^2 10^{(L/10)}}$$

This implies that:

(a)
$$P = \sqrt{4\times 10^{-10}\times 10^{(65/10)}}$$
$$= 3.6 \times 10^{-2}\mathrm{N/m^2}$$

(b)
$$P = \sqrt{10^{-12}\times 10^{(65/10)}}$$
$$= 1.8 \times 10^{-3}\mathrm{N/m^2}.$$

Example 11.4 For the problem in Example (11.2), calculate the annual amount of water withdrawn by the plant if the plant has a water intensity of 2 m^3/MWh under a carbon emission limitation of 300 kg CO_2 eq/MWh.

Solution

Let W be the amount of water withdrawn, then:

$$W = 8767\ W_i(f_s X_s + f_n X_n),$$

where W_i is the plant water intensity.

Optimization yields $X_s = 15.62$ and $X_n = 34.38$

$$
\begin{aligned}
W &= 8767 \times 2 \times (0.60 \times 15.62 + 0.2 \times 34.38) \\
&= 284{,}892 \text{ m}^3.
\end{aligned}
$$

Example 11.5 A solar gas turbine plant with a capacity of 20 MW is installed in an isolated community. The total area of the land used by this plant is 350,000 m^2, and it produces 61,000 MWh per year. Estimate the land use intensity of the plant.

Solution

Land use intensity, $I =$ Area/annual production of electricity.

Therefore,

$$
\begin{aligned}
I &= 350{,}000/61{,}000 \\
&= 5.74 \text{ m}^2/\text{MWh.year}.
\end{aligned}
$$

Nomenclatures

A Area (m^2)
f Capacity factor (dimensionless)
I Land use intensity (m^2/MWh.year)
L Carbon emission limit (kg/MWh)
P Pressure (N/m^2)
t Time (s)
X Capacity (MW).

References

Abdallah T (2017) Sustainable mass transit. Elsevier

Abdelaziz EA, Saidur R, Mekhilef S (2011) A review on energy saving strategies in industrial sector. Renew Sustain Energy Rev 15(1):150–168

Abdelsalam TI, Tian Z, Robinson A (2023) Directly irradiated liquid metal film in an ultra-high temperature solar cavity receiver. Part 1: concepts and a quasi-steady-state analysis. Sol Energy 255:355–368

Adebiyi FM (2022) Air quality and management in petroleum refining industry: A review. J Environ Chem Ecotoxicol 4:89–96

Alarcón VM, Ochoa GV, Castilla DV (2024) Energetic, exergetic, and carbon footprint performance assessment of a combined renewable simple and recompressed Brayton S-CO_2/ORC system. Energy Rep 12:3114–3133

Al-Hadhrami LM, Shaahid SM, Al-Mubarak AA (2011) Jet impingement cooling in gas turbines for improving thermal efficiency and power density. In: Benini E (ed) Advances in gas turbine technology. InTech

Al-Sulaiman F, Atif M (2015) Performance comparison of different supercritical carbon dioxide Brayton cycles integrated with a solar power tower. Energy 82:61–71

Amponsah NY, Troldborg M, Kington B, Aalders I, Hough RL (2014) Greenhouse gas emissions from renewable energy sources: a review of lifecycle considerations. Renew Sustain Energy Rev 39:461–475

Amy C, Seyf HR, Steiner MA, Friedman DJ, Henry A (2019) Thermal energy grid storage using multi-junction photovoltaics. Energy Environ Sci 12(1):334–343

Arancibia-Bulnes CA, Peña-Cruz MI, Mutuberría A, Díaz-Uribe R, Sánchez-González M (2017) A survey of methods for the evaluation of reflective solar concentrator optics. Renew Sustain Energy Rev 69:673–684

Asfand F, Palenzuela P, Roca L, Caron A, Lemarié C-A, Gillard J, Turner P, Patchigolla K (2020) Thermodynamic performance and water consumption of hybrid cooling system configurations for concentrated solar power plants. Sustainability 12(11):4739

Ashby M (2011) Materials selection in mechanical design. Elsevier

Azoumah Y, Ramdé EW, Tapsoba G, Thiam S (2010) Siting guidelines for concentrating solar power plants in the Sahel: case study of Burkina Faso. Sol Energy 84(8):1545–1553

Bellos E, Tzivanidis C, Antonopoulos KA (2017) Parametric analysis and optimization of a solar assisted gas turbine. Energy Convers Manag 139:151–165

Bhattacharyya SC (2011) Energy economics. Springer, London

Blake WK (2017) Mechanics of flow-induced sound and vibration, vol 1. Elsevier

Blattert C, Lemm R, Thees O, Lexer MJ, Hanewinkel M (2017) Management of ecosystem services in mountain forests: review of indicators and value functions for model based multi-criteria decision analysis. Ecol Indic 79:391–409

Bošnjaković M, Tadijanović V (2019) Environment impact of a concentrated solar power plant. Tehnički Glasnik 13(1):68–74

Brenniman GR (1999) Potable water. In: Environmental geology. Encyclopedia of earth science. Springer, Dordrecht

Cheng F, Luo H, Jenkins JD, Larson ED (2023) The value of low- and negative-carbon fuels in the transition to net-zero emission economies: lifecycle greenhouse gas emissions and cost assessments across multiple fuel types. Appl Energy 331:120388

Chipperfield MP, Dhomse SS, Feng W, McKenzie RL, Velders GJM, Pyle JA (2015) Quantifying the ozone and ultraviolet benefits already achieved by the Montreal Protocol. Nat Commun 6(1):7233

Čuček L, Klemeš JJ, Kravanja Z (2015) Overview of environmental footprints. In: Klemeš JJ (ed) Assessing and measuring environmental impact and sustainability. Elsevier, pp 131–193

Cumpsty NA, Marble FE (1977) Core noise from gas turbine exhausts. J Sound Vib 54(2):297–309

Dadkhah-Aghdash H, Zare-Maivan H, Heydari M, Sharifi M, Lucas-Borja ME, Naidu R (2022) Air pollution from gas refinery through contamination with various elements disrupts semiarid Zagros oak (Quercus brantii Lindl) forests. Iran Sci Rep 12(1):284

Daellenbach KR, Uzu G, Jiang J, Cassagnes L-E, Leni Z, Vlachou A, Stefenelli G, Canonaco F, Weber S, Segers A, Kuenen JJP, Schaap M, Favez O, Albinet A, Aksoyoglu S, Dommen J, Baltensperger U, Geiser M, El Haddad I, Jaffrezo J-L, Prévôt ASH (2020) Sources of particulate-matter air pollution and its oxidative potential in Europe. Nat 587(7834):414–419

Dehkordi MM, Nodeh PZ, Dehkordi SK, Salmanvandi H, Khorjestan RR, Ghaffarzadeh M (2024) Soil, air, and water pollution from mining and industrial activities: sources of pollution, environmental impacts, and prevention and control methods. Results Eng 23:102729

Dincer I (2001) Environmental issues: II-potential solutions. Energy Sources 23(1):83–92

Dunham MT, Iverson BD (2014) High-efficiency thermodynamic power cycles for concentrated solar power systems. Renew Sustain Energy Rev 30:758–770

Egerer U, Dana S, Jager D, Xia G, Stanislawski BJ, Yellapantula S (2024) Wind and structural loads data measured on parabolic trough solar collectors at an operational power plant. Sci Data 11(1):98

El Baraka A, Bouhafra O, Jorio A, Khaldoun A (2022) Improvement of WET cleaning technique of CSP reflector mirrors using novel liquids. Mater Today Proc 53:332–335

Environmental Protection Agency (2024) New source performance standards for greenhouse gas emissions from new, modified, and reconstructed fossil fuel-fired electric generating units; emission guidelines for greenhouse gas emissions from existing fossil fuel-fired electric generating units. United States of America. Available at: https://www.federalregister.gov/documents/2024/05/09/2024-09233/new-source-performance-standards-for-greenhouse-gas-emissions-from-new-modified-and-reconstructed

Fernández-García A, Álvarez-Rodrigo L, Martínez-Arcos L, Aguiar R, Márquez-Payés JM (2014) Study of different cleaning methods for solar reflectors used in CSP plants. Energy Procedia 49:80–89

Fthenakis V, Kim HC (2010) Life-cycle uses of water in U.S. electricity generation. Renew Sustain Energy Rev 14(7):2039–2048

Gallardo B, Bacher S, Barbosa AM, Gallien L, González-Moreno P, Martínez-Bolea V, Sorte C, Vimercati G, Vilà M (2024) Risks posed by invasive species to the provision of ecosystem services in Europe. Nat Commun 15(1):2631

Gargalo CL, Yu H, Vollmer N, Arabkoohsar A, Gernaey KV, Sin G (2024) A process systems engineering view of environmental impact assessment in renewable and sustainable energy production: status and perspectives. Comput Chem Eng 180:108504

Geissinger A, Laurell C, Öberg C, Sandström C (2019) How sustainable is the sharing economy? On the sustainability connotations of sharing economy platforms. J Clean Prod 206:419–429

Gong Y, Zha J, Guo Q, Guo G (2024) A new indicator for estimating the degree of mining-induced land subsidence: the overburden's average GSI value. Sci Rep 14(1):332

Hasii O, Gasii G (2024) Coal mining and water resources: impacts, challenges, and strategies for sustainable environmental management. IOP Conf Ser: Earth Environ Sci 1348(1):012017

Hasrod T, Nuapia YB, Tutu H (2024) The application of transfer machine learning to predict and impute missing sulphate levels in different acid mine drainage treatment plants. Clean Water 2:100029

He YL, Qiu Q, Wang K, Yuan F, Wang WQ, Li M-J, Guo J-Q (2020) Perspective of concentrating solar power. Energy 198:117373

Hernandez RR, Easter SB, Murphy-Mariscal ML, Maestre FT, Tavassoli M, Allen EB, Barrows CW, Belnap J, Ochoa-Hueso R, Ravi S, Allen MF (2014) Environmental impacts of utility-scale solar energy. Renew Sustain Energy Rev 29:766–779

Ho CK, Sims CA, Christian JM (2015) Evaluation of glare at the ivanpah solar electric generating system. Energy Procedia 69:1296–1305

Hopland AO, Kvamsdal SF (2023) Drinking water contamination and treatment costs. Water Resour Econ 43:100225

Hu Y, Man Y (2023) Energy consumption and carbon emissions forecasting for industrial processes: status, challenges and perspectives. Renew Sustain Energy Rev 182:113405

Intergovernmental Panel on Climate Change (2011) Renewable energy sources and climate change mitigation: summary for policy makers. Cambridge University Press, New York

Intergovernmental Panel on Climate Change (2023) Climate change 2021—the physical science basis. Cambridge University Press, New York

International Energy Agency (2023) World energy outlook 2023. Paris

Jacuzzi G, Kuehne LM, Harvey A, Hurley C, Wilbur R, Seto E, Olden JD (2024) Population health implications of exposure to pervasive military aircraft noise pollution. J Expo Sci Environ Epidemiol 2024. https://doi.org/10.1038/s41370-024-00670-1

Jain S, Wienold J, Eandi C, Gisselbaek S, Kawasaki A, Andersen M (2023) Influence of macular pigment on the sensitivity to discomfort glare from daylight. Sci Rep 13(1):18551

Jiang J, Zhang L, Wang Z, Gu W, Yang C, Shen Y, Zhao J, Han W, Hu Y, Xue F, Chen W, Guo X, Li H, Wu P, Chen Y, Zhao Y, Du J, Jiang C (2024a) Spatial consistency of co-exposure to air and surface water pollution and cancer in China. Nat Commun 15(1):7813
Jiang S, Tong X, Yu K, Yin P, Shi S, Meng X, Chen R, Zhou M, Kan H, Niu Y, Li Y (2024) Ambient particulate matter and chronic obstructive pulmonary disease mortality: a nationwide, individual-level, case-crossover study in China. eBioMedicine 107:105270
Joglekar SN, Khadke PN (2023) Sustainability assessment of a fruit and vegetable waste biorefinery: environment-social-economic aspects. In: Mandvgane SA, Chakravarty I, Jaiswal AK (eds) Fruit and vegetable waste utilization and sustainability. Elsevier, pp 213–226
Jordan P (2014) Solar energy markets: an analysis of the global solar industry. Elsevier
Kanwal H, Raza A, Zaheer MS, Nadeem M, Ali HH, Manoharadas S, Rizwan M, Saleem MK, Ahmad U, Ikram K, Waheed M, Riaz FR (2024) Transformation of heavy metals from contaminated water to soil, fodder and animals. Sci Rep 14:11705
Kartsan I, Kartsan P, Matyunina M, Zhukov A, Kolosov A (2023) Economic mechanisms of environmental protection. E3S Web Conf 420:08008
Khan MI, Asfand F, Al-Ghamdi SG (2022) Progress in technology advancements for next generation concentrated solar power using solid particle receivers. Sustain Energy Technol Assess 54:102813
Khorampoor N, Farahani SD, Mosavi A (2020) Modeling the efficiency and emissions of a hybrid solar-gas power plant. Eng Appl Comput Fluid Mech 14(1):790–804
Kowalska-Pyzalska A (2024) Diffusion of innovative energy services. Elsevier
Krumina G, Volberga L, Ikaunieks G, Naumovs L (2021) The development of brigthness evaluation method for digital billboards. IOP Conf Series Mater Sci Eng 1202(1):012035
Kumar A (2022) Ecosystem-based adaptation: approaches to sustainable management of aquatic resources. Elsevier
Kundzewicz ZW (2008) Hydrosphere. In: Jørgensen SE, Fath BD (eds) Encyclopedia of ecology. Elsevier, pp 1923–1930
Kyule B, Wang X (2024) Quantifying the link between industrialization, urbanization, and economic growth over Kenya. Front Archit Res 13(4):799–808
Li C, van Donkelaar A, Hammer MS, McDuffie EE, Burnett RT, Spadaro JV, Chatterjee D, Cohen AJ, Apte JS, Southerland VA, Anenberg SC, Brauer M, Martin RV (2023) Reversal of trends in global fine particulate matter air pollution. Nat Commun 14(1):5349
Linares JI, Arenas E, Cantizano A, Montes MJ, Rovira AJ, Porras J, Pérez-Domínguez JR (2023) Innovative integrated solar combined cycle: enhancing dispatchability with a partial recuperative gas turbine and supercritical CO_2 bottoming cycle, coupled with an ORC. Sol Energy 264:112075
Ling L, Li X, Xue H, Sun Q, Liu L, Du C, Li K (2024) The effect of acid mine drainage on the properties of an all-solid waste paste backfill body based on oil shale residue. Constr Build Mater 425:136017
Li N, Aksoy M, Nutakki TUK, Singh PK, El-Shorbagy MA, Dahari M, Alkhalaf S, Alotaibi KO, Ali HE (2024) Development of a modified gas turbine-based sustainable power generation and water treatment system; Economical/environmental considerations and data-driven optimization. J Clean Prod 450:141904
Löfving B, Billger M, Thaung J (2015) Visualization of disability glare due to veiling luminance. Energy Procedia 78:735–740
Lu X, Zhang L, Shen L (2021) Tropospheric ozone interacts with weather and climate. In: Gao M, Wang Z, Carmichael G (eds) Air pollution, climate, and health. Elsevier, pp 15–46
Luo Z, Wu S, Shi W, Hu H, Lin T, Zhao K, Hou G, Fan C, Li X, Chen G (2024) Combined effects of cadmium and simulated acid rain on soil microbial communities in the early cultivation of Populus beijingensis seedlings. Ecotoxicol Environ Saf 280:116583
Mahdi MS (1987) Essay book review. Int J Environ Stud 30(2–3):233–235

Mahmoudi A, Mousavi SA, Darvishi P (2021) Greywater as a sustainable source for development of green roofs: characteristics, treatment technologies, reuse, case studies and future developments. J Environ Manag 295:112991

Malico I (2024) Biomass for power production and cogeneration. In: Gonçalves AC, Malico I (eds) Forest bioenergy. Green energy and technology. Springer, Cham

McAuliffe D, Morlock H, Oran F (1963) What to do about gas-turbine noise. In: Aviation and space, hydraulic, and gas turbine conference and products show. The American Society of Mechanical Engineer. Available at: https://www.asme.org

Md Islam T, Huda N, Abdullah AB, Saidur R (2018) A comprehensive review of state-of-the-art concentrating solar power (CSP) technologies: current status and research trends. Renew Sustain Energy Rev 91:987–1018

Meldrum JR, Macknick JE, Heath GA, Nettles-Anderson SL (2013) Life cycle water use for electricity generation: implications of the distribution of collected estimates. In: Proceedings of the ASME 2013 power conference volume 2, 29 July to 1 August 2013, Boston. American society of mechanical engineers (ASME), Atlanta

Merchán RP, Santos MJ, Heras I, Gonzalez-Ayala J, Medina A, Hernández CA (2020) On-design pre-optimization and off-design analysis of hybrid Brayton thermosolar tower power plants for different fluids and plant configurations. Renew Sustain Energy Rev 119:109590

Merchán RP, Santos MJ, Medina A, Hernández AC (2022) High temperature central tower plants for concentrated solar power: 2021 overview. Renew Sustain Energy Rev 155:111828

Misthos L-M, Krassanakis V, Merlemis N, Kesidis AL (2023) Modeling the visual landscape: a review on approaches, methods and techniques. Sens 23(19):8135

Mousavi SM, Dinan NM, Ansarifard S, Sonnentag O (2022) Analyzing spatio-temporal patterns in atmospheric carbon dioxide concentration across Iran from 2003 to 2020. Atmos Environ X 14:100163

Münzel T, Hahad O, Lelieveld J, Aschner M, Nieuwenhuijsen MJ, Landrigan PJ, Daiber A (2024) Soil and water pollution and cardiovascular disease. Nat Rev Cardiol. https://doi.org/10.1038/s41569-024-01068-0

Murphy E, King EA (2022) Environmental noise pollution: noise mapping, public health, and policy. Elsevier

Muthu SS (2020) Assessing the environmental impact of textiles and the clothing supply chain, 2nd edn. Elsevier

Nathan GJ, Jafarian M, Dally BB, Saw WL, Ashman PJ, Hu E, Steinfeld A (2018) Solar thermal hybrids for combustion power plant: a growing opportunity. Prog Energy Combust Sci 64:4–28

Nwankwegu AS, Zhang L, Xie D, Onwosi CO, Muhammad WI, Odoh CK, Sam K, Idenyi JN (2022) Bioaugmentation as a green technology for hydrocarbon pollution remediation. Probl Prospects. J Environ Manag 304:114313

O'Leary K (2022) Fracking linked to adverse birth outcomes. Nat Med (2022). https://doi.org/10.1038/d41591-022-00049-5

Ojovan MI, Lee WE (2005) An introduction to nuclear waste immobilisation. Elsevier

Okoroigwe E, Madhlopa A (2016) An integrated combined cycle system driven by a solar tower: a review. Renew Sustain Energy Rev 57:337–350

Otieno GA, Loosen AE (2016) An analysis of key environmental and social risks in the development of concentrated solar power projects. AIP Conf Proc 1734(1):160012

Ozgen S, Cernuschi S, Caserini S (2021) An overview of nitrogen oxides emissions from biomass combustion for domestic heat production. Renew Sustain Energy Rev 135:110113

Panagiotopoulou VC, Paraskevopoulou A, Stavropoulos P (2024) Urban and industrial energy usage and consumption. In: Rahimpour MR (ed) Encyclopedia of renewable energy, sustainability and the environment. Elsevier, pp 21–32

Petavratzi E, Kingman S, Lowndes I (2005) Particulates from mining operations: a review of sources, effects and regulations. Miner Eng 18(12):1183–1199

Pfenninger S, Hawkes A, Keirstead J (2014) Energy systems modeling for twenty-first century energy challenges. Renew Sustain Energy Rev 33:74–86

Piao C, Zhu B, Jiang J, Dong Q (2024) Research on prediction method of coal mining surface subsidence based on MMF optimization model. Sci Rep 14(1):20316

Poberezhna L, Poberezhnyi L, Shkitsa L, Hrytsanchuk A, Korol O, Berezhenko B, Berezhenko E (2024) Predicting environmental risks in case of oil and gas equipment failures. Procedia Struct Integr 59:739–744

Prasad MNV (2024) Bioremediation, bioeconomy, circular economy, and circular bioeconomy—strategies for sustainability. In: Prasad MNV (ed) Bioremediation and bioeconomy: a circular economy approach. Elsevier, pp 3–32

Pryor SC, Crippa P, Sullivan RC (2015) Atmospheric chemistry. In: Reference module in earth systems and environmental sciences. Elsevier. https://doi.org/10.1016/B978-0-12-409548-9.09177-6

Reddy MT, Pandravada SR, Sivaraj N, Kamala V, Sunil N, Dikshit N (2017) Classification and characterization of landscapes in the territory of Adilabad district, Telangana, Deccan region, India. Open Access Libr J 04(07):1–39

Rijsberman F (2006) Water scarcity: fact or fiction? Agric Water Manag 80:5–22

Saikku L, Rautiainen A, Kauppi PE (2008) The sustainability challenge of meeting carbon dioxide targets in Europe by 2020. Energy Policy 36(2):730–742

Samset BH, Lund MT, Fuglestvedt JS, Wilcox LJ (2024) 2023 temperatures reflect steady global warming and internal sea surface temperature variability. Commun Earth Environ 5(1):460

Serpa S, Ferreira C M (2019) Society 5.0 and sustainability digital innovations: a social process. J Organ Cult Commun Confl 23(2):1–14

Shangguan J, Zhou S, Hu X, Xu A, Li Z (2024) MiRNA-mRNA joint analysis reveals the response mechanism of Ruditapes philippinarum to CO_2-driven ocean acidification. Aquac Rep 37:102272

Siddiqi A, Anadon LD (2011) The water-energy nexus in Middle East and North Africa. Energy Policy 39(8):4529–4540

Singh G (1987) Mine water quality deterioration due to acid mine drainage. Int J Mine Water 6(1):49–61

Speight J (2020) Natural water remediation. Elsevier

Tang B, Zhang L, Liu S, Bai X, Chen G, Shang Z (2024) Calculation of noise field in main transformer room of indoor substation based on thermal-acoustic coupling. Energy 297:131046

Torrubia J, Valero A, Valero A (2023) Energy and carbon footprint of metals through physical allocation. Implications for energy transition. Resour Conserv Recycl 199:107281

Tupov VB, Taratorin AA (2013) Influence of climatic factors and the ground surface on the required noise abatement from power equipment. Therm Eng 60(7):493–498

Tupov VB, Taratorin AA (2020) Features of noise radiation from gas turbines. J Phys Conf Series 1683(4):042089

Turchi CS, Ma Z (2014) Co-located gas turbine/solar thermal hybrid designs for power production. Renew Energy 64:172–179

Turconi R, Boldrin A, Astrup T (2013) Life cycle assessment (LCA) of electricity generation technologies: overview, comparability and limitations. Renew Sustain Energy Rev 28:555–565

United Nations (1998) Kyoto protocol to the United Nations framework convention on climate change. United Nations, New York. Available at: https://www.un.org

United Nations (2015) Paris agreement. United Nations, New York. Available at: https://unfccc.int

United Nations (2024) Climate action, what is climate change? Available at: https://www.un.org/en/climatechange/what-is-climate-change

Unver U, Ozkara G, Bahar EM (2018) A Design approach for cooling gas turbine intake air with solar-assisted absorption cooling cycle. In: Nižetić S, Papadopoulos A (eds) The role of exergy in energy and the environment. Springer, Cham

Vallero DA (2019) Air pollution calculations: quantifying pollutant formation, transport, transformation. Elsevier, Fate and Risks

Van Vuuren DP, Nakicenovic N, Riahi K, Brew-Hammond A, Kammen D, Modi V, Nilsson M, Smith KR (2012) An energy vision: the transformation towards sustainability—interconnected challenges and solutions. Curr Opin Environ Sustain 4(1):18–34

Varanasi P (1993) Thermal infrared absorption of atmospheric radiation by water vapor and its effect on climate. In: Williams FA, Oppenheim AK, Olfe DB, Lapp M (eds) Modern developments in energy, combustion and spectroscopy. Elsevier, pp 125–146

Wang J, Azam W (2024) Natural resource scarcity, fossil fuel energy consumption, and total greenhouse gas emissions in top emitting countries. Geosci Front 15(2):101757

Wang Y, Yazdi MD, Wei Y, Schwartz JD (2024) Air pollution below US regulatory standards and cardiovascular diseases using a double negative control approach. Nat Commun 15(1):8451

Wang X, Gao Y, Jiang X, Zhang Q, Liu W (2021) Analysis on the characteristics of water pollution caused by underground mining and research progress of treatment technology. Adv Civ Eng 2021(1):9984147

Wiesmeier M, Urbanski L, Hobley E, Lang B, von Lützow M, Marin-Spiotta E, van Wesemael B, Rabot E, Ließ M, Garcia-Franco N, Wollschläger U, Vogel H-J, Kögel-Knabner I (2019) Soil organic carbon storage as a key function of soils—a review of drivers and indicators at various scales. Geoderma 333:149–162

World Health Organization (2021) WHO global air quality guidelines: particulate matter (PM2.5 and PM10), ozone, nitrogen dioxide, sulphur dioxide and carbon monoxide. World Health Organization, Geneva. Available at: https://www.who.int/news-room/fact-sheets/detail/ambient-(outdoor)-air-quality-and-health

Yan H, Li Q, Feng K, Zhang L (2023) The characteristics of PM emissions from construction sites during the earthwork and foundation stages: an empirical study evidence. Environ Sci Pollut 30(22):62716–62732

Zambrano-Monserrate MA, Carvajal-Lara C, Urgilés-Sanchez R, Ruano MA (2018) Deforestation as an indicator of environmental degradation: analysis of five European countries. Ecol Indic 90:1–8

Zhang X, Jiang H, Jin J, Xu X, Zhang Q (2012) Analysis of acid rain patterns in northeastern China using a decision tree method. Atmos Environ 46:590–596

Zhang Q, Mi J, Shen H (2019) Green labeling and sustainable development. In: Filho WL (ed) Encyclopedia of sustainability in higher education. Springer, Cham, pp 1–7

Zhao J, Cao C, Chen X, Zhang W, Ma T, Irfan M, Zheng L (2024a) Source-specific ecological risk analysis and critical source identification of heavy metal(loid)s in the soil of typical abandoned coal mining area. Sci Total Environ 947:174506

Zhao Q, Yu L, Chen X (2024b) Land system science and its contributions to sustainable development goals: a systematic review. Land Use Policy 143:107221

Zhou Z, Wu W, Chen Q, Chen S (2008) Study on sustainable development of rural household energy in northern China. Renew Sustain Energy Rev 12(8):2227–2239

Appendix A
Units of Measurement

Basic units		Conversion of units
Length	metre (m)	Pressure (Pascal Pa ≡ N/m^2)
Mass	kilogram (kg)	1 atmosphere atm = 101.325 kPa
Temperature	Kelvin (K)	1 bar = 1 x 10^5 Pa
Time	second (s)	1 mm H_2O = 9.80665 Pa
		1 mm Hg = 133.3 Pa
Conversion of units		1 Pascal Pa = 1 N/m^2
Length (m)		1 psi = 6.894 kPa
1 ft = 0.3048 m		
1 in = 25.4 mm		Energy (Joule J ≡ Nm)
1 mile = 1.609 km		1 Joule J = 1 Nm
		1 kWh = 3.6 MJ
Area (m^2)		1 Btu = 1.055 kJ
1 ft^2 = 0.09290304 m^2		1 kcal = 4.1868 kJ
1 $mile^2$ = 2.590 km^2		
1 ha = 10^4 m^2		Power (Watt W ≡ J/s)
		1 Btu/h = 0.2931 W
Volume (m^3)		1 hp = 0.7457 kW
1 L (L) = 10^{-3} m^3		1 W/ft^2 = 10.76 W/m^2
1 ft^3 = 28.32 L		1 Btu/ft^2.h = 3.155 W/m^2
1 U.K. gal = 4.546 L		
1 U.S. gal = 3.785 L		Viscosity (Pa s ≡ N.s/m^2)
		1 cP(centipoise) = 10^{-3} Pa s
Mass (kg)		1 lbf h/ft^2 = 0.1724 MPa.s
1 lb = 0.45359237 kg		

(continued)

A. Madhlopa and R. Nkhoma, *Principles of Solar Gas Turbines for Generating Electricity*, Green Energy and Technology, https://doi.org/10.1007/978-3-031-84096-8

(continued)

Basic units	Conversion of units		
1 oz = 28.35 g	Important prefixes		
	Pico	P	10^{-12}
Volume flow rate (m^3/s)	Nano	N	10^{-9}
1 cubic foot per minute (cfm) = 0.4719 L/s	Micro	μ	10^{-6}
	Milli	M	10^{-3}
Mass flow rate (kg/s)	Kilo	K	10^{3}
1 lb/h = 0.0001256 kg/s	Mega	M	10^{6}
	Giga	G	10^{9}
	Terra	T	10^{12}
Speed (m/s)	Density (kg/m^3)		
1 ft/min = 0.00508 m/s	1 g/cm^3 = 1 x 10^3 kg/m^3		
1 mile/h = 0.4770 m/s	1 lb/ft^3 = 16.02 kg/m^3		
Force (Newton N ≡ kg.m/s^2)	Temperature conversion T = c + 273.15, where c is temperature in degree Celcius (°C) and T is temperature in Kelvin (K)		

Appendix B
Selected Constants

Avogadro constant, $N_A = 6.02252 \text{ x}10^{23}$ molecules per gmole.

Boltzmann constant $= 1.38054 \text{ x } 10^{-23}$ J/K.

Speed of light $= 2.9979 \text{ x } 10^{8}$ m/s.

Universal gas constant $R = 8.\ 3143$ kJ/kmole.K.

A. Madhlopa and R. Nkhoma, *Principles of Solar Gas Turbines for Generating Electricity*, Green Energy and Technology, https://doi.org/10.1007/978-3-031-84096-8